Eco-facts & Eco-fiction

When does recycling harm the environment? What evidence is there that health food is healthier? Is there a "safe dose" for toxic chemicals? Is the electric car a solution to air pollution or does it merely transfer the pollution from the car's exhaust to the power plant?

Fear and pseudo-science have created perceptions about environmental and human health that are wrong and in conflict with current medical and biological observations. Biodegradable waste is perceived as "environmentally friendly" but its decomposition in landfill sites can contribute to the greenhouse effect. Many "natural" and healthy foods are also rich in naturally occurring carcinogenic chemicals.

Eco-Facts & Eco-Fiction examines these and other aspects of the environmental vocabulary that obscure an understanding of the environmental crisis. The author replaces the obsolescence of earlier perceptions with up-to-date environmental information. Fear and misunderstanding of science and technology are removed through informal explanations of basic ideas and observations.

Examining a wealth of examples of local and global problems, this book leads to a more informed understanding of environmental issues. It brings a common language to the environmental debate that promotes communication between environmental science and environmental decision-making. That communication is essential for reaching effective solutions to environmental problems.

William H. Baarschers is Professor Emeritus (chemistry) and Senior Advisor to the Resource Centre for Occupational Health and Safety, Lakehead University, Thunder Bay, Ontario, Canada.

Eco-facts & Eco-fiction

Understanding the Environmental Debate

William H. Baarschers

London and New York

First published 1996
by Routledge
11 New Fetter Lane, London EC4P 4EE

Simultaneously published in the USA and Canada
by Routledge
29 West 35th Street, New York, NY 10001

Routledge is an International Thomson Publishing company

Typeset in Palatino by Solidus (Bristol) Ltd, England
Printed and bound in Great Britain by
T.J. Press (Padstow) Ltd, Padstow, Cornwall

British Library Cataloguing in Publication Data
A catalogue record for this book is available from the British Library

Library of Congress Cataloging in Publication Data
Baarschers, William H., 1932–
Eco-facts & eco-fiction / understanding the environmental debate / William H. Baarschers.
p. cm.
Includes bibliographical references and index.
1. Environmentalism. 2. Environmental education. 3. Human ecology. 4. Pollution. I. Title.
GE195.B33 1996
363.7–dc20 95-25369
CIP

ISBN 0–415–13020–4 (hbk)
ISBN 0–415–13021–2 (pbk)

To Robert, Alison, Emily, Brian and Alissa,
and the world they will live in

Contents

Figures

Tables

Preface

In almost any library catalogue a search for "environment" and "environmental" quickly returns several thousand entries. A few thousand more appear for related keywords like "ecology" or "pollution." Many words have been written about the environment. Why do we need another book on this topic?

Retrieving just a few of these thousands of records shows that there is a great deal of disagreement about environmental issues. Some of that disagreement is within small groups of experts who interpret the same data differently. They usually agree to disagree and search for more data to settle their differences. Much of the disagreement exists outside this small circle of experts. That disagreement follows from where we get our information, how well we trust our sources of information, and who we decide to believe.

Environmental information suffers from the same problem as much of the other merchandise in the information trade. There is a general lack of quality control on the traded product. Much traded information is either incomplete, or out of date, or outright wrong. Therefore environmental issues are a continuing source of debate. Several general questions and issues form the substance of this debate. What are we doing to this planet? Will our species survive? Are we destroying other species with which we share the planet? Could it be that we have already gone too far?

There is another general issue that has not received much attention in the environmental literature. That is the language we use in this debate. In the environmental context we use many terms and concepts with the almost automatic assumption that their meaning is clear to everybody. We assume we all understand the buzzwords. Widely held perceptions about the relationship between pollution and cancer, nuclear energy, recycling, and toxic chemicals are shaping the debate. And even if there is little consensus around the table, we do create laws and regulations that affect our society in many ways.

Where do our ideas come from? Most of our perceptions are based on two general lines of reasoning. One of these is the hypothesis that for many toxic agents, particularly radiation and carcinogenic chemicals, there is no safe level of exposure. The slightest amounts are a risk to personal and environmental health. The other line follows from the way we extrapolate. To describe toxic action we look at what happens to laboratory animals and we look at humans that are *occupationally* exposed to a toxic agent. Then we try to figure out what these things mean for the general public. This last step involves many assumptions and estimates. This is where disagreements begin. Ironically, this is also where the regulatory process finds its starting point.

In this book I will take you, the reader, on an exploration of the origin of these ideas, assumptions, and perceptions. We will retrieve some of the underlying experimental observations and see whether or not they support the basis on which we have built the environmental language. Sometimes this exploration will take us to perhaps unknown territory. We will also encounter some issues that some would consider controversial. Therefore it is essential that I back up some of the vistas on our tour with the appropriate observations. Of necessity, many of these observations have been reported in the science literature and this book contains many references to that literature. I do not expect my readers to follow up on all these references. They are there for those whose agreement or disagreement with what they read here is strong enough that they feel the need to "check up on me."

I hope that this brief exploration of the eco-language will make a small contribution to improving the effectiveness of the environmental debate. If we can handle this part of our collective challenge, the next steps may still come in time so that our children and grandchildren will forgive us.

William H. Baarschers
Thunder Bay, July 1995

Note
The numbers we find for global releases, production, or consumption of materials of environmental significance are usually fairly rough estimates. Therefore I have made no attempts to convert or to distinguish between the US short ton (about 900 kg), the British long ton (a little over 1000 kg), and the metric tonne (1000 kg).

Acknowledgements

Many people helped to shape this book. They were the classrooms full of students and adult education groups that so patiently listened to me while I was riding my environmental hobby-horse. They not only listened, they also asked many questions and showed their genuine concern with the issues at hand. I wish to thank them all.

I greatly appreciate the efficient help of Sarah Lloyd and Matthew Smith at Routledge. Their interest in the project and the technical and moral support were indispensable.

I must especially mention the efficient assistance from Ginnie Taylor and her colleagues in the interlibrary loan department of Lakehead University's Patterson Library. This project would hardly have been possible without their help. My thanks also go to Dr. John Sutherland of New Brunswick Power, who provided helpful comments on an early draft.

I must mention a few special people for their long ago contributions. They are Joke Middelberg and her late husband Lex, who encouraged me to study, and Denis Horn, who shared his love for chemistry with me early in my career.

Of course, this venture could not have been completed without the support and encouragement from my family and friends.

All illustrations are my own except Figure 12.1 which was supplied by Prof. James Kayll, Faculty of Forestry, Lakehead University, and Figure 13.1, which is courtesy of Fisons Instruments, Canada, Inc.

CHAPTER 1

Introduction

As soon as I had gotten out of the heavy air of Rome and from the stink of the smoky chimneys thereof, which, being stirred, poured forth whatever pestilentious vapours and soot they had enclosed in them, I felt an alteration of my disposition.

Seneca, AD 61

How it started

In 1798 Thomas Malthus published his famous *Essay on the Principle of Population.* Around the 1930s several biologists started to voice their concerns with population growth, pollution, deforestation, and soil erosion. The names of Julian Huxley, William Vogt, Karl Sax, and Edward A. Ross come to mind with respect to the literature of that time. Malthus, one of the first to raise the issue, was out of phase with the trends in agriculture and technology and he did not find much of an audience. We will encounter Malthus again in a later part of our discussions. Also the more recent voices from the 1930s did not receive much attention from the scientific community.

Then by the late 1950s, a change became noticeable. Garrett Hardin, a biologist we will also meet again later, showed how that change became evident from the number of press clippings recorded on the topic of population growth.[1] It was not just population growth alone that concerned these early writers. They were deeply disturbed by the associated problems of pollution, waste, and the living conditions of humans, plants, and animals.

It was not until the 1960s that the general state of the planet reached the attention of larger numbers of people. The wake-up call came from biologist Rachel Carson when she published *Silent Spring* in 1962. Environmental concerns became common property and a number of buzzwords started to permeate our language.

An "environmental chronology" starts with the ordinances passed by the Plymouth

Colony in 1626, regulating the cutting and sale of timber. It ends with the oil spill in Prince William Sound, Alaska, in 1989. Early entries in this sequence are spaced years apart; by the late 1800s several events were recorded each year.[2] Although this historical record is almost exclusively concerned with events in the USA, the message does take us beyond the borders of that one country. Although the awareness of environmental degradation mainly related to a major industrialized country like the USA, the international entries in the record illustrate the global significance of these events.

The publication of *Silent Spring* signalled the point at which an active environmental movement began to gather speed. The literary masterpiece, written by a biologist, made it clear to industrialized societies around the world that we could not go on doing what we did at that time. The subsequent appearance of Ehrlich's *The Population Bomb* set the scene for the organization of environmental groups like Greenpeace and Friends of the Earth in 1969. Barry Commoner's *The Closing Circle* was another leading influence on the direction of environmental thinking of the late 1960s and 1970s. These books were the beginning of a long literature list for the environmentally interested reader. This literature is an important starting point for examining the environmental language of today.

Literary background

There are several similarities among these three authors. All three were biologists, scientists-turned-authors, who addressed not their fellow scientists, but the public. That was unusual because at the time scientists were generally content to talk to each other as if the public at large did not exist. With respect to their environmental concerns, all three authors were becoming impatient with the conservative scientific community. Their message was urgent and they wrote with the missionary zeal necessary to convince.

All three were practicing scientists who knew where to find the scientific information they needed. They also knew how to use their literary skills to write a compelling story. From their own observations as biologists all three were acutely aware of the environment. The urgency of their message provided them with the justification to step outside their fields of specialization from time to time. Over the years all three have been ruthlessly criticized by some, and valiantly defended by others.

Although Carson was a biologist, the problems on which she focussed contained much chemistry. In 1962 the environmental impact of some synthetic chemicals, particularly the chlorinated pesticides like DDT, had in her opinion reached a crisis point. When she referred to these chemicals as "the synthetic creations of man's inventive mind, brewed in his laboratories, and having no counterparts in nature," she appealed to the emotions of her readers. She knew that chemists don't "brew chemicals" and she must have been aware of the occurrence of many natural chemicals that are at least as toxic as many synthetics. But references to "extenuating evidence" would have weakened the urgency of her message.

Commoner, also a biologist, was the inventor of the term "the technological flaw." He used this expression to describe the fixing of one environmental problem with a "solution" that then became a problem in its own right. He also formulated what he called "the laws of ecology." The third biologist, Paul Ehrlich, focussed more indirectly on environmental problems by addressing, at least in his earlier books, what he saw as the problem of overpopulation. His message was that there are too many people and not enough resources, especially food. He warned us that "If the optimists are correct, today's level of misery will be perpetuated for perhaps two decades into the future. If the pessimists are correct, massive famines will occur soon, possibly in the early 1970s, certainly by the early 1980s."[3] Ehrlich's predictions were wrong, considering that present local famines in Northern Africa are largely due to tribal wars and the misguided "help" of Western countries.[4] However, concerns about a world food shortage are being raised again today.[5]

Beyond the scientific arguments of these writer-scientists, there was the voice of another crusader, heard in the late 1960s. Ralph Nader and his consumer advocacy group entered the literary field in the early 1970s. Nader managed to focus public environmental thinking on the "behaviour of corporate North America," and away from individual responsibility. In doing so he made it quite clear that the environmental debate is, to a large extent, a political debate. He introduced the idea of "the Polluters," the large corporations that are willing to sacrifice the entire planet for the sake of profit.[6] Nader approached the environmental debate from the political angle and paid little attention to environmental science.

Also on that side of the political spectrum we find the views of Rudolf Bahro, a founding member of Germany's Green Party. Bahro is quoted as saying that "people should live in socialist communities of no more than 3,000, consuming what they produce and not trading outside the community."[7] Vance Packard, author of *The Waste Makers*, also addressed the political aspects when he contrasted the "business-Republican-conservative coloration" with the "liberal-Democratic-labour coloration" of the players in this field.[8]

On the other side of the political fence there were initially few counterweights. The best known are probably the books by philosopher Ayn Rand, in defense of capitalism and individuality.[9,10] However, after the Carson, Commoner, and Ehrlich sequence, the late 1970s and early 1980s saw the appearance of a different kind of literature that quickly resulted in a reaction, both from the political right and from the scientific community.

A public concerned with the environmental issues raised by the early biologists now faced the fear spilling out of books like Regenstein's *America the Poisoned*, Grossman's *The Poison Conspiracy*, and Samuel Epstein's *The Politics of Cancer*, and many more with similar titles. Although these books probably had a limited readership, their message spread in many directions. Abstracts appeared in the popular press, and the authors became regular guests on radio and television talk shows. An issue of *Time*, devoted to the Regenstein book, illustrates the general tone of this literature. The author of the cover story, writing about chemicals, uses a collection of words like "concoctions, alchemists, brew, festering, fouling, oozing."[11] The use of this language was presumably intended to capture the public's attention. It also created fear.

Reactions came from economists and scientists. Already in 1977 Claus and Bolander published their *Ecological Sanity*. Julian Simon, an economist, defended the view that the resources of the planet, including energy, are truly unlimited in his book *The Ultimate Resource*. Elizabeth Whelan, an epidemiologist, criticized the fear-raising literature in her book *Toxic Terror* and Edith Efron examined the cancer issue in her book *The Apocalyptics*. The Isaacs examined the political aspect of the environmental debate in *The Coercive Utopians*. More recently, Dixy Lee Ray, a marine biologist, added her voice to the science side of the debate. In the concluding chapter of her book *Trashing the Planet* she described the large majority that lives between these two political extremes. As part of her views of the environmentalist movement she wrote: "This is not to suggest that everyone who supports more responsible policies for cleaner air and water ... is a wild-eyed extremist, ... they are honest, honourable supporters of a good, clean environment and responsible human actions." She then distinguished "the *political* environmentalists" from "the rest of us."[12]

In the meantime, yet another kind of writing became part of the environmental mosaic. To the *observations* of the biologists, the *emotions* of the catastrophists, and the *rebuttals* by the cornucopians was added the *environmental ethic* of the philosophers. It was not just the question of whether the observations and the interpretations were right or wrong, the basics of right and wrong of the human relationship with nature needed to be examined. An early entry in this category was Aldo Leopold's *A Sand County Almanac*, published in 1949, which is now a classic in its field. It is still a cornerstone of modern environmental ethics.[13] Leopold's thinking can best be summarized in his own, often quoted words:

> A thing is right when it tends to preserve the integrity, stability, and beauty of the biotic community. It is wrong when it tends otherwise.[14]

The result of Leopold's writing was the evolution of what became the land ethic. The land ethic teaches respect for nature based on ethical arguments. Quotes from Leopold's writings are found in many essays on environmental ethics. The late 1970s saw the founding of a professional journal *Environmental Ethics*. The collection *People, Penguins, and Plastic Trees*, edited by VanDeVeer and Pierce, introduces some important names in environmental ethics such as Tom Regan, Peter Singer, and others.

The earlier Malthusian issues on population control also found a place in the ethics literature. In 1969 Garrett Hardin, a biologist, edited *Population, Evolution and Birth Control, A Collage of Controversial Ideas*. Hardin's own articles, "The Tragedy of the Commons" and "Living on a Lifeboat," also date from those years. They appeared again later in *Managing the Commons*, a collection of essays on population and resources.[15]

This then is the tapestry of literature that forms a background for today's environmental debate. Clearly, a wide spectrum of opinions and convictions comes to the debating table with a varied company of debaters. If this varied group wants to reach a consensus, there is little doubt that a common language among them is extremely important. A casual perusal of the relevant literature shows that the evolution of this common language has not made much progress.

Facts, factoids and chemophobia

Along the wide political spectrum of environmental interests, the catastrophist activists occupy the left, the cornucopian optimists are on the right. A bewildered public and beleaguered governments huddle in the middle. Many organizations, groups, and establishments have an interest of one kind or another in environmental matters. Post secondary educational institutions offer hundreds of courses dealing with the environment. An entire industrial sector, "the environmental industries," is making a handsome living in many industrialized countries. Yet, much of the environmental language and thinking is still stuck in the 1960s and 1970s, albeit expanded with the terrifying details provided by the popular literature and the media.

These earlier ideas are often inconsistent with more recent developments. Much of the environmental thinking of the 1960s is now hindering our attempts to bring consensus to the debating table. Some progress has been made on the political scene. There are international agreements such as the Montreal Protocol that addressed the potential problems with stratospheric ozone. The warnings of Ehrlich did find an echo in the 1994 international conference on human population in Cairo. However, the perception that pesticides are *always* persistent and that synthetic detergents are *not* biodegradable have a firm hold on the public mind. That corporations, the "polluters," cause the environmental problems is a common perception for many people. Scientists who work for these corporations have little credibility with the public.[16] The fear of radiation and radioactivity inhibits the use of nuclear energy while the environmental impact of coal-fired power plants is ignored. On the cornucopian side the rhetoric also remains much the same: "there is an unlimited supply of resources and technology will fix all our problems."

This book is not concerned with the moral judgements of whether these beliefs and attitudes are right or wrong. Rather, it is concerned with a basic understanding of the ideas and terminology, the language needed for making these judgements. It is necessary to examine these beliefs and attitudes, to understand them, and decide if we need to retain, change, or abandon them.

The early literature is now over thirty years old. One unavoidable effect of this decades-long time lapse is that much of the factual part in the early literature is now obsolete. For example, Commoner, in one of his essays, tabulates percentage increases in the use of many industrial chemicals, several of which have now been phased out.[17] Carson's ideas about the unnatural organochlorines have been overtaken by the discoveries of thousands of these chemicals made naturally by plants and microorganisms. Our present understanding of natural radioactivity must enter into the discussions on the world's energy supply.

We must also be aware of one specific aspect of the environmental language as it is found in the literary background. Rachel Carson knew perfectly well that scientists do not "brew" things in their laboratories, but the association with evil, the image of a "witches' brew" was a tool she needed in order to be heard. Similarly, Commoner's references to "the technological flaw" were an effective way of making a point. At the time there was perhaps a justification for this dramatic language. The message had to get across. However, its continued and indiscriminate usage in subsequent literature has resulted in many environmental myths that have survived the decades.

Norman Ehmann, writing in the *Journal of Environmental Health,* used the term *gnosophobia* as the fear of the unknown to derive the idea of *factoids*.[18] This fear, presumably better labelled as agnosophobia, is fed by "factoids, untruths which are repeated so often that they begin to take on the aura of truth." The perception that "90 percent of all cancer is environmentally caused and is therefore preventable" has created such a fear of chemicals that the term *chemophobia* would equally well describe this fear of the unknown. We can recognize agnosophobia, or chemophobia, with the help of a few common factoids as examples. We will discuss several others in more detail later.

Laboratory experiments show that chemicals called PCBs, polychlorinated biphenyls, can induce cancer in various animals. A very tentative *causal* link between human *occupational* exposure to PCBs and cancer has been suggested.[19] Although there are reports of other health effects and effects on wildlife, there is *no link* between PCBs and cancer in the general population. Nevertheless, our language now contains the general perception that PCBs "are linked to cancer" or are "cancer causing chemicals." The result is complex legislation in many jurisdictions that makes transport of even small amounts of PCBs very difficult and expensive. A well-maintained, properly operating, PCB-filled transformer in the school yard has become a serious liability.

The main reason for developments like the general perception of the PCB situation is a failure to distinguish between occupational exposure to a potentially toxic agent and the exposure of the public. As a result, the quite different issues of *occupational health* and *public health* easily get confused, or worse, they are seen as being the same. This also explains the situation with respect to asbestos.

In 1965 Irwin Selikoff published the results of a study on asbestos-related disease in insulation workers in the New York area.[20] That report was an important step in asbestos-related studies that began in the 1930s. In several subsequent studies Selikoff and his co-workers established that there is a causal link between asbestos exposure and lung cancer in *occupationally exposed* workers. Extrapolation took the asbestos problem from the occupational area to the public health area. This extrapolation did not take into account that we all inhale asbestos fibres, from mineral dust and car brake linings, on every street corner. This is how we have come to believe that asbestos in *any* building is threatening the life of everyone in that building. Extensive legislation has now made asbestos into an expensive burden on the public purse and a liability for many property owners.

Probably one of the best factoid examples is in the food department. It almost fits Carson's description of "chemists brewing" things in their laboratories and it is found on the cover page of the *New Yorker* magazine of November 1, 1993. It is a beautiful, full colour picture of a witch, complete with pointed hat, warts on her nose, and a black cat for company. She is stirring her evil brew with a wicked smile on her face. Foul clouds of vapour curl out of her cauldron. On her table lie the packing materials of the ingredients of her evil concoction: empty cartons of full-fat milk, paper wrap of a pack of butter and the shells of several eggs. Foods that are unattainable luxuries in many parts of the world are seen as a witch's brew, killer chemicals of the affluent society.

Medicine, biology, and chemistry have shown that many of these perceptions are wrong. Yet, in today's everyday language words like radiation, radioactive, and chemical, have become synonymous with bad, toxic, and permanent. A paralyzing fear of science and technology is creating an anti-science, anti-technology atmosphere that seriously hinders the very efforts aimed at dealing with the problems. It imposes a substantial financial burden on the economies of industrialized countries. Agnosophobia, the fear of the unknown, greatly affects our choice of priorities for the debate on the environment.

Where we are going

An important statement in one of Barry Commoner's books emphasized that:

> If power is to be derived from the will of the people, as it should be in our democracy, then the people need to have the new knowledge – about strontium-90, DDT, herbicides, smog, and all the other elements of the environmental crisis – that must be the source of the grave new judgements and sweeping programs this nation must undertake.[21]

With minor modifications the "need to have the new knowledge" is more urgent than ever. While the urgency of DDT and strontium-90 has decreased considerably, other "elements of the environmental crisis" have been added, global climate change for example. This underlines the need for *new* knowledge and cautions us against using the limited knowledge of thirty years ago.

The language of the environment

This book is not an environmental science textbook in disguise. Instead, it deals with the issues of the new knowledge versus the old, and it is written in the belief that these issues are ripe for a productive discussion. Developments leading to this "new knowledge" have added many different ideas and words to the environmental debate. Many of these words have escaped prematurely from the laboratory, others have been "processed" by the media and the popular press. A re-examination of the original meaning of this "environmental language" could help us find consensus in the environmental dialogue.

In our highly technological society specialization has created complex terminologies, different kinds of "professional jargon," that almost take on the status of a separate language. The "language" in workshops where our increasingly complex automobiles are repaired, is unintelligible in the office environment. The "language" of the country's laws is understood only by lawyers. Speaking and understanding these specialized forms of communication is often equated with "literacy." Someone who works with computers must be "computer-literate." Parents concerned with television programming for their children have coined the term "media literate."

Extending the literacy idea to environmental issues leads us to *environmental literacy*. However, the analogy leads to a peculiar paradox. It is obvious that a computer-illiterate person cannot hold a job as a computer sales representative or an instructor. A medically illiterate person could hardly function as a nurse or a physician. A parent who is media-illiterate cannot control media influence on children.

In contrast, a lack of environmental literacy does not affect our ability to earn a living or to hold public office. Concern with environmental issues does not need a diploma or certificate. The only requirement for having environmental interests is that we live on this planet. Some 80 percent of North Americans have expressed themselves as being interested in environmental issues.[22] This means that our collective knowledge of the language of the environment can have important implications for "the democratic process."

The right to be concerned for the environment is a universal right. That concern may be just the self-centred worry for clean air, water and sufficient food. It may also be a more general perspective that includes all living creatures on the planet. Paul Ehrlich, writing about population problems, likes to list the eleven inalienable rights of humanity.[23] The right to eat well, for example, and the right to educate one's children and the right to have grandchildren. Perhaps we could add a "qualifying clause" to these rights. Particularly North Americans are fond of emphasizing their rights without paying too much attention to the *responsibilities* associated with these rights. If we make Ehrlich's eleven inalienable rights subject to an *obligation* to be informed on the subject it could only improve the discussion. In the baker's dozen of chapters in this book we will explore the traps, tricks and tips of the environmental language. Perhaps we can remove some confusion, save ourselves some frustrations and improve our chances to come to a consensus.

CHAPTER 2

The three Es: earth, ecology, and environment

> From a galactic perspective, Earth is not terribly impressive ... Earth is, to be frank, an inconceivably insignificant mote in the Universe.

So wrote Anne and Paul Ehrlich in their book *Earth* in 1987. A galactic perspective is not a privilege of many people, perhaps that is why the Ehrlichs completed the description by adding "... but it's all we have."[1]

What makes earth unique is the presence of life. So far, we don't know of other planets inhabited by living beings. The relationships between the different kinds of living organisms, and between them and their environment, are the subjects of a relatively young branch of science called ecology.

A global picture

A message about these relationships came from the first manned spaceflight to the moon in 1968. Besides being unique, the planet earth is also very small. Spectacular pictures and the buzzwords of spaceflight technology inevitably led to the idea of "Spaceship Earth." Some attribute the analogy to the famous American architect, the late Buckminster Fuller. Others give the credit to economist Kenneth Boulding.[2]

The space travel picture, familiar from the science fiction literature, strongly appealed to a public that had just been shaken by the call-to-arms from Carson, Ehrlich, and Commoner. Paul Ehrlich, with Richard Harriman, used the spaceship analogy for a book on environmental issues.[3] The metaphor reinforced the implication that "we are all in this together." It gives the comforting but misleading impression that "somebody is at the wheel."

In strong contrast with the spaceship is the lifeboat of Garrett Hardin.[4] This biologist-turned-philosopher asks us to view humanity as living in several lifeboats. Rich countries represent lifeboats full of rich people who have enough space and supplies on board. Poor countries represent overcrowded lifeboats with no supplies for the occupants. As some poor people fall overboard, they swim around and hope to get a place in one of the rich lifeboats. What should the people in the rich lifeboats do? Hardin's logical argument then leads to his *lifeboat ethic*: any attempt to help or rescue the people in the poor lifeboats is tantamount to suicide.

The spaceship metaphor was a good tool for manipulating public sentiments regarding the environment. It quickly helped to lead environmental discussions in an emotional direction. For an understanding of earth and the environment and the study of ecology it is not a very productive analogy. The lifeboat ethic, on the other hand, conflicts with the other side of the ethics coin. It flies in the face of religion, compassion, and humanitarianism. Earth, ecology and environment are important words in the vocabulary we use for the environmental language.

Gaia is dead, long live Gaia

In his book *The Dragons of Eden* Carl Sagan presents the history of the universe as a chronological list of important events.[5] According to astrophysics theory, it is about fifteen billion years since a gigantic explosion, the "Big Bang," led to the formation of the universe. Sagan compresses this enormous time span into one of our own calendar years and does the arithmetic to place events on this scale. This way every billion years of universe history corresponds to about 24 days. The entire solar system to which earth belongs came into existence only when the universe had already completed some three-quarters of its existence. That was about in September of this *cosmic year*. Humans did not appear on the planet until December 31. The first cave paintings were made around one minute before midnight, where midnight on that last day represents today. The two thousand years of our Western calendar took place in about four seconds. The human species is very much a newcomer on the planet and in this short time we became what Blaise Pascal called "The Pride and the Refuse of the Universe."[6] For our environmental perspective we need to concentrate on the recent part of this history, the part after the formation of our planet.

James Lovelock, a British scientist and inventor, proposed another view of the planet. In the late 1960s Lovelock was involved in a NASA project concerned with the question of life on other planets. He started by examining what made earth so uniquely suitable for living organisms and his observations led him to view the earth as one gigantic, living organism. In *The Ages of Gaia* Lovelock explained his ideas.[7] He extended Darwin's theory of evolution with the idea that the evolution of the *species* and that of their *physical environment* are interdependent. Living organisms are influenced by the conditions in their surroundings. In turn, they affect their environment, altering it, and changing the conditions. Lovelock's discussion of these environmental feedback loops illustrates this interdependence. Living organisms and their physical environment evolve as *a single system*. The picture of earth as a huge, living, self-regulating organism then unfolds easily as the result of Lovelock's engaging, storytelling style. Studying the functioning of such an organism is not unlike studying the functioning of an animal or plant. Logically, Lovelock coined the term *geophysiology* for the study of the organism he called Gaia.

The Ages of Gaia was a sequel to earlier thoughts on the subject, published in 1979 as *Gaia: A New Look at Life on Earth.*[8] The preface to that first book contained an important caveat: "Occasionally it has been difficult to avoid … talking about Gaia as if she were known to be sentient." In a series of radio interviews, *Religion and the New Science,* broadcast by the Canadian Broadcasting Corporation in 1985, Lovelock was one of the contributors.[9] The interviewer, David Caley, demonstrated how easily Lovelock's caveat was ignored when he said: "Nature as a whole therefore appears as a single living organism, able to coordinate the activities of its constituent processes and in doing so, *it displays what we can only call intelligence*" (emphasis added).

The idea of earth as a sentient, even intelligent, organism was too hard to swallow for much of the scientific community. The new hypothesis, a novel approach to studying global links between living and non-living components, was not taken seriously by many scientists. However, the timing of the book suited the environmental movement. Just when the spaceship picture was wearing a little thin, here was an attractive new metaphor, a reincarnation of Gaia, the Earth Goddess. Environmental writers quickly adopted the new image. One popularized environmental science book, lavishly illustrated with colour graphics and photographs, appeared under the title: *Gaia, An Atlas of Planet Management.*[10] A Dutch public health official wrote *Gaiasophy* and explained that the barriers between the mechanistic approach and the spiritual view must be broken down so that "gaiasophy shows us Mother Earth as our Cosmic Body."[11] *Dharma Gaia, A Harvest of Essays in Buddhism and Ecology,* takes us still further in the religious direction.[12]

Besides the pseudo-religiousness, many scientists, in a superficial interpretation and ignoring the earlier caveat, viewed the Gaia hypothesis as a too optimistic view of environmental problems. The idea of a complex of interacting living organisms actively maintaining optimum living conditions for itself, despite destructive human activity, was seen as counterproductive to environmental science, although that was never the intention of the hypothesis. Nevertheless, much of the scientific community continued to ignore the issue.

Then, in 1988, the American Geophysical Union sponsored a major international conference on the Gaia hypothesis.[13] That meeting still pitched defenders against opponents. Some of those present were vocal in their rejection of the hypothesis, others were moving over, indicating how their own research had been influenced by Lovelock's ideas. The image of a purposeful, willful being moved into the background and a new Gaia took centre stage. Despite continuing opposition, a scientific hypothesis emerged that required study and testing, and offered many challenges for geophysicists, geologists, and climatologists. Already in 1988 studies of a global nature were underway that had been directly or indirectly inspired by the concept of a self-regulating system.

There was another Gaia conference in 1994. As Lovelock himself describes it in the second edition of *The Ages of Gaia,* it has become clear that the attitudes of the scientific community toward Gaia are slowly changing and the earlier gap between the physiology of Gaia and Neo-Darwinism is growing smaller.[14]

In 1859 Charles Darwin published his ideas on evolution by natural selection in his famous *The Origin of Species*. His observations led him to describe the survival of those organisms that were best adapted, because of random genetic variations, to their changing environments. Unknown to Darwin, this vague notion of random genetic variations, a weak point in his proposal, was subsequently supported by the work of the Austrian botanist Gregor Mendel. Mendel's work lay forgotten until 1900, when it was rediscovered, too late for Darwin to know.[15] Later, in 1924, the anthropologist Jennings wrote: "what they (organisms) do or become depends both on what they are made of, *and on the conditions surrounding them*" (emphasis added).[16]

The hypotheses of Darwin, Mendel, and Jennings slowly evolved into scientific principles that have become generally accepted. Adding to these principles the tremendous influences of the oceans, marine organisms, and multiple atmospheric events seems to call for some sort of umbrella theory. The Gaia hypothesis could well lead to such a theory. Gaia's supporters will need many more years and experiments to turn this hypothesis into generally accepted science. The critics must wait, and in the meantime they may contemplate the history of other hypotheses about the planet, like the ideas of Copernicus and Galileo on planetary motion, or, more recently, the theory of plate tectonics.

In traditional biology the available evidence readily demonstrates the concept of the evolution of the species by the principle of natural selection. The appearance of insects that are resistant to a pesticide is a modern example. When a population of insects is treated with an insecticide, nearly all the insects die. Maybe a few individuals with slightly different genes survive. They are the fittest. Maybe they can detoxify the insecticide, or perhaps they just secrete it. If they get a chance to pass this genetic advantage on to their offspring, a new population of pesticide-resistant insects has evolved.

Almost separate from the debate on evolution *per se* is the intriguing question of the origin of life. Many scientists believe that life just happened, that it arose spontaneously on this planet because the conditions happened to be just right. This approach is called *scientific materialism*. Of course, both evolution itself and scientific materialism were challenged. Even before the publication of Darwin's *Origin*, William Paley published his *Natural Theology*, based mainly on the now familiar issue of the designer and the design.[17] More recently, Wilder Smith also questioned the notion that the highly ordered structures of life could arise purely by accident.[18]

Most arguments in this arena deal with the "complex structures" of living organisms, which eventually just pushes the starting point of the argument back to the molecular level. Then the questions of order and design become more appropriate with respect to the genetic code itself. This approach of looking at smaller and smaller detail is an example of *scientific reductionism*, one of the methods of science that is severely criticized in some circles. Reductionism is a tool of science, somewhat like the use of a microscope for looking at small details. It does not "make the original macroscopic phenomenon invisible" as some critics claim.[19] It just moves the macro world aside temporarily so that we can study a detail of it. Suggestions that science and reductionism are more or less the same are wrong.

Several examples show the other face of science, the holistic face. One of these is the Gaia hypothesis, another is the global approach to climate change. A further example, also relevant to the debate on origins is the rapidly developing field of chaos theory.

Mathematical theories of chaos and chaotic behaviour have seen a steady development during the past few decades.[20] Chaos theory appears to herald the end of scientific reductionism. It offers some intriguing insights to the question of how order is an integral part of chaos, albeit a hidden part. Reductionism and holism both have their place in scientific inquiry. Chaos theory neatly combines the two approaches. Incidentally, chaos theory is important for other reasons as well. An understanding of chaotic behaviour is part of the problem of how to control it. It has been suggested that the healthy states of the heart and the brain are chaotic states. Controlling chaotic behaviour is one approach to controlling diseases of these organs.[21]

The late Isaac Asimov, a convinced scientific materialist, provided us with some numbers that are useful to the debate on origins. They are also relevant to the issue of environmental degradation. The numbers deal with the composition of the crust of the planet.[22] This crust consists primarily of oxygen (46.6 percent), silicon (27.7 percent) and aluminum (8.1 percent). A few elements occur in smaller quantities such as iron (5 percent), calcium (3.6 percent), sodium (2.8 percent), potassium (2.6 percent) and magnesium (2.1 percent). Those are percentages by weight. The other eighty-plus *elements* make up the remaining 1.5 percent. This means that most of the elements essential for life, carbon, hydrogen, nitrogen and phosphorus, are mere *trace impurities* in the crust of the planet.

In the theological debate this raises the question of how these sparsely distributed "impurities" came together, by accident, to form the complex molecules of life. In the environmental context these numbers clearly show that the layer of life on the planet is very thin.

The spontaneous appearance of life, the fundamental laws of nature, information theory (Lovelock calls it control theory), and the mathematics of chaos, together form a mixture of which controversy is made. It is an important mixture, because it underlines the many different backgrounds and beliefs brought to the table during the environmental debate.

The gospel of ecology

Ecology is a relatively young science, generally considered the domain of biologists. The *Concise Oxford Dictionary* describes ecology as "a branch of biology dealing with the relations of organisms to one another and to their surroundings."[23] Ernst Haeckel coined the term ecology in 1866. His meaning of the word was almost identical to the modern dictionary definition. Although Haeckel was a biologist, ecology was originally more philosophy than biology. In this philosophy "Nature" played an important role. The appearance of *Silent Spring* in 1962 had two results. It started the revival of ecology as a philosophy, except it was now called a movement, the environmentalist movement. John Maddox traces the beginnings of the environmental movement back further to the 1950s and the fear of nuclear testing.[24] In any case, Carson's book did intensify the scientific investigations of living organisms in their environment.[25] Ecology as a branch of science was moving to centre stage.

However, ecology has not become the exclusive domain of biologists. Ecology and ecosystem have become a part of the environmental language. The words appear in many different contexts and they mean different things to different people. Biology is nevertheless a good starting point for a brief look at ecology.

Different aspects of ecology address questions about how organisms exchange materials, food and waste, and energy with their immediate environment, how organisms and groups of organisms interact with each other, evolutionary aspects, population growth, the nature of communities, and related topics. All these interactions are interdependent, and together they form one large interwoven system, an ecosystem. Since the whole planet with its many species is so large and complex a system, subdivision into smaller areas of study, localized ecosystems, is common. Part of ecology is of a descriptive nature and another part consists of the complex mathematics of population dynamics. Several pioneers of the mathematics of chaos came from the field of mathematical ecology, or geographical ecology as it is also called.[26]

Within the biology context, ecology consists of many parts, indicating a great deal of specialization. Smaller study compartments contain quite specific, limited aspects of the relationships between organisms and their environment. Some biologists deal with ecology as it relates to different kinds of *landscapes*, or landscape types, such as grassland ecology, mountain ecology or estuarine ecology. Others study the way specific *animals* relate to their environment. They specialize in honeybee ecology or earthworm ecology. On a larger scale highlands ecology or California ecology, for example, have to do with geographical areas, highway design, and city planning. Ecology, in all its widely varying forms, has become a well-established scientific discipline among the traditional sciences. A useful general ecology text at the introductory level is Nisbet's *Leaving Eden*.[27]

As part of our relationships with other organisms, humans greatly influence the environment. It is only logical that ecological issues and environmental issues became so closely connected. The terms also started to develop a much wider meaning. For example, several dictionaries define the science of anthropology as a study of humans and their relationship with their surroundings. Only a slight shift in emphasis shows anthropology as another name for *human ecology*.

Another contribution to the environmental language came from modern philosophy in the form of environmental ethics, the rights and wrongs of our relationship with the planet and the other organisms that live on it. VanDeVeer and Pierce describe the distinction between "shallow ecology" and "deep ecology."[28] Shallow ecology is concerned with issues such as pollution, overpopulation, and resource depletion. Because it deals with vital human interests, it is characterized as an *anthropocentric* or human-centred attitude. Deep ecology holds that nature has value in its own right, independent of the interests of humans.

Nash, in his anthology of environmental history, arranged several essays on "The Force of Public Awareness" and "The Debate over Growth" under a heading "The Gospel of Ecology."[29] This is how ecology has become an attitude, a world-view or, as the Germans would say, a *Weltanschauung*, often regarded as *synonymous* with environmentalism.

Ecology also became the territory where activists, politicians, and scientists meet. As politics gets to play a larger and larger role in environmental matters, it becomes increasingly important to understand who means what with those words. It is easy to become confused by apparently scientific terms when they are not used in a scientific context.

The three Es: earth, ecology, environment

For our approach to environmental problems a more precise use of words is preferable. Therefore, we will stay with the original meaning and in our discussions ecology will mean a specific branch of biology and an "ecosystem" a physical space in which the relationships "between organisms and their surroundings" are played out.

Environment, the third E

For centuries humans must have seen the earth around them, their environment, as virtually unlimited. The thought that this unlimited expanse of land and sea was only a small part of a sphere, most of which was invisible, appeared some two thousand years ago. The Greek philosophers, particularly Pythagoras (about 580–497 BC), arrived at the idea of a spherical earth. Much later Marco Polo (1254–1324), Christopher Columbus (1451–1506) and others proved it. Scientific support for a round planet came from the contributions of Copernicus (1473–1543) and Galileo (1564–1642) and their contemporaries. Then images of earth resulting from early spaceflight brought detailed knowledge of the appearance of the planet. In a few decades we became much more familiar with our surroundings, our environment, and what was happening to it.

Initially this information did not include mention of global warming and stratospheric ozone. Nevertheless, the information from space was supplemented by the studies by historians, geographers, and anthropologists. This pool of information enormously improved our understanding of global systems. More recent studies added information about the potential for an enhanced greenhouse effect and the questions about ozone in the upper atmosphere.

It is important to emphasize that the factors that control the environment operate on a global scale. When James Lovelock suggested that the large organism he called Gaia could well look after itself, he did not suggest that pollution could go on unpunished. There was no suggestion that earth had the ability to absorb unlimited insult. Earth can probably survive whatever damage humans can dish out, but there could be drastic changes in the conditions for life. Such changes could be such that the human species becomes extinct. The planet would surely survive and a new, possibly different equilibrium would follow.

A good part of the ideas about the formation of the planet and the life on it are of a speculative nature. Nevertheless, even our present limited knowledge should have a sobering effect on notions about "managing the earth" or "saving the earth." The notion that humans could either manage or save the earth appears not much more than a delusion of grandeur. Frequently, saving the earth means saving our own comfortable environment. The viewpoint is completely anthropocentric.

Human endeavours

Whatever opinions we have on the state of the environment, the human species is clearly responsible for some of the things that are happening to that environment. Discussing these events in terms of "deep ecology" is important, but in terms of practical solutions that approach may not be productive for some time. It is a safe assumption that most of the human species would prefer to discuss current environmental issues in anthropocentric terms. The three Es are the background for this kind of human ecology.

Researchers in many disciplines have presented evidence that human activity can cause serious harm to the planet. Some of that harm is of a localized nature, such as the damage done to ecosystems like the Great Lakes or the Mediterranean basin.[30,31] Other discussions focus on the global effects of human activities.[32] We will examine some specific examples in due course.

The pre-industrial environment

On history's time-scale, most of the human existence took place on a flat surface that seemed to stretch into the distance forever. Scarcities of food and water were mostly dependent on climate, but there was space to move around. There was no end to the trees to be cut for fuel. Enough water was available for drinking and washing and for carrying away waste. There was space to throw away things that were no longer needed. What was thrown away was either natural mineral objects or biodegradable animal or plant parts. Nevertheless, even the hunters and gatherers could harm the environment. Prehistoric hunters set large field fires as part of their hunting strategy, and caused substantial changes in vegetation as early as 11,000 years ago. Early agriculture that led to soil salinization because of irrigation occurred in Iraq over 4000 years ago.[33] However, the population of the world was small. Environmental degradation remained insignificant in a global sense.

The transition from a nomadic lifestyle to agricultural settlements increased the human impact on the environment. A more reliable supply of food and drinking water, provided by agriculture, allowed for a faster population growth. Agriculture required cleared land, and wood was the only available source of energy. Thus, the early effects of population growth on the environment became visible as deforestation. The associated processes of soil erosion and desertification followed unavoidably. These early effects of human activity are described in detail in many formal textbooks of anthropology and archeology. There is historical information on the extensive deforestation around the Mediterranean in the early years of our calendar and, even earlier, of the Nile delta. Forests in North-Western Europe followed quickly. Initially the driving force for the development of the British coal mining industry was the need for coal as a fuel after the deforestation of Britain was almost complete.

Much of this deforestation took place before the large scale development of technology and industry. The romanticism of the wood-burning fireplace has a close link to deforestation. The large scale burning of wood was the first step towards environmental degradation. Even now wood is the only form of energy available to many people in the poorer countries. Incidentally, there is more to fireplaces than romanticism. We will have occasion to return to the wood stove shortly.

Some historians and philosophers seek the reasons for the present environmental problems in the attitudes of earlier generations instead of in the direct effects of growing numbers. According to Hargrove the attitude of the Greek philosophers to the natural world was a major obstacle to the perspective needed for building a conserving society.[34] The historian Lynn White, writing in *Science*, puts the blame for "a mood of indifference to the feelings of natural objects" at the door of Christianity.[35] White was strongly criticized by Moncrief, who also examined the cultural basis of the environmental crisis.[36] This author pointed out the contradictions in White's reasoning. For example, White wrote that "man has been dramatically altering his environment since antiquity." He also mentions prehistoric fire-drive hunting and the human part in the flooding of the Nile basin. These influences are clearly much older than Christianity. Moncrief sees the influences of democracy, technology and urbanization, and increasing individual wealth as the most direct causes of current environmental degradation. Furthermore, neither can we blame the wholesale deforestation of early China, also resulting from the harvesting of fuel wood, on Christianity.

The uncovering of events in the former Soviet Union on the other hand, seems to argue against the influence of increasing personal wealth. What happened in that country looks more like a typical illustration of what Hardin called "the tragedy of the commons."[37] There are many variations of the destructive influence of humans on the environment.

Another view is the picture of earlier civilizations "living in complete harmony with nature."[38] Many of these earlier populations were hunters and gatherers that had not embraced the agricultural mode of living that allows a more rapid population growth. As various anthropologists explain, the nomadic existence of such groups was the *primary constraint* on population growth.[39] If four-legged food tries to run away and traveling with children means carrying them, it is hard to have a big family. A simpler explanation for these people controlling their numbers must be the obvious limitations on the number of mouths that could be fed. The methods used to control these numbers were not particularly benign by today's standards. Infanticide and abortions were common means of population control in many pre-industrialized societies.[40]

The anthropological literature contains the numbers to support this view. The population of North America has been estimated at about ten million at the time the Europeans arrived by the end of the 1400s.[41] That number is hardly large enough to cause an environmental impact of global proportions, or to generate a global vision of the limitations of the land. In contrast, there are also speculations that population densities in agricultural central Mexico had reached critically high levels long before Europeans arrived on the continent. Of course, none of these observations precludes the existence of religious or spiritual beliefs associated with the human relationship with the environment.

In summary, the environmental impact of earlier human population groups was slight because their numbers were limited. As humans learned to manipulate their surroundings, the population began to grow rapidly. These growing numbers were largely ignorant of the effect of their actions on the environment. When we search for solutions to the predicament, we have to look beyond Greek philosophers, Christians, or the mythology of earlier civilizations.

Industrialization

Environmentalism began in the USA primarily in terms of conservation. Legislation for conservation of forests, animals, and ecosystems, and the establishment of parks and conservation areas form the early part of Nash's chronology.[42] Early events of a chemical nature were the Oil Pollution Control Act in 1924, a major air pollution problem in Pennsylvania in 1948, and the London Killer Smog of 1952. The Environmental Defense Fund was created in 1967 and the Santa Barbara oil spill occurred in 1969. The first Earth Day was in 1970. That year also saw the appearance of three US Federal Acts, including the Resource Recovery Act, which was amended into the Resource Conservation and Recovery Act (RCRA) in 1976. The amended Act also included the promotion of domestic waste management. In 1972 the United Nations Conference on the Human Environment in Stockholm attracted worldwide publicity. These were the reactions, the causes started much earlier.

By the mid-1650s, on an island off the coast of Western Europe, an energy crisis was underway. The cutting of wood for fuel was leading to deforestation, and restrictions were placed on cutting the long straight trees that were needed for making the masts for the tall ships. When the British ran out of wood they started to dig underground for fuel. For the few million people on that big island there was a seemingly endless supply of coal under their feet. The problem was that the coal mines, in a country with a wet climate, tended to fill with water. Manually pumping that water out was hard work. Thomas Savery and Thomas Newcomen invented a steam-driven pumping mechanism to take the water out of the mines. The machine worked for almost fifty years, then the better known steam engine of James Watt replaced it in the late 1700s.[43]

People quickly realized that the steam engine could do much more than just pump water. It could drive all sorts of other machinery. All that machinery, including the steam engine itself, was made from iron, and the manufacture of the iron, in turn, consumed part of the coal. The "industrial revolution" was under way. Rapid technological developments increased industrial activity at a tremendous rate. Wealth increased, populations increased, and so did the production of waste and pollution.

Whether this industrial revolution occurred because of Christians "subduing the earth," or whether it was because of the profit motive of a budding capitalism is again a question of historical interest. We must focus on the total environmental impact of this industrialization process. Incidental to the notion of subduing the earth, we note that only some seventy years ago, long after the Industrial Revolution, the anthropologist Clark Wissler wrote: "It would seem then, that the dependence of man upon nature must be taken as one of the normal conditions of human life."[44]

The early view of earth as a limitless sink for the waste of human activity had already ended by the early 1600s, well before technological development really took off. About that time early legislation relating to the environment began to emerge. As was the case in Britain, early legislation in the New World also dealt with keeping tall trees for making ships' masts. There is also early state legislation limiting the hunting of some animal species. By the mid-1800s the idea of nature parks and similar forms of conservation appeared. As population numbers kept growing, the flow of legislation aimed at protecting the environment became a steady stream.[45]

Population, resources, and consumption

The population worries of Paul Ehrlich, and before him of Thomas Malthus, focussed attention on the world's food supply.[46] Malthus did generate some sort of popular consensus, with followers on both sides of the Atlantic, but his message was out of phase with world events.

As Hardin explained, in a static environment the size of a population is controlled much like the temperature of a room is controlled by a thermostat. Upward changes in population are counteracted by increases in mortality because of food shortage. Population decreases are then compensated for by natural exponential growth. Hardin calls this the operation of a *demostat*.[47] The carrying capacity of the habitat determines the point at which the demostat is set. Malthus brought his message at a time when the "set-point" of the demostat was actually moving up because of agricultural development and the Industrial Revolution. As a result, his message, which was not a pleasant one anyway, was relentlessly attacked. It is interesting to note that Malthus and his contemporaries were almost exclusively involved with food. Issues of energy and waste were not part of their concerns.

In his 1968 book *The Population Bomb* Ehrlich wrote, "sometime between 1970 and 1985 the world will undergo vast famines, hundreds of millions are going to starve to death." These predictions, repeated in a 1969 essay in *New Scientist*, initially had a better reception.[48] The book saw nineteen reprintings between May 1968 and May 1970. However, the eventual outcome was not unlike Malthus' message. The prediction was

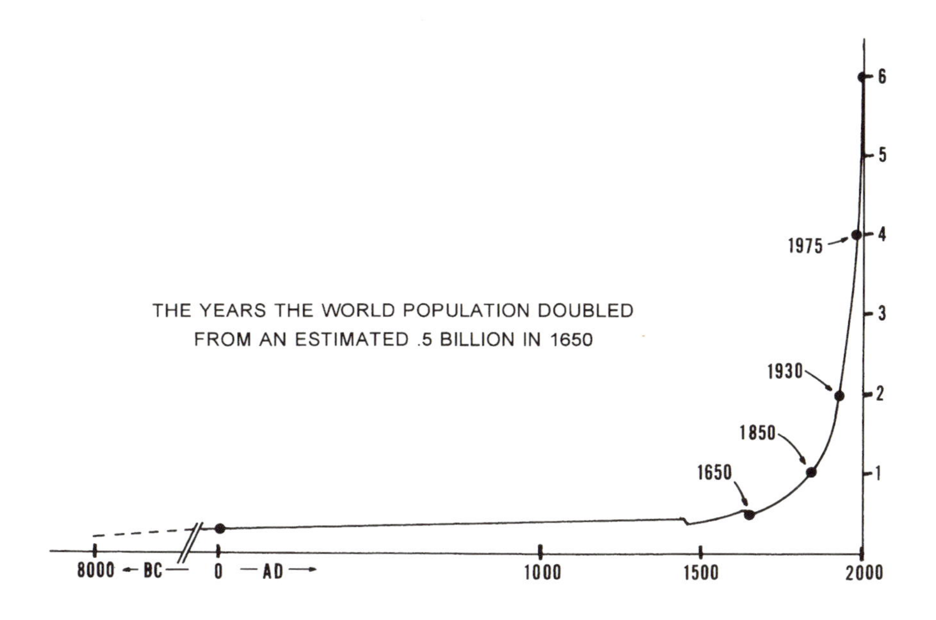

Figure 2.1 Exponential human population growth
Source: Number estimates according to Ehrlich and Ehrlich, *Population, Resources, Environment* (2nd edn), San Francisco CA, W.H. Freeman & Co., 1972

quickly shown to be incorrect, and international worries for a worldwide food shortage did not materialize.

Many books on general ecology, social geography, and history show the familiar graphical picture of how the world population has grown over the centuries. The graph, shown in Figure 2.1, ran as an almost horizontal straight line for many centuries. It reached a modest level of five million around 8000 BC. The appearance of agriculture around that time allowed a more rapid increase and the estimated world population by the year 1000 was about fifty million. Then the graph becomes a curve and begins to rise steadily. Two pronounced dips occur during the fourteenth and seventeenth centuries, when bubonic plague epidemics wiped out an appreciable part of the population of Europe. By the year 1700 the curve resumes its upward trend. By 1850 the world population reached one billion (one thousand million). Considering that we are presently approaching six billion, it may now seem odd that even the modest number of one billion in 1850 was enough to fire up the Malthusian Leagues.

Ehrlich was wrong on the food issue, but in *How to be a Survivor* he began focussing on other aspects of population growth, particularly the perception of resource scarcity.[49] In the early 1960s the production of consumer goods increased rapidly in the industrialized world, and questions began to surface on how long resources like petroleum and coal could last. While questions about resources were being debated, the disposal of waste also became an issue. Space for landfilling became less readily available and incineration raised questions about air pollution. It turned out that the only "resource" that was really getting scarce was the space for storing our wastes. Initially only the proliferation of industrial waste was a problem, but it did not take long before household waste was added to the quandary.

Is the sky the limit?

A tremendous upsurge in industrial activity followed the end of the Second World War. This was industry for the purpose of rebuilding, as opposed to the immediately preceding industrial effort aimed at war and destruction. From an environmental viewpoint the consequences were the same. Both efforts were urgent and left little room for environmental concerns. The research and the thinking about how to "save the earth" were not very pressing priorities then.

Limits to growth

It was probably *The Limits to Growth*, published in 1972 under the auspices of a group called The Club of Rome, that for the first time placed the complexity of the problem before a general audience. *Limits to Growth* was a report on "The Project on the Predicament of Mankind." The project was to examine a whole complex of interconnected problems: poverty, environmental degradation, uncontrolled urban spread, inflation, economic disruptions, and other symptoms of a society in trouble. The resulting message was simple: if we continue on this disastrous road of overconsumption and overpopulation the human race may not survive. The study turned out to be too complex an assignment.[50]

The research team of seventeen was based at the Massachusetts Institute of Technology. The reputation of the MIT name and the seriousness of the conclusions were factors in the wide distribution of the book. Over three million copies were sold worldwide.[51] Despite the marketing success, the scientific community and many economists severely criticized the book. A detailed critique, published in 1973, contained a collection of essays addressing the various aspects of *Limits*.[52] We will return to some aspects of the Meadows' study and its recent sequel in a later part of our discussion.

A second warning, this time by the World Commission on Environment and Development, appeared in 1987 as another report, *Our Common Future*.[53] This book is sometimes referred to as the Brundtland Report, after Gro Harlem Brundtland, former prime minister of Norway, who was the driving force behind that project. The book is best known as the source of the concept of "sustainable development." It suggests that the wealthy industrial nations must play a leading role in the control of over-consumption and overpopulation. On a per capita basis their consumption of resources and their contribution to environmental degradation are vastly larger than is the case in poor nations. Yet, population control *per se* remains of major importance for many countries in Asia, Africa, and South America.

The international population conference in Cairo, in September 1994, resulted in a promising approach to population control. A significant point was driven home by a network of women's organizations. A "new paradigm" for population control is moving in the direction of better reproductive health services and the reproductive rights of women.[54] The political problem of human population control is getting some benefit from applied sociology and biology.

In the past two decades, part of the environmental debate has shifted to include the well-being of all species, animals and plants, that live on the planet. The approaches to both sustainable development and population control did not address this issue. The authors of *Limits* did not mention the predicament of other species, and the Bruntland Report discussed "species and ecosystems" from the viewpoint that "Conservation ... is crucial for development."[55] On the international level the concern with conservation and biodiversity is an anthropocentric issue.

We should point out that the opponents of *Limits to Growth* have a strong tendency to overlook some crucial issues. In arguing the abundance of natural resources on the planet, Julian Simon, author of *The Ultimate Resource*, does not address the problems of domestic waste disposal. Also, his solution to the management of nuclear waste takes no account of the social problem of the public fear of nuclear waste. Simply arguing that this fear is unfounded does not make it go away.

Global consequences

By far the highest consumption of petroleum products is in the industrialized countries in Western Europe, North America, the former Soviet Union, and Japan. Vast quantities are burned as fuel and the rest feeds a petrochemical industry with a vast appetite. This explains why, after the early deforestation problems, most of the current difficulties with the environment appear to be concentrated in those industrialized areas. However, the resulting effect on the atmosphere, whatever that effect may turn out to be, is a global effect. Similarly, refrigeration is mostly a privilege for the wealthy, but problems with the ozone layer, whatever these may be, would have an impact on the whole planet. Therefore, the entire world population is exposed to the consequences of the lifestyles of a few rich countries. In the meantime, manufacturing and other industrial activities have been moving to the Pacific Rim and China. Since environmental degradation is an integral part of industrial activity, this means that the associated problems are also moving into those areas.

When Rachel Carson stated that "this pollution is for the most part irrecoverable, and the chain of evil it initiates ... *is for the most part irreversible*" (emphasis added), she was wrong, *at that time.*[56] As we will see, there are natural mechanisms for the self-purification of rivers and lakes, and natural processes for removing pollution from the earth's atmosphere. These mechanisms operate on a global scale. As the understanding of these global processes continues to improve, it is becoming increasingly doubtful if they are sufficient to handle the problems created by human activity. It may well be that Carson's point of irreversibility is coming closer now.

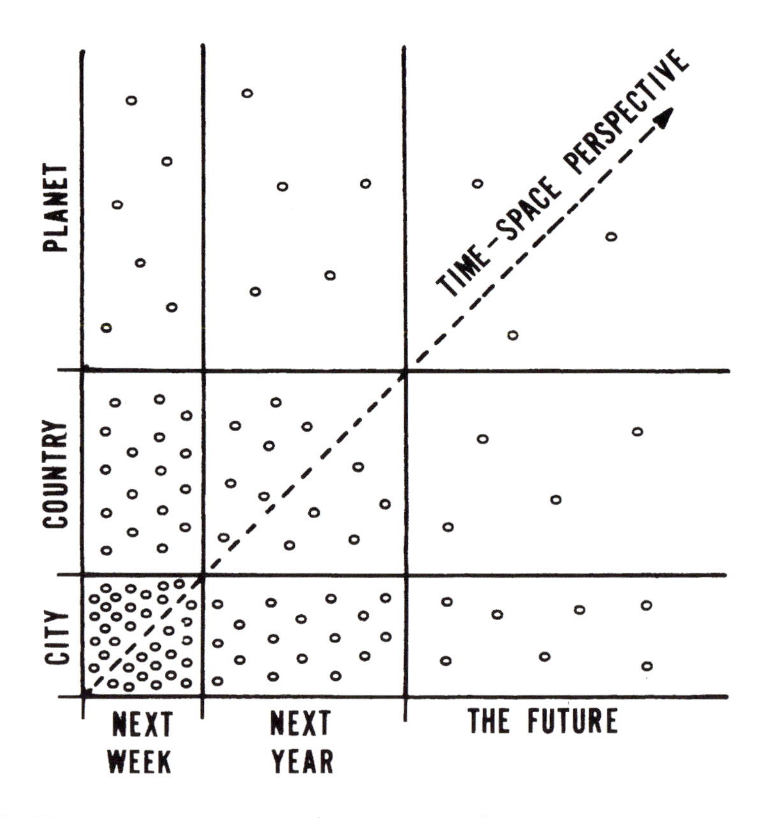

Figure 2.2 Time–space perspective of environmental concerns
Source: Meadows, Meadows, Randers, and Behrens, *The Limits to Growth*, New York, Universe Books, 1972, p. 19

The three Es: earth, ecology, environment

In *The Limits to Growth* the authors presented a simple diagram that places human concerns in a context of time and space.[57] We can readily create an environmental version of that diagram such as depicted in Figure 2.2. Most people think and worry about local problems, like the polluted well on their own property, or the derailed tank car in their city, on a short time-scale, like weeks or months. A few others extend their concerns to problems like their county's food supply or the extinction of an animal species on another continent, over a span of years. Very few have a global perspective that extends into a future well beyond their own lifetime, a perspective that includes problems that affect the whole planet and stretch into the distant future. When we add to this mixture of attitudes the problem of fear, agnosophobia, we can readily understand the battle between the "act-now-regardless-of-cost" proponents and the "more-research-is-needed" crowd.

Sustainable development?

In the enlightened 1990s it is of course unthinkable to deny poorer countries the right to the same level of wealth and comfort that the industrialized countries have enjoyed for so long. This line of thinking led the World Commission to coin the term "sustainable development". This group categorically stated that "Humanity has the ability to make development sustainable."[58] Sustainability is then defined as "to insure that it [the development] meets the needs of the present without compromising the ability of future generations to meet their own needs." The word found tremendous response with all those concerned about the environment. Sustainable development and sustainability rapidly became buzzwords in all parts of the environmental literature.

According to Ontario Hydro International, the People's Republic of China intends to build 200 medium-sized power plants by the year 2000.[59] Because of its sheer size, a development like this will have substantial effects on the environment, particularly if these power plants were to be coal-fired. There could well be similar developments in several other countries in the near future. The question is, is this development indeed possible "without compromising the ability of future generations to meet their own needs"?

The "catching up" process in developing countries is not the only impact on sustainability. In North America and Europe there is a strong emphasis on resources like forests and fish stocks. We have recently seen the sustainability question raised in the context of the fishery industry off the Canadian East coast. There is a danger that this emphasis could obscure major global sustainability issues like the possibility of climate change, drinking water supplies and hunger. We must be careful that the story of sustainability does not begin to sound like Hans Christian Andersen's fairy tale of the emperor's new clothes that turned out to be invisible.

We can now identify a few principal factors that play a role in dealing with the environmental predicament and the problems of human ecology. One factor that can contribute to the solutions is an elementary understanding of the planet and an awareness of the global character of natural processes. A preoccupation with "unspoiled nature areas" and promoting "ecotourism" must not disguise the global effects of industrial activity.

A second part of the problem is population growth and population density, which must be considered in the context of carrying capacity. The graph in Figure 2.1 cannot possibly continue its present course. Per capita consumption must be part of the arithmetic used to describe population. The wealthy countries in the world cannot point to the numbers in poorer countries without adding their own consumption habits to the picture.

A third part of the solutions is the understanding of the many different languages in use for the discussion. We must acquire an elementary understanding of all the languages spoken by biologists, economists, chemists, anthropologists, toxicologists, philosophers, and whoever else is participating in the debate.

CHAPTER 3

Science and the scientists

> Even in science, those who make the first and often sensational claim, get much wider attention and are credited with more credibility than those who come later with calm facts.[1]

Science and technology

Initially, human existence depended on hunting and gathering, and later on the manual labour of farming. When industrialization appeared, society welcomed the new scientific discoveries and technological advancements. The change from the drudgery of manual labour to comfort and affluence was seen as progress, made possible by science and technology. The discovery of blood typing and transfusions earned Karl Landsteiner a Nobel prize in 1930. Synthetic fibres like rayon and nylon, available by the end of the Second World War, were a marvel of technology. Penicillin, one of the first antibiotics, was a triumph of science. Thomas Hobbes' picture of "the life of man, solitary, poor, nasty, brutish, and short," faded as science offered a better life.[2] Then the picture changed again. Rachel Carson published *Silent Spring*.[3] Barry Commoner invented the "technological flaw."[4] Science and technology were in trouble.

Perspectives

Humans survived so successfully because they could manipulate their surroundings. The ability to control fire and use tools dramatically affected the human impact on the environment compared to that of other animals. That impact was very small for centuries because the number of humans was small.

As culture evolved, the need to understand the physical world became important. Early records show that these efforts nearly always included magic, mythology, and religion. Around the beginning of the Western calendar the Greek philosophers departed from this track and approached the problem strictly from the basis of human intellect. They questioned, reasoned, and reached conclusions. They did not experiment.

The idea of manipulating materials in the experimental sense came with the development of alchemy.[5] Although there was a distinctly philosophical side to early alchemy, the principal objectives were materialistic. The "Philosopher's Stone," also called the "Elixir of Life," was supposed to give longevity. Making gold from ordinary metals would lead to wealth. "Capitalism and greed," often mentioned in one breath with science and technology, have a long history.

The attempted *transmutation* of common metals into gold was the major experimental effort of alchemy, which lasted well into the Middle Ages. It never succeeded in making gold, but it did fashion the skills for experimentation. Relevant to today's environmental concerns, it is ironic that one of the most important ingredients in the alchemists' formulations was mercury.

Two early scientists profoundly influenced the early development of science. The first was Roger Bacon (1214–92) undoubtedly one of the greatest scholars of natural philosophy. He insisted that real knowledge could be obtained only by *experiment and observation*.[6] The second was Theophrastus Bombastus von Hohenheim (1494–1541), the son of a German doctor, who, according to the fashion of the day, Latinized his name and called himself Paracelsus. Although still an alchemist, Paracelsus promoted the idea that helping the sick was more important than making gold. He made alchemy respectable. His recognition of the human body as a complex *chemical* system is still a cornerstone of modern toxicology.

Science as we know it started somewhere around 1600, the time of Galileo. At that time scholars did not consider themselves as chemists or physicists or mathematicians. They were simply *scientists* who felt quite comfortable crossing the boundaries between subjects. Several of them were at home in the physical sciences as much as in philosophy and theology. René Descartes was a mathematician and a philosopher. Blaise Pascal was a child prodigy in mathematics and physics and his religious essays are in every library. There are many examples of these multidisciplinary scholars. The details of their lives and accomplishments make for fascinating reading.[7]

These early students of nature made the first discoveries that led to an elementary understanding of physics, chemistry, and biology. The application of these sciences to medical technology brought control of infectious diseases and the development of diagnostic tools like X-rays. Surgery became commonplace and the development of drugs for a thousand diseases moved rapidly. Biochemistry and medical technology began to boost human life expectancy, particularly by reducing infant mortality. In the 1950s the worldwide human life expectancy at birth was about fifty years, in North America and in Western Europe people could expect to live until about age 65.[8] Notwithstanding pollution and industrial wastes, life expectancy continued its upward trend. By the mid-1980s people in industrialized countries lived on average until age 72, and the world as a whole made it to 65. These numbers are still increasing.

Although usually mentioned in one breath, there are not only similarities but also significant differences between science and technology. A historian, writing in *Science*, described a "marriage of science and technology" as a "union of the theoretical and the empirical approaches to our natural environment."[9] This perspective, although it sounds poetic, is quite wrong.

Science, in any specialization, is systematic and formulated knowledge and the pursuit thereof.[10] Natural science is described as "science dealing with material phenomena and based mainly on observation, *experiment*, etc." (emphasis added). To equate science with "the theoretical" clearly shows a lack of understanding. Similarly, technology is not an empirical approach, standing alone, waiting to be "married to science." Technology is the *application* of science. In this context, the scratch-plough of earlier civilizations is not technology but tool-making. This little dispute between the historian and the non-theoretical scientist is a small example of the communication barrier between today's disciplines.

Early advances in science were followed by the development of technology now known as the Industrial Revolution. In this move away from manual labour and toward the machine, technology provided the tools made possible by science. The steam engine technology was made possible by an understanding of the science of the combustion process and the behaviour of gases under pressure.

The popularized environmental literature of the 1960s and 1970s had a dramatic result. Carson's "elixirs of death" and Commoner's "technological flaw" changed the public image of science and technology. Synthetic fibres and space-age building materials became associated with polluting industries. The "Green Revolution" that helped feed the world turned into food "laced with pesticides." The marvels of manufactured consumer goods became "chemicals in the environment." Governments were under pressure and scientists and engineers were on the defensive. A gap between "science and technology" and the public began to grow.

Several years ago A. M. MacKinnon, then president of Ciba-Geigy, was quoted as saying: "As chemistry has advanced, we've left a lot of people behind in the dark – alone with their fear, afraid of what they don't understand."[11] Considering the general perceptions concerning radiation and nuclear power, the same is true for physics. In the same address, MacKinnon also referred to companies having to function in "a business environment polluted by fear."

The increasing gap between science and technology on one side and a concerned and worried public on the other, is of growing concern to the scientific community. Unfortunately, when scientists and engineers address these problems, they have a preference for speaking to the converted. They write in their own professional literature, mostly read by other scientists and engineers. Articles of this kind regularly appear in the news portion of journals like *Nature* and *Science.*

However, it is the public that, through the tax system, funds a substantial portion of scientific research. It is also the public that besides enjoying the benefits of that research, also fears the effects science and technology may have on personal and environmental health. So the legitimate question is, who speaks to the public, or conversely, who speaks for science? Several writers of environmental books have asked "where are the scientists?"[12]

Max Perutz was a refugee from Austria in the 1930s. He won a Nobel prize in 1962 for his work on haemoglobin. In addition to a very active scientific career Perutz also has an ongoing interest in promoting the public understanding of science. His book *Is Science Necessary?* is most important reading in today's climate of hostility and mistrust of science.[13] It provides a much more comprehensive picture than the limited space in the next few paragraphs can provide. For the moment we will concentrate on some of the environmental aspects.

Science and the environment

The term environmental science calls up a picture of scientists in white lab coats or decontamination suits, wearing gas masks and carrying instruments. In the classroom on the other hand, environmental science is synonymous with ecology, the gospel of ecology, that is. Even ecology as a branch of biology is presented against the backdrop of the environmental predicament. Classroom environmental science specifically does not include white coats or decontamination suits. Often it does not recognize either chemistry or economics or physics.

As science developed, researchers faced a rapidly increasing amount of knowledge. Boundary lines formed between chemistry and physics and biology and mathematics. Scientists no longer studied theology or philosophy; philosophers no longer pursued scientific inquiry. A divergence between the sciences and the humanities appeared. Today's scientific community is made up of specialists who concentrate on anthropology, biology, chemistry, or economics or mathematics. One result of this specialization process was that all these practitioners developed their own language. The words of a plant physiologist are different from those of an entomologist and an inorganic chemist does not speak the same language as a biochemist. Even larger are the language differences between physics and political science, or between chemistry and philosophy.

Concerns with the environment had an interesting effect on this trend toward scientific specialization. The environment involves chemicals, so chemists are expected participants in environmental science. Chemicals foreign to a living organism can cause biological damage or disease. That means the involvement of toxicologists, physicians, biologists, and physiologists. Environmental problems nearly always involve numbers of people, animals, or plants. That is why we need mathematicians and statisticians to crunch the numbers and epidemiologists to interpret the results. Engineers and planners are needed for preventive or remedial action. All these specialists are involved with the scientific aspects, the technical side of environmental issues. Together, they practice a truly interdisciplinary craft called environmental science. But there is more.

Questions about social and psychological aspects of industrial development bring sociologists, anthropologists, and psychologists into the picture. Environmental ethics has brought the philosophers on board. Environmental issues need enormous amounts of money, for prevention, research, and for the clean-up afterwards. Economists, insurers, and bankers are vital to the mix.

There is a curious educational problem with this collage of expertise. The "non-science" part of environmental science is usually ignored by the scientific community. This means that traditional scientists often have a poor understanding of the social, ethical, and economic aspects of the environmental problems they are involved with. That is unfortunate, but the other side of the coin looks very similar.

The technical part of environmental science is frequently ignored in the social sciences and humanities. The teaching of environmental law or environmental ethics and the application of economics to environmental problems often proceeds smoothly in the absence of basic scientific principles. This is why lawyers and politicians and philosophers often have a poor understanding of the environmental science side of the environment.

On each side of this field we find some die-hards who believe this is as it should be. There are those in the social sciences and humanities who see science and technology as "guilty." Their solution to environmental problems is based on democracy, politics, and activism. Similarly, in the scientific community we find those that hold to the dogma that understanding the environment needs a degree in science. When these two halves of the team don't understand each other, don't even talk together, solutions for environmental problems come very slowly. Fortunately, in the middle of the field, where the talking takes place, the crowd is getting larger. Maybe solutions to this curious problem are closer than we think.

People in science

If, while travelling along the educational road map, a person's innate curiosity is directed at the natural world, the road can lead to a science study. Examples of the pursuit of science for its own sake, to satisfy personal curiosity, are found throughout the scientific community. Niels Bohr was fascinated by the structure of the atom, and Marie and Pierre Curie devoted their careers to radioactivity. They contributed to our understanding of the atomic structure of matter, radioactivity, electromagnetic radiation and other important questions about the world we live in. Applications, like atom bombs and radiation therapy, were far from their minds.

Others recognized that applied science, scientific study with the specific objective of developing new technology, is a powerful tool in the continuing process of industrialization. Benjamin Thompson's (1753–1814) main ambition in life was to place scientific research at the service of the community. He believed that this would result in more industry and more jobs for his fellow citizens. Similarly, William Perkin (1838–1907), a chemist, achieved his famous place in history because of his spectacular commercial success in the development of the dye industry.

Practicing and non-practicing scientists

Apart from the need for making a living, the basic reason for choosing a science career has not changed appreciably. Curiosity about the physical environment is still a good reason for most science students. Sometimes this is just part of the individual's character, for others the interest was aroused in school. Many students acknowledge these influences of their early school years.

The British physicist Stephen Hawking knew early in life what he wanted. A popularized description of this famous scientist's work gives a glimpse into Hawking's early days: "by the time he [Hawking] was eight, he knew he wanted to be a scientist." It is a miraculous coincidence that Hawking decided to specialize in theoretical physics before he became ill with amyotrophic lateral sclerosis (Lou Gehrig's disease). His incredibly brilliant mind continued to function within an otherwise almost useless body.[14]

Francis Crick, co-discoverer of the genetic code with James Watson, describes a different experience. About to become a scientific civil servant, Crick wanted to switch directions, but wasn't quite sure which way. Talking to friends about parts of science that he was interested in but didn't know much about, made him realize that he was only *gossiping* about these topics. He discovered that "what you are really interested in is what you gossip about." He called this important personal discovery "the gossip test."[15]

Study in the sciences normally starts with a three- or four-year university degree. The doctorate takes another three or four years and a further one or two years of postdoctoral work adds up to about ten years. Similar time frames apply to studies in the humanities and social sciences. Starting at a fairly early age, this ten-year period constantly exposes young people to their mentors and their peers *in the same discipline.* By the time they start a permanent career, their biases and opinions, and their use of language are well established.

There are enormous variations in the nature of the work that scientists do. Some of it is, or at least appears to be, quite esoteric. Some is obviously very much applied to everyday life. Although the scientific importance of the Crick–Watson work on the genetic code is now generally appreciated, this was not always the case. In their early days, the Watsons and Hawkings of this world were immersed in high-level research that had no obvious everyday significance. Beyond this high powered science, there are thousands who apply their craft to many problems for many reasons.

Some scientists still search for solutions to questions that seem esoteric or even frivolous. In considering such work we must remember that a good deal of the things we now take for granted, was once labelled esoteric. At the time Pierre and Marie Curie pursued basic questions about radioactivity, nobody could foresee the later developments of nuclear bombs and nuclear medicine.

Others work on problems in medical science or agricultural technology. And yes, many scientists apply their skills and knowledge to environmental problems. The key word here is "apply." All these examples are about scientists that *do* experiments in chemistry or physics or biology. Others *measure* things in biology or economics. Some scientists also *teach* their skills to others. What they all have in common is that they are actively pursuing scientific activities, continuously honing their practical and theoretical skills. They are *practicing scientists.*

Not every science student becomes a scientist and sometimes trained scientists change direction later in life. In other words, there are many people who have a science background, but who are no longer practicing science. Several of these non-practicing scientists are important players in the environmental arena.

One group contains persons in leadership positions in all aspects of modern society, such as politicians, media personalities, and writers. Several leaders in the environmental movement in this group told their stories in a series of interviews.[16] Some started their career with a science-oriented study. The interviewer of the "green crusaders" tells about his own studies in biology. Others studied science to confirm some preconceived ideas about science and scientists.[17] Most of these people came to pursue their present activities when they became disenchanted with "the scientific establishment." A favourite expression coming out of this kind of background is "how easily science can be co-opted to serve selfish interests." True of course, but meaningless, since it applies to every kind of human endeavour.

Few of these crusaders kept up to date with their original field of study. Mowat still believes that biology consists of "the killing of animals to study their physical mechanisms." However, things have changed since the 1930s. Advances in molecular biology, modern genetics, and modern biochemistry did not just come from "the killing of animals." Of course, developments in highly technological surgery did rely on killing animals. Perhaps the best time to judge the right or wrong of that use of biology is when we are on the way to the operating room for heart surgery or a hip replacement. In the meantime, Mowat's status as a prominent writer and environmentalist lends credence to his views of biology.[18]

Another non-practicing scientist, David Suzuki, tells the interviewer that he had to "walk away from his scientific colleagues." Yet, in the same interview Suzuki states:

> A physicist can describe oxygen and hydrogen in their most intimate details. He knows with absolute certainty that if he takes two atoms of hydrogen and combines these with one atom of oxygen, he will get a molecule of water. But if he (the physicist) is asked: On the basis of what you know about those individual parts, what will be the *property* of water? he won't know.[19]

As critics of reductionism have already indicated, this view is not uncommon among ecologists and philosophers who prefer the holistic approach and the idea that the "indivisible whole" or the "unifiedness" are representations of genuine knowledge. The example of the water molecule shows a flaw in this reasoning. It is precisely from the properties of the elements oxygen and hydrogen that we *can* describe the properties of water. The details of the atoms and their properties explain quite accurately why water is a liquid and why the closely related hydrogen sulphide is a gas. Seeking to predict properties of chemicals, eventually including toxicity, by applying theoretical principles consists of a mathematical approach to finding a *quantitative structure–activity relationship* or QSAR. Research activity in the field of QSARs is recent, exciting and very promising.[20]

The example of the water molecule and the physicist shows how quickly scientific information is overtaken by new discoveries and observations. A decade of development in theoretical chemistry rendered this particular example obsolete. Because of Suzuki's stature as a media personality and environmentalist, his ideas are readily accepted. Clearly, when we listen to the social wisdom of these non-practicing scientists, we must be very cautious with their views of science.

Another group of non-practicing scientists specifically aims at educating the public on environmental issues. They prepare review articles, summaries or reports that contain the original scientific information in a condensed form. Since these literature abstracts are meant for general readers, they are translated into lay language. The quality of that translation depends on the up-to-date knowledge and the integrity of the writer.

Other scientists stop practicing their profession when they become civil servants. They then have to take their orders from the politicians and protect their empires. Their statements and actions have significant effects on environmental policy making. This can lead from the merely annoying to the absurd. For example, Michael Belknap, who was a director of New York's Council of the Environment, wrote in an article in 1971 that "although proof of the cause and effect of pollutants and disease is not established, the evidence is overwhelming." In a similar enlightening observation Julius Richmond, then the US surgeon general, said in 1980: "toxic chemicals are adding to the disease burden in a significant, although not precisely defined way."[21] If proof is not established, what is the overwhelming evidence about, and if something is not precisely defined, how can we know that it is significant?

Who pays the piper?

The pursuit of science is expensive. Several early scientists were either independently wealthy or were protégés of wealthy nobility. Antoine Lavoisier (1743–94), the pioneer of measurement in chemistry, was the son of wealthy parents. Amedeo Avogadro (1776–1856), who was the first to understand the relationship between atoms and molecules, was an Italian Count. This is probably why historian Lynn White generalized that "science was traditionally aristocratic."[22] However, Joseph Priestley (1733–1804), a Presbyterian minister-turned-scientist, was financially supported by the Earl of Shelburne. John Dalton (1766–1844), who uncovered the numerical relationships in chemistry, was the son of a poor weaver and Michael Faraday (1791–1867), the inventor of electrochemistry, was the son of a blacksmith. These not-so-aristocratic scientists did have a hard time financing their studies.

Today science research is funded by private foundations, industry, and the granting agencies of various levels of government. The distinctions are important. In many walks of life, it can be important to know who signs a person's pay cheque. Science, including environmental science, is no exception.

Government granting agencies and private foundations usually fund scientific research at a variety of educational institutions. Such funds have few strings attached except the requirement for *scientific merit*, a concept we will examine shortly in further detail. To establish such scientific merit is important to the career of the researcher. A list of "grants received" is almost as important as the list of scientific reports published when it comes to promotions or an appointment to a senior position. Most academics develop the skills of "grantsmanship" early in their career.

For governments, the reason for scientific or technical studies is often the need to solve a specific problem. The funding is a matter of budgets and budget management. The scientific merit of the project is not always decided by disinterested parties. Because the work is often urgent, results may be released prematurely. Mistakes affect the careers of senior civil servants or politicians. Government scientists may not always be free to express their own opinions.

In industry, a scientist's freedom of expression is often limited by contract. Results of scientific research can be of major significance for the company's competitiveness, may involve patents, or affect the company's financial affairs. If environmental issues are involved, there can be severe restrictions on what a scientist can or will say in public.

For scientists working in a private organization, the political bias of that organization is important. It may strongly influence the direction of research, the nature of a published report. How selectively or exhaustively the literature was searched, and how the scientific information was translated into lay language, all depend on the bias of the author or the employer.

The methods of science and "the scientific method"

Fernand Seguin, who wrote in the now defunct *Science Dimension*, argued that the very reason for the public's perception of science is "the lack of understanding of the scientific method."[23] Many books have been written on this important, somewhat pompous sounding topic. Much of this literature is stored in library departments containing the "Philosophy of Science" or "History of Science" books. However, this mystic scientific method can be briefly and simply stated.

The scientific process starts with making and recording *observations*. These may lead to the formulation of a general *hypothesis*, which must be verified by further *experiments*. Methods used for testing the hypothesis have to satisfy certain criteria, such as reproducibility or the use of control experiments. When a sufficient number of experiments support the original hypothesis, a more general *theory* can follow. Such a theory will be valid until further experimentation suggests its rejection. Sometimes further observations and experimentation may lead to the formulation of *universal natural laws*. In all these situations the critical questions are about the experimental evidence and its agreement with the theory and with earlier observations.

Henry Bauer, a chemist, presented a different viewpoint. In his book *Scientific Literacy and the Myth of the Scientific Method*, he explained in detail how science works without having to resort to "some magical scientific method." Bauer's approach is to look at common work practices of scientists. Rather than using philosophical or historical arguments, following a few simple steps readily explains the way science works.[24]

The methods of science

Bauer begins by making a distinction between frontier science and textbook science. By frontier science he means the large collection of not yet very cohesive discoveries, theories, hypotheses, and preliminary "findings" that are still in a volatile state. To this he applies a "knowledge filter," a continuing process of further experimentation, discussion, and revision. The knowledge filter effectively removes errors, inconclusive results, and other questionable parts. This process gradually leads to textbook science which may then be assumed, reasonably safely, to be correct.

One pillar of science methodology is the experiment. Bauer's terminology relates to two kinds of results obtained from two kinds of experiments. Mixing oxygen and hydrogen in the right proportions gives a mixture that on ignition will burn to form water. A clear description of the results, a written report, follows careful observation. This allows other scientists to repeat that experiment and get the same results. When several repetitions confirm the outcome of the initial experiment, the result becomes a scientific fact, or as Einstein called it, a scientific truth, which then finds its way into the textbooks.[25]

On the other hand, if an experiment cannot be confirmed after several further attempts, the knowledge filter temporarily stops the process. The researchers will be criticized and the results are not acceptable to the scientific community. An example of such a situation was a report of a nuclear fusion reaction that was achieved without the extremely high temperatures normally required for such reactions.[26] Confirmation of these results would have been literally of world shaking importance, but several attempts at other institutions to repeat the results failed. Much publicity and criticism were levelled at this so called "cold fusion." However, the original researchers observed *something* unusual. This work still continues; the knowledge filter is still busy.[27]

The formation of water and the cold fusion experiment are at the opposite ends of a long scale. In the middle are many situations with more or less certainty or ambiguity. Experiments with uncertain or ambiguous results may need considerable interpretation. Such situations are the testing of a new drug or toxicity measurements on a chemical. Large scale studies on human populations, so-called epidemiological studies, frequently give rise to ambiguities. We will discuss these methods later in more detail, but the results are important in terms of Bauer's "frontier science." The uncertainties that are inherent in frontier science mean that scientists can change their minds when new information becomes available. Often a given chemical or drug or food additive that was initially considered relatively non-toxic is later found to have unacceptable side-effects. Conversely, estimated risks or hazards may well be scaled down on the basis of new information

Even the best possible assessment of uncertain results will contain *some* subjectivity. Frontier science often means that the initial report includes a certain amount of the scientist's opinion. It is quite normal to find scientists or groups of scientists that have different opinions about the same experimental observations. The opinions of scientists are very much a part of experimental science. There is an important caveat, however.

Scientists can hold an opinion, about interpreting experimental results for example, on the basis of their scientific expertise, a truly scientific opinion. They can also, as persons who happen to be scientists, hold a personal opinion based on their ideological or religious beliefs. When it comes to political or economic arguments, it can be very difficult to distinguish between these different kinds of opinions. In other words, in the environmental debate it is almost impossible to distinguish between ideology and science. This problem is often at the root of bitter divisiveness within the scientific community.

Stephen Hawking is a typical example of a modern *theoretical scientist* who uses the purely intellectual approach to answer questions that are not suitable for experimentation. The principal tool for his work is mathematics, usually applied with the help of computers. Hawking asks questions about the origin of the universe, and why the universe is here in the first place. This may seem pretty esoteric stuff, but some of the answers do have a bearing on environmental problems, particularly global issues.

Similarly, the theoretical work of Albert Einstein, Erwin Schrödinger and Werner Heisenberg formed a foundation that was essential for the development of modern science. These studies are examples of a scientific approach that initially does not involve experiments. Often, the theoretically obtained results can eventually be tested and confirmed by experiments and observations.

A variation of this theoretical approach has become of increasing significance, particularly with respect to environmental problems. That approach is based on computer modeling. This method can be useful in situations where some limited experimental results are available, but larger scale experiments are not feasible because of size, cost, or time limitations. In computer modeling researchers attempt to describe a particular system by means of a set of mathematical formulae. Making small changes in the variables and recalculating the new picture, gives information about how gradual changes would affect the original system. Typical candidates for the computer modeling approach are the global warming problem, stratospheric ozone depletion, and the kind of "what-if" problems of large scale disaster scenarios. A well-known example was the prediction of the "nuclear winter" by Carl Sagan and his group.[28]

In modeling studies, the knowledge and expertise of the researchers controls the quality of the information the computer can work with. The scientists' assumptions and preliminary data are the *input* for the computer, and the quality of the input decides the quality of the *output*. The importance of the quality of the input follows from a term computer modelers use for this relationship. They call it: "garbage in, garbage out."

The textbook science of Bauer's model is, of course, influenced by the writer of the textbook. Writers cannot avoid a stamp of their personal biases and opinions on their writing. Therefore the fact that a given scientific observation has made it to the textbook stage is, by itself, not a guarantee of scientific truth. A textbook can contain vast amounts of information and, notwithstanding careful checking, errors can and do creep in. Furthermore, the author of the book also had a part in applying the knowledge filter. The writing of science is not as easily subjected to quality control as the practicing of science.

A recent study has shown...

In any case, after experimentation or computation is complete the results must be put in writing: a formal report has to be submitted to the research community. There are good reasons for this necessity. Science is cumulative, that is, each new experiment is built on previous work. The present study or the current project nearly always relates to earlier studies in the same field, and a written report keeps everybody up to date. A report is also important for planning the next move. That further move may be a next phase or a new study, not necessarily by the same people. When scientific results are published, they become public. Anyone is free to check up, follow up, re-investigate, or continue the work.

Publishing the results of a scientific study also has to do with questions of evidence and scientific integrity. Does the study give evidence to support the hypothesis or theory? Does the study confirm earlier observations, strengthening existing evidence? Or does the study present strong evidence to reject earlier viewpoints? Coincidentally, the scientific report may also present evidence in support of further funding, in support of career advancement, or in support of a political decision.

In more general terms, the written report on the experiment or study provides the evidence needed to judge the value of the work that is reported. This evaluation is a complex issue, but it is very important. Bauer presents it as the puzzle analogy that he in turn borrowed from Michael Polanyi:

> Science is like a jigsaw puzzle put together by many workers, each in full view of all the others. Science stays on track not because of any magic scientific method, but just because any piece that does not fit cannot be used.[29]

It is important to see how all these scientists stay "in full view of all the others."

Peer review

The most common method for reporting scientific results follows the same pattern as in the social sciences and the humanities, that is, publication in a professional periodical, a journal. The researcher submits the report to the editor, who sends it to other scientists and asks for their comments. These other scientists, usually unknown to the author, have expertise in the same subject area. The reviewers, the author's peers, will recommend acceptance and publication or they will suggest other courses of action to the editor. Other options can be suggestions for additional experiments, demands for re-examining an interpretation, or an outright rejection of the report.

An alternative to publication in a journal is a presentation to a professional meeting or conference. Peer review also applies here. For most conferences the contributor must submit a written version prior to the conference. Acceptance of the presentation and its subsequent publication in the conference proceedings are subject to the reviewers' approval. The conference presentation has the added feature that there is often a discussion or question period. The audience, the author's peers, can ask questions, show disagreement, or suggest alternative approaches.

Reviewers base their opinion on a few basic criteria: the methodology of experiments, the validity of computations, the interpretation of results, and the relationship with other work in the same area. This evaluation of the *scientific merit* of the study ensures that experiments are well designed and described, can be reproduced by others, that calculations were correct, and that the interpretation is reasonable. Good science also means that other existing studies on the topic have been acknowledged and considered in context. The peer review process is a crucial part of Bauer's knowledge filter. It is how scientists solve the puzzle "in full view of each other."

Sometimes there is the occasional attempt to "make one of the pieces fit" while others are not looking. Or somebody merely makes an accusation of such an attempt. Then the normally peaceful process becomes perturbed. Scientists are just like electricians and clerks and auto mechanics and physicians, there are good ones and bad ones. Notwithstanding peer review, some bad science does find its way into the literature. It is important to understand that in scientific literature, as in other literature, the fact that a piece of information appears in print is not a guarantee that it is correct.

There are several case histories on record to show that *corrective action* is part of the quality control process. For example, in the USA there exists an Office of Scientific Integrity, a part of the National Institutes of Health, that carries out formal investigations into official complaints in the health sciences. One such investigation involved a longstanding argument about low level lead poisoning in children. Two psychologists attacked a published report on this issue by a fellow scientist, psychiatrist Robert Needleman, accusing him of professional misconduct. The dispute eventually made it to an open hearing before this authority.[30,31]

Investigations of this type are usually quite distasteful affairs and are fortunately not common, but they are part of a system that aims at protecting scientific honesty and integrity. Arnold Relman, chief editor of the *New England Journal of Medicine*, presented the useful distinction between deliberate fraud and just sloppy science.[32] The "cold fusion" events at Texas A & M University in 1989 also resulted in a lengthy investigation that in the end showed that there was "no direct evidence of fraud."[33]

While deliberate fraud is rare, sloppy science, particularly at the reporting end, is not uncommon. Poor science can involve insufficient or inadmissible statistics, or the omission of significant "negative" results. In the rush to stake out territory or to obtain research funding, results of the frontier science variety can be released at the wrong time and to the wrong audience. The well-known "study" that showed that oat bran reduced blood cholesterol is a typical example. The results had not been confirmed and were released in a time of general preoccupation with diet fads. Common sense almost immediately made it clear that any fat-free food that replaces fried eggs for breakfast will reduce the amount of fat in the blood, including cholesterol.

The self-policing mechanism of the academic community, the peer review, is not perfect, but it does provide a reasonable level of protection for the "consumer" of scientific information. Therefore it is common to regard reports and opinions that were not subjected to peer review as secondary in importance compared to those in peer reviewed journals and conference proceedings. This is not because of the scientists' arrogance, or because of a contempt for non-scientists. It is simply part of a process designed to yield *good science.*

The information trade

An economic development started sometime in the 1980s that is based on a trade in information. Company A buys a mailing list from company B to use for its own advertising by mail. Company C compiles a directory of suppliers of widgets and sells copies of it. Information has become a commodity that is bought and sold like crude oil and pork bellies. However, there is a significant difference. If wheat contains dirt or vermin, or if pork is of poor quality, it cannot be sold. Quality control decides whether a commodity or manufactured product is acceptable for sale. The quality of traded information in mailing lists, reports, magazines, or books is not so obviously subjected to quality control.

The methods used by scientists are nearly always in direct conflict with the practices followed by the media. In science, hearsay or anecdotal evidence is not evidence. Unconfirmed, casual, and verbal information has no place in scientific research which is a systematic analysis of direct observations. In contrast, the anecdotal stories distilled from interviews and from off-the-cuff comments by scientists are the very lifeblood of the news media. So-called scientific findings reported in newspapers and on television are wrong in an alarmingly large number of cases. Quality control, or rather the lack thereof, is a serious problem in the trade in scientific information.

Quality control and quality assurance

Receiving two copies of the same junk mail, although common, is merely annoying. The mailing list with the error was sold without proper quality control. In science the so-called primary literature, the original reports in the scientific literature, is subject to quality control by peer review. Not a perfect solution perhaps, but perfection is hard to find in an imperfect world. The "offspring," the reviews, summaries, and digests of this literature constitute *second generation* information. Here quality control is a different story. This second generation information is of two kinds, one stays within the scientific community, the other is intended for a much wider audience.

The science literature contains many *review* articles. Such literature reviews are written by practicing scientists who summarize the progress in a given area of scientific research. Scientists prepare these reviews for the benefit of their peers, to help them find their way in the original, or primary literature. Several scientific periodicals are exclusively devoted to these literature reviews. They are of great value and they help us cope with an enormous amount of professional literature. Therefore it is important that a literature review is up to date and complete, so that the reader can rely on the information. Review articles of this kind are nearly always subject to peer review.

Notwithstanding various quality control tools, it is impossible to exclude bias, particularly in reviews of politically controversial topics. An example from each end of the spectrum illustrates this. Economists George Reisman and Clark Wiseman, in separate essays, argue for the abundance of natural resources and the wastefulness of recycling. The first author stated that the world is "solidly packed with natural resources," and that the earth possesses a "practically infinite supply of energy." He sees no problem in the production of waste.[34] Wiseman looked at recycling as an economic waste and concluded that there is nothing wrong with using landfills.[35] Both authors ignore problems with the global drinking water supply, neither mentions the potential of climate change or the serious societal problems with the "infinite supply of energy." Both are clearly biased toward the cornucopian viewpoint.

On the other side we find a similar example in an essay on pesticide use. The authors spent four pages describing "some chemicals found in plant foods, ... are known to cause human illness or even death." Then they describe the potential risk to humans from pesticides in over six pages of almost entirely *occupational exposures*. Nearly three pages of *calculated* risk estimates from pesticide residues in food are also part of their views. These authors showed their anti-pesticide bias by the very careful selection of literature they included in their essay.[36] Evidence of bias is not hard to find at either end of the spectrum.

A different quality control problem is found in adult education. Specialized training courses and workshops have become commonplace in the 1980s and 1990s. Many private companies have been created that specialize in this short-term education in health and safety, environmental technology, money management, human resources management, and similar areas. They produce a profusion of video tapes, instruction manuals, overhead slides, quiz sheets, and course notes. Many of these training manuals and handouts are third or fourth generation material, frequently modified, paraphrased or just copied from a previous course. These materials contain more than their fair share of errors. Errors keep being perpetuated and neither instructor nor trainee is aware there is something amiss. Quality control was lost and factoids are the result.

The literature review format, in a popularized version, is also chosen by a variety of special interest groups. The word "study" exclusively means a literature study in these situations. Few of these popularized literature reviews are peer reviewed. The writers of these reports identify themselves by name but they rarely give details of their credentials. They simply call themselves "researchers." There is little or no quality control. The reader does not know if the author has actually studied the topic or has merely paraphrased what others have written about it. The reader is generally not equipped for "checking up" on the writer, because that requires access to, and understanding of the primary literature. The reader of these reports is therefore wholly dependent on the writer's expertise, bias, and honesty.

State-of-the-art reports on scientific topics are common in government service. These are often also of the "translated" kind, since they are targeted at politicians, civil servants, and the public. The objective of such reports from scientists in government agencies is to lay the groundwork for regulations and legislation. Several authors have harshly criticized some of these reports.[37,38] These criticisms are a legitimate part of the peer review process and form an important counterweight to the regulatory literature.

Another important factor in the quality control aspect of scientific and technical information is the recognition and prevention of obsolescence. The nature of scientific investigation causes a continuous change in the status of knowledge. This particularly applies to Bauer's frontier science. Although elemental carbon is today just as black as it was forty years ago, the knowledge about the behaviour of carbon atoms has increased manifold. The knowledge of forty years ago is no longer sufficient to discuss today's problems. That also applies to the forty-year-old information in the early environmental literature, or a forty-year-old science education that was not kept up to date.

The lack of quality control in the information trade has serious consequences for environmental decision making. Policy decisions, environmental remediation, and personal decisions concerning health or diet based on un-science can be expensive. After the actual cost of the decision, when the remedy does not work, there is the additional cost of corrective action. We will encounter several examples of this particular form of "decide now – pay later" type of policy making. This "un-science," created by poor quality information, carries a hefty price tag.

In addition to the political and economic consequences there is a more fundamental problem with the scientific misinformation. That is Norman Ehmann's agnosophobia, the factoids, and the fear of the unknown. We must now also add the additional personal fear of criminal prosecution that many industrial executives have to live with. Appeasing the electorate, politicians have found a way to protect themselves from political pressure. Many jurisdictions now consider certain environmental events as "crimes" for which company executives and boards of directors are held legally accountable.

Perceptions

The credibility gap between the scientific community and the public is clear from a collection of opinion polls published some years ago.[39] One of these suggests that on a given environmental issue only 7 percent of the public will believe a scientist from a major corporation, and only 21 percent will trust a scientist working for the (US) government. Another poll found that only 18 percent of those surveyed would accept risk estimates from academics. *Zero percent* would accept such estimates from a scientist working in industry. In contrast, these same polls also show that when it comes to believing a scientist, some 66 percent of the public would put their trust in a scientist from an environmental organization.

These numbers are readily explained. The people who answered these polls are constantly exposed to a truly bewildering mixture of different biases. The arrogant disregard of scientists, the radical zeal of environmentalists, the defensive indignation of company executives, and the suave talk of politicians are not a good information database. Add to this the "translations" produced by the media, and attempts to form a rational opinion become a frustrating experience.

Several attempts to explain this communication gap have been made, and solutions have been offered. Ottoboni suggested that the public often sets its expectations of science too high.[40] She based that suggestion on an essay by Alvin Weinberg.[41] Weinberg explained with a few examples how it happens that apparently scientific questions cannot be answered by experimentation. His example of the low level radiation problem illustrates his point and is important for a later part of our discussion.

To double the spontaneous mutation rate in laboratory mice requires quite a high level exposure to radiation. We can measure what that radiation level must be. We can *calculate* what radiation level would increase the mutation rate by *one half of 1 percent*, but only if the response in the mice decreases with decreasing radiation *in a linear fashion*. However, we don't know if the response and the radiation level have a linear relationship. Suppose we insist on *measuring* the small one half percent change, instead of calculating it. Weinberg then shows that, in order to measure the half percent change with a 95 percent confidence level, the experiment would need 8,000,000,000 laboratory mice. Even if we would be satisfied with a mere 60 percent confidence level, we would still need 195 million mice. In other words, the question about this low mutation rate cannot be answered by an experiment.

There are several other examples, such as "the probability of extremely improbable events," to explain how many interactions, between society on one side and science and technology on the other, cannot be answered by either calculation or experiment. Weinberg proposed to call these *trans-science questions*, since they are questions of fact, can be stated in the language of science, but science cannot give the answer. Because of the complexity of the multiple interactions in an ecological system, the study of environmental issues frequently runs into the trans-science barrier. We will encounter many more examples as part of our discussions.

Attempts to make headway with the removal of this credibility gap can also benefit from another tool. Several scientific journals, specifically *Nature, Science,* and *Environmental Science and Technology* are laid out in two sections. The rear section contains the usual, peer reviewed reports on scientific research. The front section of these publications contains extensive editorials, scientific news, and a detailed correspondence in the format of letters to the editor. These letters are a valuable form of debate among scientists on controversial topics. It is not unusual to find these letters to the editors containing extensive references to earlier literature. They are an important part of "frontier science." Particularly the news items and the editorials are usually presented in plain language.

Another useful tool is the textbook on environmental science. Although textbooks are subject to peer review just like research publications, there is more room for errors, omissions, or bias. A full size, thousand page, advanced level textbook represents a gigantic task of editing and proofreading. Using environmental science textbooks also needs a cautious look at the biases of the authors. Chemists usually pay little attention to the ecology part of environmental science, biologists tend to ignore the chemistry.

For example, one such text contains some two hundred and sixty references to the literature. Only *one* of these points the reader to *Environmental Science and Technology,* an eminent publication in the field.[42] Another textbook suggests that "synthetic organics are highly soluble in fat or fatty compounds." Since thousands of synthetic chemicals are freely soluble in water, this is an unacceptable generalization.[43] In contrast, from the environmental science textbooks written by chemists we get the impression that a comprehensive chemistry background is needed to understand environmental issues. Several books sporting the title "Environmental Chemistry" are primarily chemistry textbooks aimed at the chemistry student who is interested in environmental problems.[44] All these books are useful for building some background in the subject of environmental studies, provided we use them with a healthy dose of common sense and critical thinking. Let's close this chapter with an example of not so critical thinking.

There are still a few unspoiled pieces of nature left on the planet. Supposedly, the growing popularity of eco-tourism is offering hopes for saving some of these natural ecosystems. Before we get carried away about the benefits of eco-tourism, it would be interesting to consider how much fuel is burnt or how much carbon dioxide is produced to bring thousands of eco-tourists from central Europe, for example, to the Brazilian rain forest. And we should also look at the other end.

An eco-travel story by a Dutch journalist, written with a slightly cynical slant, is interesting as a story, but the most important part in it is a summary of the numbers of eco-tourists. According to the World Tourist Organization about 50 million people travel to developing countries for an "exotic holiday," and the number grows by 10 to 20 percent annually.[45] Many of the targeted countries are quickly realizing the economic potential and are building hotels and cashing-in any way they can. Improper planning for such tourism activity can lead to serious environmental degradation.[46] It would appear that eco-tourism is just tourism with a little green paint.

CHAPTER 4

Chemistry, chemicals, and the chemists

> The poisoning of America by chemicals did more to galvanize the environmental movement than any other demonstration of man's failure to understand ... the natural order.[1]

This is the opinion of Victor Scheffer when he discussed the development of the environmental movement in the United States of America. There are, according to a rough estimate, some nine million known chemicals on the record.[2] We are adding more at a rate of about a hundred thousand a year. Considering that virtually all these chemicals have the potential to cause adverse health effects, the additional comment that "only an estimated 76,000 are in daily use" hardly sounds reassuring. Does this mean, by implication, that chemistry, the business of making chemicals, is a demonstration of our failure to understand the natural order?

Chemistry

The word chemistry has a special place in the language of the environment. In much of the literature "chemicals" are simply bad. They cause disease, they poison the planet. There is the suggestion of an assumed basic understanding of chemistry and chemicals, but it has a negative connotation. The implication is that "everybody knows that chemicals are the cause of the problem." It would seem logical, therefore, to examine first the words, and then look at what chemistry is about. At least one dictionary of the environmental language does not apply that logic and contains no entry under the word chemistry.[3] This is curious enough to merit looking at the dictionary problem more closely.

About dictionaries

An entry in another book that has the format of a dictionary reads:

> Chemical: a substance made of natural molecules, either found or produced by refining, mixing or compounding, these processes described as chemistry (see artificial).[4]

This description is internally contradictive and it contains fundamental errors. It is contradictive because it contains the improbable suggestion that something "made of natural molecules" is at the same time "artificial." The term "natural molecules" is meaningless, mixing and compounding are the same thing, and chemicals are rarely produced by mixing. They are either *isolated* from a natural source, or they are *synthesized* in a laboratory or in a chemical plant. Refining means purifying. Refining does not produce chemicals, it merely removes impurities from an impure chemical. The main problem with this description is the notion that chemistry only means *making* chemicals. It ignores analysis, or the study of the properties of a chemical, like solubility or toxic effects. The reference to "artificial" has no meaning other than its psychological impact. A chemical cannot be artificial. We often find "artificial" in juxtaposition with "natural" or "organic." The implication is that somehow natural or organic are "good" and artificial is inherently "bad." This approach is also found in another book. This formal "dictionary of environmental science," does not explain the word chemistry *per se*, but contains the following:

> agrochemical: any inorganic, artificial or manufactured chemical.[5]

And, indicative of the real problem there is the entry:

> detergent: an inorganic substance used to remove dirt. Earliest detergents were soaps made from alkaline salts and fatty acids.[6]

There is nothing artificial about agricultural chemicals. Although some agrochemicals are indeed inorganic, a very large number of them are organic chemicals. Detergents are made from petroleum. They are most certainly not inorganic. These are serious errors that raise questions about the language used in the environmental discussions. What does this language mean, how did it develop? We clearly need to examine some of these words and their origins in some more detail.

About chemistry

When the alchemists pursued their search for the Philosopher's Stone, the chemicals they used were exclusively of mineral origin, chemicals we would call inorganic today. Gold, silver, and mercury were the main ingredients for this elusive and miraculous "stone," and the acids, sulphuric and hydrochloric acids, used to dissolve these ingredients are still called "the mineral acids" today.

Even as alchemy changed, during the 1700s and 1800s, into chemistry as we know it today, most experimental work still dealt with the transformations and properties of mineral, inorganic substances. Robert Boyle studied the behaviour of gases using air, oxygen and nitrogen. John Dalton used mainly metals and minerals to discover the chemical rules of ratios and proportions. When Antoine Lavoisier did his work on the importance of accurate measurements with his self-made instruments, the mineral materials available at the time were quite sufficient for his needs.

In terms of chemical make-up, the entire universe is built from about one hundred kinds of building blocks called *elements*. Not "the elements" like in "weather," but chemical substances, the chemical elements. An element is a substance that consists of fundamental particles called *atoms* that are all of the same kind. A chunk of the element iron contains only iron atoms and a piece of the element carbon contains only carbon atoms.

This "handful" of elements does not explain the tremendous variety of chemicals we recognize around us, some nine million of them. There is a lot more to it. The next step is the notion that atoms can combine into larger particles, *molecules*. Molecules can be built up from atoms of the same kind, or they can contain many different kinds of atoms, atoms of different elements. A collection of molecules represents an amount that we can see and weigh, of substances we call chemical *compounds*. All the molecules of the compound water are built from one atom of oxygen and two atoms of hydrogen. All the molecules of the compound cholesterol consist of twenty-seven atoms of carbon, forty-six atoms of hydrogen and one atom of oxygen.

We can use a simple form of shorthand to describe these molecules. We use the first, or the first and another letter of the Latin name of each element to describe it. For example, we use C for carbon, Co for cobalt, and Cd for cadmium. Then we add subscript numbers to indicate how many atoms there are of each kind. The combination leads to the idea of chemical formulae. So, in chemistry shorthand, water becomes H_2O and cholesterol is $C_{27}H_{46}O$. In the case of water this formula is enough to describe the molecule, for cholesterol it is not. For cholesterol we have to add a drawing of the geometry of the molecule, the arrangement of all the atoms in space. Such a drawing that represents the architecture of a molecule is called a *structural formula*. A properly drawn structural formula is a sufficient and complete description of that molecule.

Berzelius (1779–1848), a Swedish chemist, divided chemicals into two large categories. One contained those chemicals that were directly derived from living things, either from plants or animals. He postulated that a "vital force" was essential for the formation of these chemicals from living organisms. Naturally, Berzelius called these chemicals "organic chemicals." All other chemicals, not derived from living organisms, were *therefore* called "inorganic." Berzelius' mysterious vital force was the subject of much debate at the time, until Friedrich Wöhler, a German, settled the argument with a landmark experiment. It is interesting to see how today's "New Age" type of thinking is returning to these ideas of vital forces and "inner energy."

Wöhler started with an inorganic chemical, ammonium cyanate, and, in the laboratory, converted it into urea, a clearly recognizable organic material. Urea is a water soluble chemical that contains a few nitrogen atoms in addition to carbon and hydrogen. It is the form in which animals secrete their excess nitrogen in urine. Because of its high water solubility urea is an excellent fertilizer to supply nitrogen to plants and micro-organisms. Wöhler showed conclusively that there was no need for a "vital force" to make organic materials. Today urea is a common, very much organic agrochemical, that is manufactured on a large scale. This manufactured agrochemical is completely identical to the natural urea in the urine of the cattle in the next field.

The continued study of mineral materials established inorganic chemistry as a separate sub-discipline. The chemical behaviour of the heavy metals, lead, cadmium, mercury, and others is very important for environmental science. Inorganic chemistry also deals with complex minerals such as asbestos. When inorganic chemistry becomes involved with the radioactive elements like uranium and its relatives it is sometimes called *nuclear chemistry* or *radiochemistry*.

After Wöhler an entire branch of chemistry developed that deals with making (synthesizing) and studying and analyzing chemicals that are related to living or formerly living organisms. In this context related means *similar to*, not necessarily *derived from*. These chemicals have a molecular make-up based on the elements carbon, hydrogen, and oxygen, sometimes with a few atoms of other elements such as nitrogen, sulphur, chlorine, and phosphorus. Although the idea of a vital force was shown to be invalid, the term "organic" stuck. Today we use the term "organic chemistry" to describe the chemistry of the many chemicals that are based on carbon. The adjective "organic" no longer means a direct association with living organisms.

The fascinating chemical mini-factories within living cells led to another sub-discipline, *biochemistry*. The multitude of chemical events that occur in living cells is, of course, vitally important to the health of living organisms. Breakdown of food chemicals like carbohydrates and proteins to produce energy involves the biochemistry of nutrition. The interaction between, say, food chemicals in the cell and foreign chemicals that don't really belong there is the *biochemistry of toxic action*. Physiology, toxicology, medicine, and psychology are all built around the foundation of chemistry and biochemistry.

The work of Lavoisier, with a focus on the problems of accurate measurement, developed into the specialty we know today as analytical chemistry. To analyze, according to the Oxford dictionary, means "to examine minutely the constituents of." In terms of chemistry this examining can be done in different ways. To characterize the various components in a complex mixture, food for example, is a form of *qualitative analysis*. This tells us if a specific component is a carbohydrate or if a specific chemical, like cortisone for example, is present or not. *Quantitative analysis* is concerned with how much of a chemical is present in a mixture. The quantitative analysis of a soil sample tells us precisely how much of a given metal, lead for example, is present in that sample.

Analysis can also mean the actual separation of a mixture, such as food, into its individual components. We can then study the behaviour of the individual pure proteins and sugars. Analytical chemistry, in all its variations, is probably the most important branch of chemistry for dealing with environmental problems. Both in environmental chemistry and in food chemistry the "art" of precise analysis is more important than ever. It requires ever more sophisticated and expensive equipment.

Environmental chemistry

Whenever there is a spill with environmental implications the first question is *what* is it. This means qualitative analysis. The next question is *how much* is there. Here the answer depends on quantitative analysis. If we have no information on the spilled material, the first question could take a long time to answer. Literally weeks or months of laboratory work could be required to identify a completely unknown chemical. Fortunately this is rarely the case. The spilled material nearly always has a history; we will have some information on where it comes from or who made it. A qualitative analysis can quickly confirm such prior information. Even for an unregistered or illegal waste dump site some information can often be obtained to help reduce the analysis work.

The question of quantities is usually straightforward. It is a matter of collecting the samples properly, getting the samples to a suitable laboratory, and performing the analyses. Straightforward does not necessarily mean simple or fast. Samples must be properly collected so that they are *representative* of the whole. The laboratory must be equipped for the work. This often means that the lab is *accredited*, and its staff must have the right training and experience. Some analytical protocols only take a few hours, other methods can take days to complete.

The reports that subsequently come back from the analytical laboratory will contain a few buzzwords. Chemicals can be present in a sample of soil or water in very small quantities. Especially in environmentally important samples quantities are often reported in parts per million (ppm) or parts per billion (ppb). One part of salt in a million parts of water would be the same as one milligram of salt in a litre of water. Or, in a more familiar context, one part per million would be a *very* dry Martini made with one drop of vermouth (about 0.05 mL) in almost seventy bottles of gin. For these incredibly small quantities the small error caused by ignoring the density of liquids and solids becomes insignificant. Therefore a parts per million number does not need to be qualified by saying whether it applies to a liquid or a solid, it is a *dimensionless* number. In the gas phase we must be more specific. The concentration of a contaminant in air is often reported in milligrams per cubic meter of air, or mg/m^3, and we must take into account things like temperature and pressure.

The quantitative analysis of a soil sample for, say, a pesticide is only as good as the personnel doing the work and as good as the instrument that was used. While we will look at chemists later, instrumentation needs a brief comment here. Keeping in mind the economic importance of environmental chemistry, the competition for making better, more precise instruments is fierce. New equipment appears on the market frequently, and the new instrument can usually measure smaller quantities of the pesticide than the older one. Therefore the lab report will never say "there is no pesticide in this sample." Instead, there would be a statement like "the amount of pesticide in this sample is below the *detection limit*" or "no pesticide is *detectable* in this sample." In a formal report this is usually indicated as "nd" which means "not detectable."

Any statement about detectability is meaningless unless the minimum amount that can be detected, the *detection limit,* is clearly stated. The instrument is the principal limitation. The formally written analysis protocol, which is legally required and approved in many jurisdictions, usually prescribes how this detection limit must be measured. These technical details are important because there will be almost certainly a new instrument available tomorrow. Therefore the notion of "zero" in environmental chemistry has little meaning. In dealing with environmental reports or policies or regulations we must always keep in mind that what was zero today will no longer be zero tomorrow. We have already moved from measuring micrograms in the 1950s to measuring picograms in the 1980s and 1990s. That is an improvement factor of one million. At the same time, we must keep in mind that there is no relationship between toxic effects and our ability to detect a chemical. Small amounts only matter if they do affect living organisms.

An important feature of the analysis part of environmental chemistry is the problem of dealing with extremely small quantities. Parts per million and parts per billion are often bounced around with a great show of confidence as if everybody understands precisely what these numbers mean. Yet, the technical difficulties in making such measurements are truly formidable. They require very sophisticated and very expensive instrumentation and the laboratory and the glassware have to be impeccably clean. Solvents and other chemicals used during the analysis must be superpure. Any contamination of the laboratory or the equipment casts doubt on the accuracy of the results. The opportunities for error and inaccuracy are many.

Many other kinds of chemistry are important for dealing with environmental issues. There are many questions relating to how chemicals interact with each other in special circumstances. An example is the chemical behaviour of chlorinated molecules when they encounter ozone molecules in the upper stratosphere. Other questions deal with how sunlight affects industrial pollutants, the so-called photodegradation of chemicals. Still other questions deal with the biochemical interaction between living organisms and environmental contaminants, the important issues of toxicity, detoxification, and biodegradation.

At the technical level of carrying out the analyses, measuring concentrations, or synthesizing chemicals, we need highly specialized technical expertise. At the discussion level, where we have to deal with decisions on controlling and regulating and personal lifestyle, an understanding of a few basic principles and familiarity with the buzzwords is quite adequate.

Chemicals

This brief sketch of the history and some of the words of "chemistry speak" are standard ingredients of every elementary chemistry textbook. For those readers who wish to refresh an earlier, perhaps a little rusty, background in chemistry, a specific kind of introductory chemistry text is most useful. This is the textbook style aimed at students in the health sciences. These books usually start with an introduction to general chemistry, followed by discussions of organic and biochemistry with an emphasis on biology and health.[7,8,9] The book by Kruus and Valeriote is now somewhat dated, but it still contains useful information.[10] One important wrinkle in the chemistry language that we still have to deal with is the problem of "inorganic detergents" and "artificial chemicals."

Organic, inorganic, and artificial

The Collins dictionary suggested that a detergent is inorganic.[11] A detergent is a *wetting agent,* it allows the dispersion of an oil in water and vice versa. Detergent action is applied in many situations, from keeping butterfat dispersed in the milk to keeping vegetable oil dispersed in our watery stomach content. The classical examples are materials known as soap. Soaps are not "early detergents" they are just detergents. Soaps are not, and were not, made "from alkali salts and fatty acids" either. Soaps *are* the alkali salts of fatty acids. Soaps are detergents derived from natural materials, like fats and oils. When automatic washers made their appearance several decades ago, natural soaps did not work properly and synthetic detergents, derived from petroleum, became necessary. Synthetic detergents are related to hydrocarbons, and they are as organic as any chemical can be.

The idea that certain chemicals are "artificial" usually produces a smile, patronizing or otherwise, on the face of a chemist. The expression may have its origin in the food industry. We will use candy bars as our example. An early "discovery" of European settlers in the Americas was the flavouring agent vanilla. The name was given to an *extract,* a decoction, made from the dried beans of *Vanilla planifolia,* a plant of the orchid family. The characteristic flavour soon found its way into the food industry. Vanilla, the extract, is expensive. It is coloured brown because it contains a large number of other, coloured chemicals also extracted from the vanilla beans. It was soon found that the active ingredient, the chemical responsible for the flavour, has a relatively simple molecular architecture. They called that active ingredient *vanillin.* It is a pure white crystalline chemical that can easily be synthesized in the laboratory.

Synthetic vanillin is now made on a large scale from certain byproducts of the pulp and paper industry. A molecule of vanillin made in a laboratory is *indistinguishable* from a molecule of vanillin made by the vanilla plant. We will shortly return to this issue of synthetic and natural molecules in more detail. A candy bar containing an extract of vanilla beans contains *natural flavouring.* By contrast, a candy bar containing synthetic vanillin is *said* to contain artificial flavouring. It is instructive to look at a comparable situation.

Sugar and a chemical called malic acid are both crystalline materials. A solution of a mixture of these two tastes like apple juice. The taste can be considerably improved by adding another chemical, a close relative of isoamyl alcohol, also called "apple oil." This flavouring agent gives the concoction a *taste* like that of apple juice, but it is not the real taste of apple juice. The mixture contains an *artificial flavour.* But there is nothing artificial about the apple oil that was added; it is a very real chemical.

Back to the synthetic vanillin, which is a pure chemical that is not "contaminated" with coloured plant materials. For the candy bar, the flavour is caused by the chemical called vanillin. Whether that vanillin came from a vanilla bean or not is immaterial. If the vanillin came out of a laboratory, the candy bar contained a *synthetic* flavour. The erroneous label "artificial" obscured the distinction between the chemical and its origins. Ironically, as consumers our belief in the "good of nature" is so strong that pure, synthetic vanillin, which is colourless in solution, is coloured brown with caramel (burnt sugar) so as to imitate the impure, brown extract from vanilla beans.

We find another piece of the semantic puzzle in the *Green Dictionary,* cited earlier.

This book explains that the word organic is "most commonly used in the green movement with respect to farming and food." However, the word organic has been used in chemistry since Berzelius and the notion of organic gardening is only a few decades old. Of course, languages evolve and organic gardening has become part of our vocabulary. However, we must remember the context. It is wrong to argue that since organic gardening is natural and not "artificial," organic is therefore synonymous with natural. That reasoning goes somewhat like: organic is natural and the opposite of natural is "artificial." Therefore, since another "opposite" of organic is inorganic, then inorganic and artificial must mean the same thing.[12]

The confusion is further compounded by mixing the language from different scientific disciplines. Monkhouse's *A Dictionary of the Natural Environment* is clearly slanted toward geography and defines "organic" as "living organisms and their remains."[13] This reference book therefore makes the distinction between *organic weathering* (of rocks) as "the breakdown of rocks by plants" and *chemical weathering* or corrosion, which is "the breakdown of rocks by chemical agents such as water and oxygen." These descriptions are suggestive of a difference between "organic" and "chemical." The authors of the Collins dictionary just took one step too far by using this mixture of geographical and environmental terms for explaining some chemical concepts. The meanings of terms like organic, chemical, natural, and inorganic cannot be freely interchanged between geography, chemistry, and environmental science.

This inappropriate extension of "artificial" and the frequent use of "organic" have become part of the environmental language. While organic gardening may be considered as a legitimate development in our language, the "artificial agrochemical" and the "inorganic detergent" have no place in it. It follows that the environmentally aware reader should be extra cautious when reading environmental chemistry written by geographers. And also, of course, when reading environmental geography written by chemists.

Families, derivatives, and other relationships

Of the one hundred-odd substances we called elements, carbon is unusual because it can attach itself to its own kind to form molecules containing many carbon atoms. Carbon atoms can *form bonds* with other carbon atoms. Several other elements can do this, but what makes carbon atoms special is that they can form bonds with their own kind in a virtually unlimited number and in almost unlimited patterns. This is why there are so many different chemicals based on carbon, organic chemicals. On our list of nine million, by far the largest part consists of organic chemicals.

Within the molecular make-up, the usual companions of carbon are the atoms of hydrogen and oxygen. Sometimes atoms of nitrogen, phosphorus, sulphur, and a few other elements are also present. The result is the absolutely staggering number, literally in the millions, of different possible combinations. Not all these combinations have been found, made or discovered, but a very large number have. To keep track of such a bewildering variety, in terms of their names, behaviour, and toxic properties, appears a truly daunting task. Fortunately, there is a remarkable order in this apparent chaos.

We can organize chemistry by means of groups, some smaller and some very large, of chemicals that are *closely related* in terms of their molecular architecture. This means we can subdivide organic chemicals into a much smaller number of groups or *families*. Within each family we find close relatives and sometimes not-so-close relatives. Some families are very large, others are smaller.

Examples of these families are readily found in the language used for the environment, health, and nutrition. There are very familiar family names like hydrocarbons, alcohols, and organic acids (as distinct from mineral acids). Some families, such as the PCBs, dioxins, and chlorinated hydrocarbons have achieved much notoriety in the environmental context. Families like carbohydrates and proteins are important in nutrition. In biological chemistry, families of chemicals may have little relationship in terms of their chemical architecture. We use other criteria to group together families like vitamins and hormones and neurotransmitters. Their family relationships have more to do with their biological or physiological functions than with their chemical architecture.

Instead of talking about relatives, we can also use the term *derivatives*. Chemical A could be *derived* from chemical B, either just by being very similar in molecular makeup, or by actually making (deriving) A from B in a laboratory. The idea of relatives and relationships is very useful as a descriptive tool for understanding traits and properties. Both the hydrocarbons and the *related family* of their chlorinated cousins do not easily dissolve in water, but both readily dissolve in oil or in fat. This similar behaviour of both families plays a role in their environmental behaviour.

The analogy of a family relationship is very useful but it is subject to some limitations. First there is the matter of sub-families. The dioxins, for example, form a small, special group within the family of chlorinated hydrocarbons, also called the organochlorines. That does not automatically mean they behave the *same* as all other organochlorines, but they do have some things in common. A second limitation of the family analogy is the behaviour of the members within a family with respect to living organisms. Members of the same chemical family can behave in a totally different way when it comes to their toxic properties. Within the family of alcohols we find chemicals like glycerol and ethylene glycol which are very closely related chemically speaking. Glycerol is essential in every living cell, ethylene glycol, better known as antifreeze, is literally a deadly toxic chemical. We will come back later to this question of chemical relationship and toxic behaviour.

Synthesis and nature, friends or enemies?

Medical science, physiology, and biochemistry give abundant evidence for the notion that many chemicals are essential for maintaining good health. If these essential components are missing from our food, the result is malnutrition. Among many such essential ingredients the common vitamins are probably the most familiar. There are two aspects of the family of the vitamins that are important in the environmental context. These aspects relate to their chemistry and their toxicity. We will deal with the toxicity question later. A brief comment about their chemistry deals with synthesis.

These vitamins can often be isolated, extracted in pure form, from food. Thus vitamin C, also called ascorbic acid, can be obtained as a pure, white crystalline substance from citrus fruits, tomatoes, and other plants. These white crystals are "a substance characterized by a definite molecular composition," in other words, a pure chemical. Analysis tells us the exact proportion and arrangement of all the carbon, oxygen, and hydrogen atoms within a molecule of vitamin C. We can represent all these various atoms by plastic balls and build a *molecular model* of vitamin C (see Figure 4.1). Suppose we go to the laboratory and extract some vitamin C crystals from a few tomatoes. We carefully purify the crystals and place them in a container and label the container "sample A, natural source."

In the same laboratory we find a chemist who could put together carbon and hydrogen and oxygen atoms in exactly the same *proportions* and in exactly the same *arrangement* that we know applies to the vitamin C molecule. Some *building blocks* used by that chemist may be chemicals obtained from petroleum, or from coal, or from other plants or animals. This is a time-consuming process, but in the end we will get some carefully purified, white crystals. These crystals go into another container and this time the label will read: "sample B, synthetic source."

We can now use different methods, ranging from laboratory instrumentation to biological tests, to prove, beyond any doubt, that samples A and B are completely identical. The multiple functions of vitamin C in the human body *will* be performed with absolutely equal speed and efficiency by the vitamin C from either sample. Furthermore, if the samples were only identified as sample A and sample B, an analytical chemist would not be able to tell which sample is natural and which one is synthetic, except....?

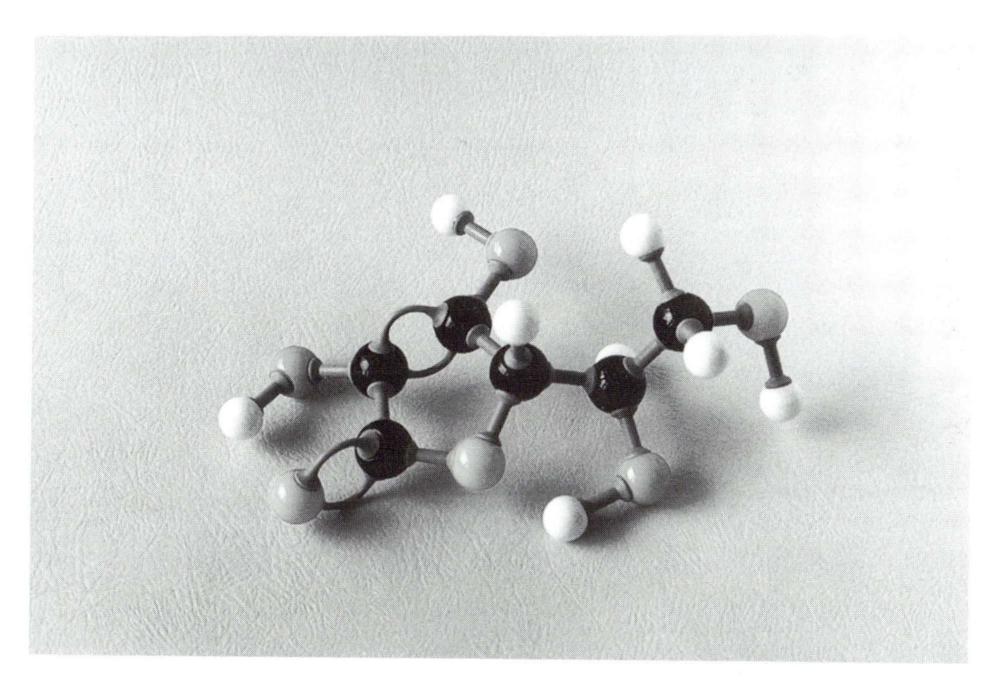

Figure 4.1 A molecular model of vitamin C – natural or synthetic, who can tell?

The synthesis involved the manipulation of starting materials, simpler molecules, to which other molecular parts were added. There were transfers between glass containers involving various solvents. Since vitamin C is an acid, also called ascorbic acid, the synthesis may involve an interaction with alkali. Sodium ascorbate, the sodium salt of vitamin C, can be used as food supplement in the same way as the acid itself. If the final synthesis product was not very carefully purified, some of the chemicals used in the process could have remained in the sample as *impurities.*

However, the removal of vitamin C from plant material also involved other chemicals such as solvents and acids and alkalis, and there would have been several transfers between containers. The tomatoes themselves contained literally hundreds of chemicals made by the tomato plant. For example, tomatidine is a chemical toxic enough to be used as a fungicide.[14] What if the vitamin C from the natural source is not properly purified? Incidentally, complex chemicals obtained from plants are frequently named after the source from which they are obtained. The alkaloid solanine was named after the plant family Solanaceae (e.g., potatoes and tomatoes). Tomatidine was first found in Rutger's tomato plant, a specific cultivar of *Lycopersicon esculentum.*

Yes, some sophisticated equipment would help our chemist to show the difference between sample A and sample B, by recognizing, in a chemical sense, the *different impurities* in the two samples. In the same way, the only difference between the natural vanilla extract and synthetic vanilla lies in the *impurities* left behind from the extraction process in one and from the synthesis in the other case. Particularly in the trade in vitamin pills, the commercial emphasis on the "natural source" product is nothing more than a play on the ignorance of an unsuspecting public. At the same time, what good is a rule if there are no exceptions?

Many chemicals that occur naturally in plants and animals have a molecular architecture *in three dimensions,* they have a definite shape in space. This three dimensional geometry works a little like a pair of hands with a pair of gloves, the right glove does not fit the left hand. In this sense some molecules have a *handedness,* and the "left-hand-molecule" looks like the mirror image of the "right-hand-molecule." Most operations in the laboratory can't "see" the handedness of a molecule, but the living cell can.

For some molecules, vitamin E for example, the synthesis gives us, inevitably, both forms in the same process. Such a mixture of the two forms is called a dl-mixture. Synthetic vitamin E is a fifty-fifty mixture of the "left-hand" form and the "right-hand" form. The body can use one of these forms more effectively than the other, and health food aficionados argue that the natural source vitamin E is better. Of course, the same "better" can be achieved by using a little more of the synthetic dl-mixture. In any case, the international standard, the IU or *international unit* for vitamin E is measured in terms of the synthetic dl-mixture.

What does all this have to with the environment? Well, in the first place food is an extremely important part of the environment in the widest sense. Second, these examples from the food sector dispel the common perception that synthetic chemicals are "alien to nature" and therefore inherently bad. Of course, all this does not exclude the possibility that synthesis can also be used to make chemicals that do not occur at all in nature. We won't open up that can of worms just yet.

Chemical factoids

We have described factoids as untruths that are repeated so often that they begin to be accepted as truth. While the world of advertizing is full of factoids, the environmental debate could well take second place. Let us look at how some factoids are born. Our examination begins with finding a better definition of the word chemistry. A formal dictionary of science describes chemistry as:

> The scientific study of the properties, composition and structure of matter, the *changes* in structure and composition of matter and the accompanying *energy* changes.

And a chemical:

> A substance characterized by a definite molecular composition.[15]

As an *adjective* the word chemical occurs in some eighty entries in this dictionary.

Molecular composition means the exact numbers of carbon atoms, hydrogen atoms, oxygen atoms, etc. which together make up a molecule. If in a substance all the molecules have the same definite molecular composition that substance is a pure chemical or a pure *compound*. The meaning of the word compound in chemistry is different from, for example, in the construction industry. In construction a caulking compound is a *mixture* of any number of chemicals. Although the word compound is the standard expression in chemical literature, we will stick with the term *chemical* to avoid this confusion.

This difference between a pure chemical and a mixture of chemicals is important. While in a laboratory we frequently work with individual chemicals, in everyday life we very often deal with mixtures of chemicals. Paints, lubricating oils, prescription medicines, and food are all mixtures of chemicals. Plants and animals are made of chemicals, water is a chemical, and the air we breathe is a mixture of chemicals. Since carbohydrates, fats, proteins, vitamins, and mineral salts are all substances characterized by definite molecular compositions, they are chemicals. Food *consists* of a mixture of chemicals, no matter how it was grown or processed. Therefore the frequently stated assertion that we should be able to eat "food without chemicals" is meaningless. A meal without chemicals is an empty plate. Of course, "food without chemicals" is an *expression* often used for food without additives or contaminants. Nevertheless, the distinction is important and, as we will see, food without contaminants does not exist.

No counterparts in nature?

It is not difficult to find examples of synthetic chemicals that can have negative effects on the environment. Chemicals like DDT and PCBs readily come to mind. The potential effects of some of these chemicals on health and environment must have provoked Rachel Carson to her sweeping statement about "the synthetic creations of man's inventive mind, brewed in his laboratories, and having no counterparts in nature."[16]

That statement may have had the dramatic appeal needed to make a point convincingly, but it was wrong.

When DDT had been banned in the early 1970s replacements had to be found. Ironically, the thiophosphate pesticides used as replacements are much more toxic and dangerous than DDT ever was. Yet, this group of pesticides, with names like malathion and parathion, readily illustrates the error of the "no counterparts in nature" idea. Within living organisms the thiophosphates interfere with an enzyme, cholinesterase, which is critical for the proper transmission of nerve impulses. They are called *cholinesterase inhibitors* or nerve poisons. In agriculture, rapid biodegradation partly compensates for their very high toxicity.

These nerve poisons indeed have their "counterparts in nature." One natural cholinesterase inhibitor occurs in the West African Calabar bean, *Physostigma venenosum*, which was also introduced into India and Brazil. This makes the Calabar bean toxic. In some societies, medicine men used it as the "ordeal bean." The pure chemical responsible for the toxic action is called *physostigmine* (by chemists) or *eserine* (by physicians). It is a valuable chemical that is useful in the treatment of glaucoma. Physostigmine was described in the literature as early as 1864 and it was synthesized for the first time in 1935, long before *Silent Spring*. Incidentally, as with vitamin C and vanillin, synthetic eserine is undistinguishable from natural eserine for medical purposes. Or to put it differently, "organically" grown Calabar beans will kill you just as dead as synthetic physostigmine.

We can readily identify the toxic, natural counterparts of many more synthetic chemicals. Particularly in the medical arena many successful drugs were developed using *natural chemicals as the model* for the synthesis. Chloroquine is a synthetic antimalarial, an organochlorine incidentally, that was modeled on the naturally occurring quinine. Many plants make a vast array of chemicals that are quite toxic to insects, to defend themselves against being eaten. When we get to the issue of food we will see that by far the largest part of pesticides in food are nature's own pesticides.

There is one general example of the natural counterpart that must be mentioned in this particular context. There is a commonly held opinion that most human cancer is caused by industrial pollution, by synthetic chemicals, or by nuclear energy. We will explore the relationships between chemicals, radiation, and cancer in much more detail shortly. At this point it is relevant to point again to the counterpart in nature. Naturally grown food plants contain a larger variety of carcinogenic chemicals than have ever been synthesized by humans. That does not mean that food causes cancer. A complete list of the many chemical factoids in circulation would be beyond the scope of our discussion; it would serve no useful purpose in any case.

Finally, although it does not relate to the "natural counterpart" argument, the notion of chemical factoids would not be complete without the story of furniture polish.

Instead of using furniture polish "containing toxic solvent" we are often advised to use a "mixture of vegetable oil and lemon juice" instead. None of these recipes explains how we are supposed to mix the oil and the juice. Lemon juice is mainly water, and water and oil don't mix. What is more relevant, vegetable oils are "unsaturated fats," and are good for our health. But furniture? On exposure to air these oils slowly oxidize and polymerize. Some oils, like linseed oil, polymerize so rapidly in air that they "dry" quickly. They were once used for making paint, real oil paint, and floor covering. The name linoleum is derived from linoleic acid, the main component of drying oils like linseed oil. However, some not-so-unsaturated oils, like most cooking oils, dry very slowly and form a very sticky coating when applied to a solid surface. Anyone who has spent time cleaning the fan of a kitchen hood, will think twice about applying such a concoction to furniture. Nevertheless, this factoid is frequently repeated from one book to the other and from one community workshop to the next.

Nine million chemicals and still counting...

The mention of nine million chemicals is a perfect emotional argument for the political part of environmentalism. What does it mean? The American Chemical Society has published a periodical called *Chemical Abstracts* since 1907. This monthly listing contains abstracts from all reports published in the chemical literature the world over, including biological chemistry, veterinary chemistry, and medical reports relating to chemistry. The Chemical Abstract Service skips very few scientific journals. This whole gigantic system is computerized and can be searched on-line or by CD-ROM. Although nine million is an estimate, it is a reasonable estimate.

From the environmental viewpoint, how can there be any hope to create a complete database on the behaviour, particularly the toxicity behaviour, of so many chemicals? Is this not an argument to put shackles on an industry out of control, a clear case against chemists and the chemical industry? Let's put these numbers in some perspective. What is the real significance of having nine million chemicals on the books?

Beyond molecular architecture, we can also classify chemicals by their origin or source. One large group of chemicals is found in plants, for example, another in animals, a third in soil. In each group there are many known chemicals that are part of the nine million inventory. Chemicals responsible for the brilliant colours of plants, flowers, and foliage are called acetogenins. As early as 1965 the literature knew a thousand chemicals in this group.[17] Nitrogen containing chemicals, the alkaloids, are made by many plants. Normally we pay attention only to a few familiar names, like morphine and nicotine. Plant chemistry literature contains thousands of alkaloids.

Similarly the animal kingdom contains large numbers of well-identified chemicals. Some twenty amino acids that occur in every living organism can combine and form thousands of chemicals called proteins. Many of them, enzymes for example, are known by name and are part of the nine million. Other chemicals, the hormones, work as messengers to start chemical processes in specific organs. Micro-organisms are particularly rich in complex chemicals. A twenty-year-old survey of only the more important chemicals made by fungi contains a thousand entries.[18]

Besides chemicals characterized in pure form, many products obtained from plants are complicated mixtures of chemicals. A wine made from grapes is mainly water and about 10 percent alcohol. Less than 1 percent of the wine is a mixture of many chemicals responsible for the colour and flavour. The essential oils, products like peppermint oil and lavender oil, are complex mixtures of many chemicals. Thousands of chemicals are a normal part of the planet's crust. On top are biodegrading plant products, next is a layer of minerals, different types of sands and clays. All these are complex mixtures of many individual chemicals. Still deeper we find more enormously complex mixtures, coal and petroleum. The individual chemicals within these mixtures, described one time or another in the chemical literature, are counted in their thousands, and are part of the inventory.

Although the numbers in these various categories are very large, we are still far from the nine million total. A huge contribution to that total results from human activity. Not an environmental problem waiting to explode, but chemical research in action. What is involved is the chemistry aimed at identifying the architectural make-up of chemicals and their synthesis, the normal process of chemical research.

Laboratory chemicals

Cholesterol is a naturally occurring chemical with great significance for human health. It was discovered in 1812 by one Michel Eugène Chevreul. The name cholesterol is from Greek, *chole* meaning bile, because cholesterol was recognized very early as the principal constituent of human gall stones. To establish its molecular architecture, to synthesize it and to find out how nature synthesizes it, its biosynthesis, became a major scientific pursuit during the first half of this century. That pursuit led to a clear picture of the whole family of the steroids. In the steroid family we find names like progesterone, cortisone, and many others related to cholesterol. The steroids are very important for human health and modern medicine. Yet the number of these naturally occurring chemicals is hardly significant with respect to the nine million grand total. Where the large numbers do become significant is right within the laboratories where this enormous amount of experimental chemistry was done.

When cholesterol is made to interact, in the laboratory, with hydrogen gas, the product is another chemical called *dihydrocholesterol.* Mild oxidation of cholesterol gives *cholestanone.* The living cell cannot use these two cholesterol derivatives and does not make them, they are not "natural" chemicals. They were needed only for the laboratory studies on cholesterol. To unravel the secrets of the chemicals in the steroid group, to establish relationships, many of these synthetic derivatives were made, described, and reported in the literature. All that is left of them now are small amounts, sitting in some laboratory's *sample collection,* but still part of the nine million inventory. It is these sample collections that make up a very large part of that inventory. Even in the early days of steroid chemistry we find over two hundred such derivatives-made-for-study from cholesterol alone.[19] That was in 1959, and we have had some thirty-five years of intensive research activity since. Whereas only a few dozen steroids occur naturally in living organisms, thousands of derivatives were made of them for studying their molecules, their chemical behaviour, and their effect on living cells. *Some* derivatives made of *some* steroids led to the development of oral contraceptives in the late 1950s.

Sample collections like these do not exist only in steroid chemistry, they have also evolved in the chemistry of alkaloids, antibiotics, petroleum chemistry, and many other specialized areas. These sample collections of derivatives are the result of the study of any chemical of any significance, particularly those relevant to health or medicine. The process is continuing. Intensive, worldwide research activity on problems like cancer and AIDS adds many thousands of specimens to the already formidable grand total. In the vast majority of cases the existence of these millions of chemicals is confined to the pages of *Chemical Abstracts* and the sample collections in chemical research laboratories. They do not enter the environment, they do not contaminate food or drinking water, they have nothing to do with environmental problems or health problems. It is true that we have problems maintaining control of the thousands of industrial chemicals coming out of factories and chemical laboratories, but they are thousands. For a useful discussion of environmental problems, the "nine million chemicals question" is utterly irrelevant.

What's in a name?

Some of the chemicals we encountered have rather unusual names. A brief look at the origin of these names is useful. Chemicals found in plants, like tomatidine and solanine, often get names that are derived from the Latin or common name of the plant. Some of the older names come from the effect a chemical has on human health. An alkaloid found in the ipecac plant is a powerful emetic and is called emetine. Synthetic chemicals, particularly those of commercial significance, are often named after the person who made the compound for the first time. Sometimes a chemical is named to honour a chemist, who may or may not still be alive. The pesticides dieldrin and aldrin were named after the German chemists Otto Diels and Kurt Alder, who made significant contributions to the development of modern chemistry.

This somewhat haphazard method of naming individual chemicals cannot possibly work with all of the nine million chemicals we "enumerated" above. Moreover, and particularly in the case of laboratory sample collections, many chemicals are *derived* from other chemicals. To some extent this still allows for naming one chemical after another, often by including some reference to the chemical process that changed the first chemical into the second. For example, carrying out a chemical process of adding two hydrogen atoms, called hydrogenation, to a molecule of cholesterol gives *di*hydrocholesterol. This can be quite confusing, considering that the removal of hydrogen atoms by a process called *de*hydrogenation would give the name *de*hydrocholesterol.

The considerable chemical difference between dihydrocholesterol and dehydrocholesterol would not be obvious. Considering that there are, literally, millions of chemicals to deal with in terms of names, there is an obvious need for a more systematic approach. This systematic approach has been worked out by an international organization called the International Union for Pure and Applied Chemistry, IUPAC for short. The system, although complex, makes it possible to give a chemical a name that gives information about its molecular architecture. A chemical that is the result of an oxidation often has "oxide," or a contraction thereof, in its name. The family of the alcohols frequently have names with the ending -ol, like alcohol (ethanol) or cholesterol.

This last example illustrates an important point. Even though the names of two chemicals may sound similar, their behaviour, health effects, or toxic properties may be vastly different. The toxicological properties of cholesterol and alcohol make this obvious. Similar sounding chemical names easily lead to errors that can be quite substantial. Such errors can involve wrong conclusions, wrong policy decisions, and, in a worst case, public alarm. Errors of this kind are not uncommon and the environmental literature contains its fair share. We will have occasion to see other examples of this nomenclature trap in further parts of our survey.

Living better with chemistry

Although the North American advertizing slogan "Live Better Electrically" has been obsolete for some time, its effects are all around us. In view of today's image of the chemical industry, it is not surprising that a similar slogan "Better Living through Chemistry" has also all but disappeared. However, in a subtle way this too is alive and well. One only has to observe carefully the television commercials and the glossy magazine advertizing for cosmetics, cleaning products, and convenience food.

In a fundamental sense all matter in the universe *consists* of chemicals, from our food to our bodies and the entire planet we live on. Now that we have unmasked factoids like "artificial" chemicals, looked at some aspects of synthesis, and dealt with the "nine million chemicals" question, let's have a second look at our chemical society.

Useful chemistry

The emphasis in this part is on *useful.* Not in a philosophical sense, questioning the "useful to whom" and "do we really need all these chemicals?" By useful we mean chemicals that we make or mine and which are *being used* by society as a whole. Somebody wants them, somebody makes them, and somebody buys them. Whether we could do without is a question for another debate.

Many pure chemicals obtained from plants are in widespread use, mainly in medicine and cosmetics, but also for making other consumer products. The huge number of chemicals obtained from coal and from petroleum plays a key role in any industrialized country. Despite the occasional cry for doing with less, and going back to nature, the reality is that a modern industrialized society is completely dependent on its chemical industries, both petrochemical and pharmaceutical. An easy illustration of this dependence is to stand up, fully dressed for going to work, in front of a mirror. Take paper and pencil and list all objects you carry on your person that are unrelated to petroleum. Remember that the "pure cotton" shirt was probably cut, sewn and pressed with oil or gas derived energy, and the cotton was grown with fertilizer and harvested with a gas or diesel burning tractor. In subsequent chapters we will have many opportunities to see more "chemistry in action." At this point it is instructive to illustrate our dependence on chemistry with a couple of case histories.

Many willow (*Salix*) species contain derivatives of salicylic acid, a chemical from which aspirin (acetyl salicylic acid) is derived. Teas and other extracts of willow contain salicylic acid and some of its close relatives, but not aspirin itself. These concoctions containing the *salicylates* were used by ancient European cultures, the Greeks for example, to relieve pain.[20] Aspirin, a wholly synthetic chemical, is now made in very large quantities from petrochemicals. In combination with other drugs like codeine and caffeine, aspirin is one of the most frequent ingredients in prescribed drugs in the industrialized world. In recent years it has been touted as having possible preventive value in some aspects of cardiovascular disease, even cancer.

The investigation of the group of chemicals known as the salicylates and the synthesis of aspirin have had, and continue to have, a major effect on modern medicine. There are many more examples of "good" chemistry that relate in one way or another to human health. The history of this medical part of our chemical heritage, although outside the scope of our discussion, is interesting reading.[21] It is very difficult to visualize modern medicine without the contributions from chemistry and biochemistry. A few representative examples are presented in Table 4.1.

Other examples come from the petroleum industry, specifically from the area of polymer chemistry. The I.G. Farben Company in Germany produced the first wholly synthetic fibre in the early 1930s. It had only limited value because it did not tolerate heat. It melted at about 85°C. However, by 1935 the first polyamides, Nylon and Perlon, became commercially available, followed by the polyacrylic fibres such as Orlon. Polyesters such as ICI's terylene soon joined a selection of synthetic fibres that changed the clothing industry forever. Particularly in sports and other outdoor activities the development of synthetic fibres has a significant influence. Without synthetic fibres the world's clothing industry would look rather different.

Similar illustrations of the chemical society are found in building materials, transportation, the food industry, and virtually every other aspect of life in the industrialized countries. We truly live in a chemical society. Probably the best way to decide if we actually "live better with chemistry," is to make a few lists, as in Tables 4.1 and 4.2, and consider what it would be like without.

Chemical waste

A few final comments on the chemical society are necessary before we attempt to put the genie back into the bottle. Many chemicals can be and are obtained from plants. Plants that produce useful chemicals are often cultivated and harvested like other crops. These are chemicals that we may consider as renewable resources. The renewability is dependent only on our ability to manage the crop and sustain the supply of the chemical.

The chemicals obtained from the petrochemical industry appear different. If we burn

Table 4.1 Some common chemicals from the medicine cabinet

Aspirin	acetyl salicylic acid	analgesic
Librium	chlordiazepoxide	tranquilizer
Tylenol	acetaminophen	analgesic
Chlor-Tripolon	chlorphenyramine[a]	antihistamine
Litarex	lithium citrate	antidepressant
Anginine[b]	nitroglycerin	vasodilator
L-Dopa	Eurodopa[b]	antiparkinsonian

Notes: [a]Polaronil in Germany, [b]and several dozen other trade names

Table 4.2 Synthetic fibres and other polymers

Product	*Made from*[a]	*Used for*
Vinyl, PVC	vinyl chloride	building materials, bottles, pipes
Saran Wrap	vinyl chloride	food packaging, cling wrap
Polystyrene	styrene	car parts, foamed pieces
Teflon	tetrafluoroethylene	non-stick cook ware
Dacron	polyester[a]	fabric
Nylon	polyamide[a]	fabric, rope
Kevlar	polyamide[a]	fabric, tire belting

Note: [a]Or family name

the products of the petrochemical industry they are gone. By the time we have burned all the natural gas, gasoline, diesel fuel, and bunker oil there will no longer be a petrochemical industry. This view is disputed by some economists who argue that the planet is so vast that we will never run out of resources. They say the finite nature of resources is an economic problem, not a physical one.

This argument about the finiteness of natural resources is of limited significance when we add the waste problem to it. The exploitation of both renewable and non-renewable resources inevitably produces waste. Fuel made from biomass still produces carbon dioxide, and cultivating plants for chemicals also produces waste. In terms of the waste problems the difference between renewable and non-renewable resources is only a matter of relative quantities. We will discuss waste and waste problems in more detail later, but there is one aspect that fits in our story at this point.

Waste is a particularly difficult problem for the chemical industry. How does chemical waste come about and why is it unavoidable? The purification of vitamin C or vanillin, either the natural products or the synthetic ones, does not only produce the pure chemicals. Purification means that the impurities are removed *from* the product we want, but where are they removed *to*? Somewhere the process yields plant materials, solvents, acids or alkalis, that were all part of the purification process.

There are often ways to minimize the volume or the weight of this waste. The search for doing chemistry with less impact on the environment is an urgent research topic. These efforts are already showing considerable promise. Research that addresses this environmentally benign approach to synthetic chemistry is becoming a regular feature on the conference scene.[22,23] However, ultimately there will be waste. This is a fundamental rule that is true for every chemical process. There *will* be chemical waste in every industry that processes chemicals, which means every industry.

This waste must either be stored or disposed of. The storage may be temporary or long-term, the latter often meaning forever. The disposal must occur "in an environmentally responsible manner." In addition to the storage problems, waste must often be transported. All these operations contain the potential for environmental damage. Accidents in the transport industry, unusual weather conditions like heavy rain, floods, or earthquakes, can cause spills. When this happens, we are vigorously protecting the environment with legislation that deals uncompromisingly with "those that are to blame." Only those responsible for the *supply* of the spilled chemicals are qualified to take that blame. Those that *demand* the use of the chemicals are exempt from punishment.

CHAPTER 5

How toxic is toxic?

Toxic agents

> A poison is any substance which, when ingested, inhaled or absorbed, or when applied to, injected into or developed within the body, in relatively small amounts by its chemical action may cause damage to structure or disturbance of function.[1]

This formal definition from *Dorland's Illustrated Medical Dictionary* attempts to be all-inclusive so that there will be no room for doubt or error. We will look at the several different parts in turn.

Chemicals

First, there is mention of a *substance*. This means that a poison can be an element (a chemical), or a pure compound (also a chemical), or a mixture of two or many chemicals. Second, it makes no difference by what pathway the poison enters the body, the "routes of entry." Inhaled and ingested are self-explanatory, absorption perhaps less so. Many chemicals can readily penetrate the intact, healthy skin and are then absorbed into the bloodstream. Such absorption can also occur through the mucous membranes of the eyes, mouth, and nasal passages. Skin absorption frequently involves liquids such as industrial solvents. Absorption through mucous membranes is common with gases and vapours.

Third, the key phrase is "by its chemical action." Living tissue consists of chemicals, such as proteins, fats, hormones, minerals, water, and carbohydrates. These chemicals can undergo chemical changes, either by interacting with other chemicals or by interacting with physical agents like heat or radiation. Fats can be made into soap, and bones into gelatin or glue. Many chemical changes can occur *within the living cell.* A living cell is a busy mini-factory full of chemical activity. Thousands of chemical changes take place, to transform energy, to make new cells, to transmit messages, and to *defend the cell* against foreign invaders. These invaders may be foreign organisms like viruses, or chemicals that do not belong in that cell. When the foreign chemical causes a chemical change in a "rightful occupant" of the cell, for example a protein, the changed protein may no longer be *biologically functional.* Then the cell no longer works properly and there is a "disturbance of function." The foreign chemical, when it interacts with body chemicals, acts as a poison, has a toxic effect; it is a toxic chemical.

What happens depends on which organ the affected cells belong to. For example, most chlorinated hydrocarbons affect the cells of the mammalian liver, and *can* cause disturbance of liver function. They are liver poisons. Other poisons may affect the kidneys, the heart, or blood cells. If the poison changes many of the chemicals in the cell, the whole cell may fall apart. When that happens with many cells, a substantial amount of tissue may be destroyed, resulting in "damage to structure." Damage to structure at the surface of skin or mucous membranes can be caused by "corrosive chemicals." In this case "corrosive" refers to tissue.

A final point about Dorland's definition is important in the environmental context. This is the possibility that the poisonous substance is "developed within the body." In addition to poisonous chemicals made by industry, nature also makes a substantial contribution of toxic chemicals. Since all living organisms have been exposed to these chemicals for thousands of years, they must have developed a way of coping with toxicity. This coping mechanism consists of the ability of a cell chemically to change the molecules of a poison. This change can have two results. The foreign chemical usually becomes more soluble in water, and it often also becomes less toxic. Improved water solubility helps to secrete the changed chemical more easily. The process is called *detoxification.*

These *in vivo* chemical changes into less toxic chemicals involve a break-up of the toxic molecule into smaller bits and pieces; it is a degradative process called *biodegradation.* Eventually the pieces are small enough and non-toxic so that other organisms can use them as food: the reverse of biodegradation is biosynthesis.

A decrease in toxicity occurs often, but not always. In some situations the biodegradation process, at least in its very initial stages, results in the formation of a breakdown product that is actually *more toxic* than the original poison. That is the Dorland meaning of the poison being "developed within the body." The organophosphate pesticides like malathion and parathion are examples of chemicals that are initially transformed into more toxic products. If the cell, or indeed the organism, survives this first step, detoxification or biodegradation would still be the eventual outcome.

Other descriptions of a poison are very similar but there may be minor variations. The McGraw-Hill dictionary adds one significant detail by defining a *toxin* as:

> Any of the various poisonous substances produced by certain plant and animal cells, including bacterial toxins, phytotoxins, and zootoxins.[2]

This indicates a distinction between a poison and a toxin. According to Melloni a toxin is only produced by micro-organisms; Blakiston specifies that a toxin is protein in nature.[3,4] All three dictionaries agree that the key difference between poisons and toxins is that the latter are produced by living organisms. In formal toxicology language the distinction is useful because it allows for more precise language. For our present discussion it is sufficient to *be aware* of the existence of bacterial toxins, for example botulinus toxin, produced by the bacterium *Clostridium botulinum*, and phytotoxins or plant toxins, like strychnine and curare. Zootoxins or animal toxins, produced by snakes and insects, are often called "venom." The McGraw-Hill dictionary omits an important class of toxins, those made by fungi, the so-called *mycotoxins*. This group is important. It includes chemicals such as the penicillins and the dangerously poisonous aflatoxins, from the fungus *Aspergillus flavus* which can grow readily on improperly stored grains, nuts and other dry foods.

In the environmental language we often hear the term "toxics." Toxic, in all medical dictionaries, is an *adjective*, a qualifier *pertaining to* ..., but not a noun. When languages evolve and new words appear from time to time, they must be explained. The new word "toxics" is usually perceived as having to do with human activity, often in a negative sense. The use of "toxics" tends to exclude naturally occurring toxic substances, and by implication refers only to industrial chemicals. As we will see, there is no basic difference between the action of industrial poisons and that of natural chemicals. For our present discussion we prefer to stay with poisons and toxins.

Poisons and toxins *are* chemicals, or mixtures of chemicals. Most chemicals that are not naturally at home within the living cells of an organism provoke a reaction in that organism. We will look at several important aspects of toxic behaviour, but our space is limited. For readers who desire more detail, Ottoboni's *The Dose Makes the Poison* and Rodrick's *Calculated Risks*, are good supplementary sources written in plain language[5,6] A useful pamphlet deals with the question: *What makes a chemical poisonous?*[7]

Physical and biological agents

When a hen applies the warmth of her body to her egg, chemical changes start to take place within the egg. Proteins are formed, fats are consumed, and after some time the content of the egg has been transformed into a chicken. An increase in temperature, i.e., heat, started a *chemical activity* in the egg. Application of much more heat to the egg, for example with boiling water, gives a boiled egg from which no chicken can be hatched. This larger amount of heat caused chemical changes within the egg different from those caused by the hen. In the boiled egg the changed chemicals are no longer biologically functional; the heat has caused a "disturbance of function." Heat is a physical agent that had a toxic effect on the proteins that were intended to form an embryo. Chemical activity, that is interaction between chemicals, is directly dependent on temperature. Placing a steak in the freezer slows down the chemical activity in spoilage-causing bacteria; keeping the steak at room temperature allows the bacteria to grow, and the barbecue heat will kill them.

Sunlight, a form of "radiation," causes changes in the chemicals on the surface of a photographic film. Light energy has been transferred to the film and the result is the photographic process with its many applications. Sunlight can also cause chemical changes in the dyes used in fabrics. This makes curtains *fade* in sunlight. When the radiation is of a kind that contains more energy, such as ultraviolet light or X-rays, other chemical changes are possible. Some of these changes, taking place at the surface of human skin, may affect the mechanism of cell division and growth, or influence the formation of melanin, the pigment associated with skin colour.

When the radiation contains enough energy to penetrate right through the tissue, as with X-rays, it can cause chemical changes *within the tissue*. Radiation like X-rays can cause cells to misbehave and form a cancer, and they can also be used intentionally to kill cancer cells. In the latter case radiation is used to cause damage to the structure of cancer cells. Heat, sunlight, ultraviolet light, and X-rays are *physical agents* that can cause a clearly defined chemical effect which has the potential to become toxic action.

Many naturally occurring and a few manufactured chemical *elements* emit radiation. When a living organism is exposed to such radiation, or ingests a large enough dose of the radioactive material, the associated radiation can have a toxic action similar to that of X-rays. These elements are called *radio-isotopes*, they are *radioactive*, and most of them occur as natural minerals in the crust of the earth. Radio-isotopes can also result from *atomic fission*, the splitting of atoms. Atomic fission occurs mostly in nuclear power stations and in some nuclear bombs. The spontaneous fission of a uranium isotope, uranium-238, is a rare natural event; rare because it has a half-life of 10^{17} years. Depending on the associated energy, some forms of radiation can travel appreciable distances through air and can penetrate through various materials.

The changes caused by physical agents are changes in the chemicals of the cell. Thus, toxic action can be caused by either a chemical or a physical agent. Although in terms of everyday English it sounds a little strange, in our present context it is quite alright to describe ultraviolet light or noise as a toxic "substance."

Electrical energy can also cause chemical changes. The common lead-acid battery is an example. In terms of its harmful effects on the (human) body, however, electrical energy usually causes direct, almost instantaneous physical changes that interfere with the body's own electrical activity. In this sense electricity in the form of lightning or electrocution is not really a toxic agent, although it can do a great deal of harm.

Chemicals made by micro-organisms were included above as chemical agents. A micro-organism can also affect human or animal health by direct entry into the host body. Then the interactions between the invading micro-organism and the host can be of several different kinds. The microbe can secrete toxic chemicals directly into the cells or the bloodstream of the host, or it can multiply using the resources of the host. When the micro-organism causes adverse health effects, it is a toxic agent. *Biological agents*, in toxicology language, include bacteria, fungi, and other parasitic organisms.

The dose makes the poison

The alchemist and physician Paracelsus was probably the first to recognize that a living organism is really a system of interacting chemicals. A very complex system to be sure, but nevertheless a *chemical system*. He was probably also one of the first to treat his patients with chemicals of a mineral nature. Until that time, treatment of disease was almost entirely limited to concoctions derived from plants. Paracelsus' success rate was somewhat questionable, and he probably lost a few of his patients. In true alchemist style he continued to use compounds of mercury, arsenic, and other minerals. These medical methods now look very primitive, but Paracelsus' careful observations have retained their importance and have developed into a fundamental concept in modern toxicology.

Paracelsus realized that the *amount* of a chemical given to a patient was of crucial importance. If a small amount of his "remedy" did not cure the patient, a large amount of the same chemical could kill. He formulated this observation as "Dosage Alone Determines Poisoning." A more detailed translation from the German actually says that everything is a poison if the dose is large enough.[8] This has become a fundamental and permanent principle in toxicology. Conversely, everything is non-toxic as long as the dose is small enough. This means that the term "toxic chemical" has little meaning.

The explanation lies in the detoxification and biodegradation we just discussed. The breakdown of a chemical takes time and that time depends on the *number of cells* available to do the job and the *amount of the poison* involved. The larger the organism, the larger the number of cells and the larger the amount of poison that can be detoxified in a given time. This leads to the important definition of *dose*, that is the amount of the toxic chemical relative to the size or weight of the organism. Dose can be measured in grams or milligrams of the chemical *per kilogram body weight* of the animal or human. The concept of dose explains why a two-hundred-pound "hulk" can "hold his liquor" while a hundred-pound "petite" can't. Two stiff drinks in the blood and tissues of a two-hundred-pound body have only half the toxic effect they would have in a one-hundred-pound body.

An individual consuming too much alcohol becomes intoxicated. In more formal language we say that a person responds to a high dose of alcohol by showing the symptoms of central nervous system (CNS) depression. The key words are again the *dose*, the *response*, that is the reaction of the body, and the *nature of the response*, that is CNS depression. It follows that the *degree* of the response is directly related to the dose. For many chemicals we can roughly estimate the dose that would be lethal to an average human, but it may be difficult to consume such a dose in a short time. For example, there is a lethal dose of solanine in a couple of hundred pounds of potatoes. There is also a lethal dose of oxalic acid in about fifteen pounds of rhubarb and a hundred aspirin tablets contain a lethal dose of acetyl salicylic acid. We will look at the important relationship between dose and response in more detail shortly.

Alcohol also illustrates other toxicological events. It can affect the human body in two ways. Besides causing CNS depression alcohol is slowly detoxified, by oxidation, in the liver. A fair amount of liquor, say several drinks consumed in a short time, will quickly produce the intoxication we just mentioned. The same amount of liquor consumed over a full day would not produce this effect, but the liver would still have to do the same amount of work. The immediate effect caused by rapid entry of a relatively large dose of poison is an acute poisoning or *acute toxic effect*. Slower entry of smaller doses over a longer time produces the other poisoning effect, stress on the liver. This is chronic poisoning or *chronic toxic effect*.

Another important aspect of toxic action deals with the time lag, or *latency period*, between exposure and the onset of disease. This time lag can be quite long. In some forms of cancer it can be as much as twenty years. The actual entry of a chemical into the body and the time factor involved both determine *exposure*. A person exposed to asbestos at age 50, *may* get asbestosis or cancer at age 70 or later. A child exposed at age

10 *may* get cancer, because of that exposure, at age 30. The age difference means a significant difference in the consequences of the exposure. Incidentally, this is one of the reasons much public health and environmental legislation includes special considerations for children.

The various biochemical mechanisms for counteracting the effects of many toxic agents, both chemical and physical are called the body's *defense mechanisms*. Particularly important in terms of these defense mechanisms are the individual's age and general state of health. In young children biochemical defense mechanisms are generally not yet well developed. The biochemical activity in an organism is programmed into the genetic material that the organism inherited from its parents. This genetic information is also one of the factors in an individual's defenses against disease, including disease caused by exposure to chemical or physical agents. There is often also a genetic aspect in other "spontaneous" diseases. The recent literature contains a steady stream of reports that points increasingly toward genetic factors in many of today's common health problems.

Measurements

The effect of a chemical on an individual can vary from a beneficial effect to something barely noticeable, to a more or less severe toxic effect. According to Paracelsus this is true for any chemical. There are estimates suggesting that for the statistical adult human male, drinking about seventeen litres of water will dilute electrolyte concentrations in body fluids to fatally low levels. In this case pure water has a toxic action.

It is often necessary to describe toxic action in precise terms. Many chemicals, prescription drugs for example, have beneficial effects at one dose level, but are dangerous, sometimes fatal, at higher doses. A well-known example is vitamin A. This chemical, also called retinol, occurs in animal oils, particularly fish oils like cod liver oil. Vitamin A does not occur in plants, but carotene and its relatives, abundant in green and yellow vegetables, are readily converted into vitamin A. This vitamin is essential to human health and a deficiency can lead to an eye disease called *xerophthalmia*. On the other hand, a chronic overdose, leads to a condition known as *hypervitaminosis A*. The symptoms are vomiting, loss of appetite, headache and skin lesions. Excessive consumption of vitamin A by pregnant women can result in birth defects in their offspring. The beneficial and toxic effects of vitamin A are *dose-dependent*. The health effects of a low dose are opposite to those of a high dose. They are determined by the body weight of the subject and the amount of the chemical.

In order to describe these effects, the response of the organism to the chemical must be measured in some way. We must examine these measurements and methods before we can productively discuss a number of controversies surrounding the effects of chemical and physical agents. Experiments make it possible to express the basic concepts of toxicology in numbers. Measuring toxicity is usually straightforward and tabulating the numbers is simple. Interpreting the meaning of those numbers is not.

How toxic is toxic?

Measuring toxicity

In a recent essay on aquatic toxicology the authors start with the assertion that: "No instrument has yet been devised that will measure toxicity."[9] This statement should be underlined and strongly emphasized, because there is a widespread belief that there are machines for measuring toxicity. In this chapter the focus is on the testing process with animals and what it involves. We will keep the questions about alternatives for later.

Imagine several laboratory animals, rats for example, that are identical in *almost* all respects. They would be nearly equally good, or bad, at biodegrading a specific toxic chemical. If these rats were given the same dose, they would *almost* all respond the same way. If we want to measure acute toxicity in terms of lethality, the response of this group might have been the death of all rats. If a second group of animals were given a smaller dose, the response may be less severe, for example, 40 percent of the second group may have survived. Note that the response is not the same for all animals in the group, the rats are *almost* identical. If a third group of rats were given a still smaller dose, possibly 70 percent of the third group could have survived. There is a numerical relationship between the severity of the response, that is the number of deaths, and the size of the dose. This is the *dose-response relationship.*

This relationship can be represented graphically. The above experiment gave three dose levels and three responses, representing three points on a graph. Repeating the test with another three groups of rats and a further three dose levels would complete a *dose-response curve* with six points as presented in Figure 5.1. Somewhere on the curve is a point where exactly half the animals survived (or half of them died, as with the half-full glass and the half-empty glass). The dose that killed half the animals is the LD_{50}, one of the best known and most often criticized tests in the toxicology testing repertoire. It is of crucial importance to recognize that for a *more toxic* chemical a *smaller amount* is needed to kill 50 percent of the rats. Of a less toxic chemical a larger dose is needed; the *lower* the LD_{50}, the *higher* the toxicity. In Figure 5.1 point A indicates the LD_{50}.

The dose-response principle is also valid for measuring chronic toxicity. If different groups of rats are given *sub-lethal* doses of the chemical over a longer time, some animals at some dose levels may get sick. The nature of the sickness depends on the action of the chemical. Some chemicals cause cancer, others may damage the liver or the kidneys. This second experiment will produce a dose-response curve for chronic toxic effects of that chemical.

A third kind of toxicity test is important to complete the picture. In addition to chemicals that are acutely fatal or cause disease at low level chronic exposure, some chemicals will, at appropriate dose levels, have no effect on the animals themselves, but will affect the offspring. Feeding the chemical to pregnant laboratory animals is part of *reproductive toxicity* testing. Reproductive toxicity can also mean measuring the effects some chemicals have on *fertility*, in both male and female subjects.

The toxic effects of chemical or physical agents can be measured with many different organisms. The mechanisms of toxic action in plants are very different from those in animals. Toxic effects in plants are important for environmental studies on landscapes, wildlife habitat, or aquatic systems. Many chemicals can inhibit the growth of fungi, algae, or bacteria. The percentage of growth inhibition, relative to a control, and the dose

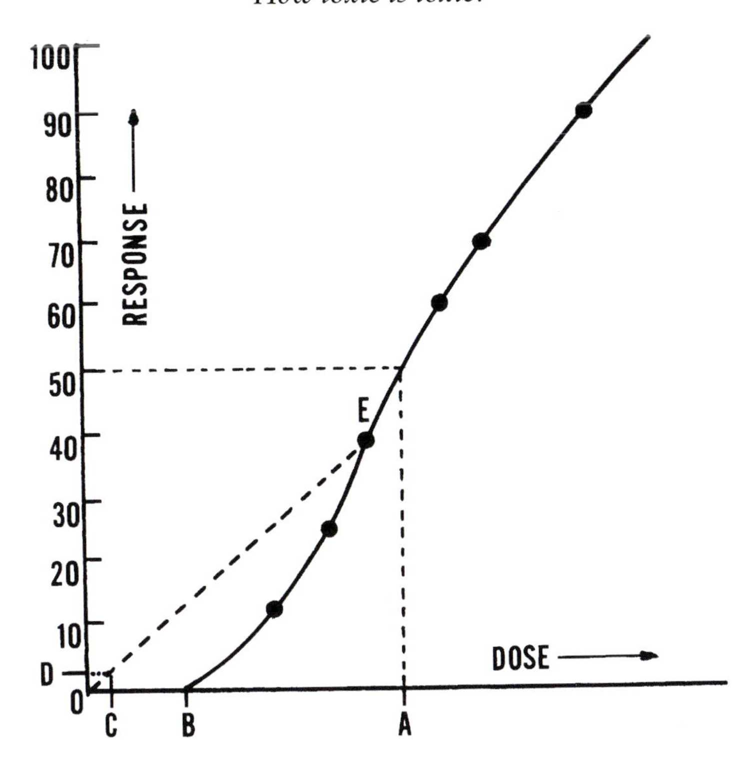

Figure 5.1 A dose-response curve showing the LD_{50}, the NOEL, and a linear extrapolation to zero threshold

of the chemical in the growth medium give dose-response curves similar to those obtained with animal tests.

The dose-response curves that are measured in all these tests are rarely straight lines. They are often curved or S-shaped. They also show some peculiarities near the origin of the axes, that is the area of low doses and low responses. The shape of the dose-response curve in that region is of critical importance in environmental toxicology.

Toxic, non-toxic, or beneficial?

If the quantities of the toxic material are small enough, and are administered gradually, it is possible that the organism can detoxify and excrete the material completely without suffering any damage. There will be no visible response, and at that point the dose is said to be below the *no-observable-effect level*, or NOEL. The no-observable-effect level can be read from the dose-response curve, it is the highest dose the test animals can tolerate without showing any observable adverse health effects. Observable includes whatever is seen when the animals are eventually killed and opened up. The euphemism says the animals are sacrificed, and then subjected to a detailed autopsy. If there are truly no adverse health effects visible, the chemical can be said to be non-toxic *at that dose level.* In Figure 5.1 the NOEL is at point B. Toxicity testing normally does not include detailed observations below dose B. As we will see, that is an unfortunate omission.

Rats, guinea pigs, mice, and rabbits are some of the most frequently used laboratory animals. To arrive at any meaningful conclusion it is essential that a variety of animal species are used. One of the variables in toxicity testing is species specificity. A well-known example comes from the plant world. The herbicide 2,4-D is toxic to broad-leaved plants like dandelions and poplars but is *almost* non-toxic to grasses and conifers. Another example is the thalidomide tragedy of the early 1960s. The drug thalidomide, briefly used as a tranquilizer mainly in Europe, was considered "safe" for pregnant women. But it had not been tested on that one particular kind of monkey that could have shown the birth defects caused by this chemical. Similar differences between species are found with many chemicals. TCDD, one of the chlorinated dioxins, is about five thousand times more toxic to guinea pigs ($LD_{50} = 0.001$ mg/kg) than to hamsters ($LD_{50} = 5.0$ mg/kg).[10] Other examples of differences between species are well known. Penicillin is lethal to guinea pigs and hamsters in amounts comparable to therapeutic doses in other species. The common anesthetic fluoroxene, which has been used on humans safely for many years, is fatal to dogs and rabbits in low doses.[11]

Particularly with respect to chronic toxicity, effects in the low dose and low response area can become impractical to measure because very large numbers of animals would be needed. This is where we enter Weinberg's trans-science territory.[12] When that happens and the lowest dose for which an effect could be reliably measured would be, for example, at point E (Figure 5.1), the *assumption* is often made that the dose-response curve in that region approaches *a straight line that goes through the origin.* This would mean that the lower part of the dose-response curve would look like the dotted line O–E in Figure 5.1. The consequences of that assumption are enormous. Relating dose and response in that way would mean that for the smallest possible dose, for example the dose at point C, there would be a response in a small percentage, D, of the subjects. It would mean that the only way we could ensure a *zero percent* response would be at *zero dose.* What we see here is the reasoning that led to the "no-safe-dose" concept. This "zero-threshold" or "no-safe-dose" hypothesis holds that even the smallest amount of a toxic agent will have an adverse health effect. This hypothesis is common in two areas of toxicology, carcinogenic chemicals and ionizing radiation. It is of key importance to recognize that the "zero-threshold" is derived from a *hypothetical* dose-response curve that is *assumed to be linear* and going through the zero dose point. It is also peculiar that the "zero-threshold" concept is rarely used for toxic agents other than carcinogens and radiation.

A peck of dirt

Let us picture the scene. On a warm summer day a little 2 year old drops an ice cream and tries to retrieve its remains from the floor. The mother's common response is that it does not matter, because the child "has to build up resistance." In a mid-1960s essay "Sufficient Challenge," H. F. Smyth refers to this attitude with an example from his own childhood days, when children were told: "You've gotta eat a peck of dirt before you die."[13] In everyday life the notion of a beneficial effect, building up resistance from a low level exposure, the peck of dirt, is not new or controversial. Let us examine this idea in terms of the dose-response curve in conventional toxicology.

Measuring toxicity toward fungi is a very simple experiment. A small piece of a fungal culture, placed in the centre of a petri dish with nutrient, grows in a perfect circle that reaches the circumference of the dish in about five or six days. A dish with "clean" nutrient is the control, several other dishes contain increasing amounts of a chemical, a herbicide for example. If the herbicide is toxic to the fungus, the cultures containing the chemical will grow slower than the control. A number for this *growth inhibition* can be obtained by measuring the diameter of the cultures with a ruler, and expressing the difference with the control as a percentage of that control. The percentage growth inhibition, plotted against the concentration of the herbicide, gives the usual dose-response curve. Here too we often find an apparent anomaly at the low end of such a curve.[14]

It happens frequently that a few of the cultures containing the smallest amounts of the test chemical *are growing faster* than the control. They grow to the full circle of the dish before the control has reached that point. In those low concentrations the chemical has *stimulated growth*; it had a beneficial effect on the fungus. Mycologists, the brand of biologists that study fungi, are quite familiar with this effect. It is the same effect we saw in our vitamin A example and which is so familiar to nutritionists.

When Smyth presented about two dozen examples of the beneficial effects of apparently toxic agents, he used the phrase "sufficient (but not overwhelming) challenge." Among his examples were the well-known story of Captain Cook's sailors infecting the people of Easter Island with measles. The island's population was wiped out because it had never been exposed to measles. It had not "built up resistance." Another example was Smyth's own observation during inhalation experiments with carbon tetrachloride on laboratory rats. The rats that inhaled 50 ppm grew better than the controls and it took 400 ppm to cause an equal but negative change in growth.

The benefits of "sufficient challenge" have been known among pharmacologists as the Arndt–Schulz Law or Hueppe's Rule. Smyth called it *hormoligosis*, a term he attributed to a 1960 paper by Townsend and Luckey. Since then, the term has changed to *hormesis*. The general applicability suggested by a "law" or "rule" was not readily accepted. A major criticism was the lack of an experimental explanation, and the Arndt–Schulz law gradually disappeared from the toxicology vocabulary. However, supporting experimental evidence has not disappeared and the issue of hormesis has also penetrated into the field of radiation. The same T. D. Luckey recently published *Radiation Hormesis*.[15] There are many hundreds of studies in the literature that describe experiments and observations that demonstrate the beneficial effect of chemical and physical agents at low doses. We will briefly look at two details.

The first deals with the shape of a dose-response curve in the low-dose and low-effect area where it shows hormesis. Since the vertical axis of our graph represents an adverse health effect in the positive (upward) direction, a beneficial health effect would be represented along the negative (downward) part of that vertical axis. A curve that shows hormesis would then have a shape as represented in Figure 5.2. The part O–P on that curve is the dose range where the organism suffers adverse effects from deprivation of the chemical (or physical agent), for example a vitamin deficiency. In the

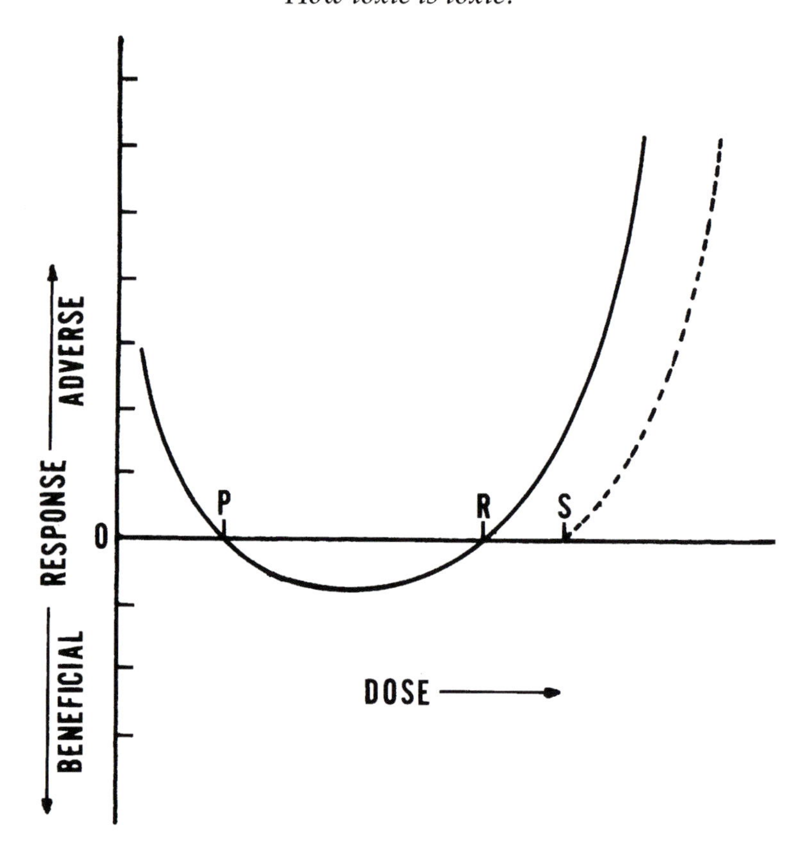

Figure 5.2 A complete dose-response curve showing hormesis and a "high dose" NOEL

dose range P–R the chemical has a beneficial effect. In the range P–R the effect is a "sufficient challenge" leading to hormesis. At point R the dose exceeds the "sufficient" range and becomes an "overwhelming challenge" and the organism suffers adverse health effects. R is the same dose that we earlier identified as the NOEL; only then we neglected to make observations between points O and R.

The second point about hormesis concerns the transition from an accepted phenomenon, i.e., the "peck of dirt" or the effects of vitamins, to a controversial environmental viewpoint. If we accept the common observations of hormesis as described above, there is no chemical or biochemical reason why hormesis should be limited to a dirty ice cream or measles or vitamins. Therefore, when we look for other examples of hormesis we are certain to find them. The place to look is in the literature.

In another, more detailed and more recent review we find almost a hundred studies that demonstrate that this stimulatory effect of toxic agents is widespread in nature. Hormesis has been observed in bacteria, fungi and yeasts, plants, algae, invertebrates and vertebrates, and in cell cultures. The agents used in these studies include metals such as cadmium and lead, solvents, and many synthetic and natural chemicals. PCBs can promote the growth of juvenile Coho Salmon and minnows. The growth of clam and oyster larvae is promoted by low level concentrations of some 52 pesticides, and chlordane and lindane have been shown to promote the growth of crickets. Low levels of ionizing radiation have been shown to stimulate growth in the blue crab (*Callinectes sapidus*) in experiments that lasted 70 days.[16]

Hormesis means that high and low doses produce opposite effects. This does not mean that we can ignore the toxic effects of physical agents and chemicals on human and environmental health. What hormesis does mean is that by pursuing zero exposure, often at great economical costs, we are also pursuing zero benefit from those agents. We will return to the issue in a later part of our discussion. First some other measurements.

Epidemics and statistics

In a conventional medical dictionary an epidemic is defined as a "sudden outbreak of infectious disease that spreads rapidly through a population."[17] In this context diseases like asbestosis and cancer, which are not infectious diseases, don't lead to epidemics. However, the study of epidemic disease, which is called epidemiology, has extended the idea to include the study of "diseases that relate to the environment or ways of life." This "study of diseases" related to environment and occupation is an important part of human toxicology.

The main objective is to find out if, within a group of people, a cohort, there is a *cause–effect relationship* between the incidence of disease and the exposure to a chemical or physical agent. When the study involves lifestyle, additional factors such as diet, exercise, and age are also considered. Epidemiology can take several different forms and is usually quite a complex process.[18]

An epidemiological study compares the health or disease patterns in a population group, or a group of workers, exposed to a certain agent, with those of an unexposed or "control" group. The numbers of the sick and the healthy and the numbers of exposed people are analyzed statistically to decide if the result is meaningful. If in an exposed group of ten thousand individuals three subjects got sick and in the control group two got sick, the difference is hardly significant. If both groups contain only ten subjects, three is significantly more than two and further investigation is clearly indicated. The numbers and the proportions have to be large enough so that the information is *statistically meaningful.*

An important aspect of epidemiological studies is the issue of *confounding factors.* A confounding factor is a particular circumstance that could influence the outcome of the study. For example, to study the effects of grain dust on the health of grain workers, one would study the respiratory health of an exposed group in comparison with a control group representing the general population. Other factors that affect the respiratory system, such as smoking, could influence the results either for the worker group or the control group, or both. In such a study smoking could be a confounding factor, and the smoking habits in both groups must be carefully documented. The statistics must include correction factors that compensate for the effect of the confounding factor, in this case cigarette smoking.

An example of how confounding factors are used is a study involving babies that were exposed to PCBs, through their mothers' diets, before birth.[19] The authors said that "over seventy confounding factors were considered" in the study that might have affected the results attributed to PCB exposure, although they didn't specify these factors. The study dealt with the mothers' consumption of fish from heavily polluted Lake Michigan. Considering all the possible chemicals in that lake and in the fish that could confound the PCB effects would be very difficult. In a follow-up study it was found that the *in utero* exposure to PCBs had a continuing effect on the mental development of these children.[20] This time the authors did list some possible confounding factors, such as DDT and lead. The list did not include potential confounders like chromium or mercury or cadmium.

An epidemiological study can sometimes be supported by evidence from related research. In this PCB case such support comes from the observation that related health effects were seen in a study with rhesus monkeys.[21] Even among experts there is usually ample room for disagreement concerning the conclusions from epidemiological studies. However, the epidemiological approach is often the only possible way for studying health effects in human subjects. The issue of confounding variables is a difficult part of epidemiology; we will encounter several more examples.

Making sense of it

With the proper equipment and trained personnel the measurement of toxicity is straightforward. After all the experimental work is done, the results are numbers, dose-response curves, and a large collection of samples in the pathology laboratory. Then we have to decide on the meaning of this information. The graphs and numbers must be interpreted before the information becomes useful. Particularly, what does it mean for the human animal?

Crunching the numbers

It has been argued that animal tests are meaningless and give no useful information. A critic of animal testing stated that the LD_{50} test fails to give useful information in "areas of great concern such as the exact poisonous dose of a substance" or "the harmful and non-harmful dose of any given substance."[22] This is of course true for the LD_{50} test, but this test is an important prelude to more complex testing for chronic toxicity or reproductive toxicity. Our brief exploration of the methods of testing shows that "the exact poisonous dose of a substance" and the "harmful or non-harmful dose" are meaningless expressions.

Special breeding techniques are used to ensure that the animals in an experiment are as similar as possible. The rats, or mice, are often identified as belonging to specially bred, numbered or named strains. Yet at the LD_{50} dose level only 50 percent of the animals die. Obviously the rats in the experiment, although very similar, are not identical. In a chronic toxicity test the toxic agent may cause a certain cancer in a *statistically significant number* of rats, but not in all. Some rats will survive the experiment without getting sick.

If different kinds of rats, from different sources, are placed in one group, the results will be much different. Some bigger, healthier animals will survive relatively high doses and some others will get sick at lower doses. In the human species, four beers into a two-hundred pound body and two beers into a one-hundred pound body represent the same dose of one beer per fifty pounds. That dose will produce symptoms of intoxication in many people, but some individuals may show an effect only at a higher dose. A smaller dose, say one beer per two hundred pounds, one-quarter of the first dose, may produce visible effects in others.

In addition to the dose, other variables influence the way a laboratory animal, or a human, reacts to a toxic chemical. Those variables are the general state of health, nutritional status, and genetic heritage. Defense mechanisms are less efficient if the organism is already stressed by other health problems. The association of health effects with the genetic make-up of the individual has another consequence. A chemical can be completely harmless for thousands of people, but eventually one individual *will be found* to suffer harmful effects from this harmless chemical. For the human species, "the harmful and non-harmful dose" of any given substance *can never be measured.*

Another important detail of toxicity testing addresses "the exact poisonous dose of a substance." A few years ago it was found that the artificial sweetener sodium cyclamate could cause cancer in laboratory rats. Somehow media coverage of the story focussed on the quantities of the chemical given to the animals. The popular calculation at the time was, for a human to consume the same dose would require several cases of diet soft drink per day. The public dismissed the studies as meaningless and a good deal of ridicule was directed at the results. What is the sense of giving lab animals such "enormous amounts" of a chemical?

One explanation suggested that "researchers testing for toxicity with laboratory animals use such high doses to speed up their experiments," supposedly to cut costs. The reader is left with the wrong impression that the scientists are cutting corners at the expense of public health.[23] High doses of chemicals are not *routinely* used, with the exception of carcinogenicity testing. A second look at the dose-response curve in Figure 5.2 readily explains how we get to these high doses.

In the dose range O–R there is no adverse health effect. In order to measure an adverse effect the dose must be increased, for example to S. Now if the dose-response curve we are about to measure is not the solid line in Figure 5.2, but the dotted line, there will still be no effect at dose S, which would then be the NOEL, the highest dose that has no observable effect. The only way to find out if this is the case is to increase the dose beyond point S. If the true dose-response curve lies to the right of the dotted line, we may have to use dose levels so high that they start to interfere with the animal's normal food intake. That means the experiment cannot be continued.

However, when no toxic effects are found the question always remains what would have happened if the dose had been increased one more time? In other words, the absence of visible effects does not mean that the chemical is non-toxic. It only means that a toxic dose level has not yet been reached. There is always the possibility that a higher dose, if the test had been continued, would have shown toxic action. Therefore the fundamental rule in experimental toxicology is that *a chemical can never be proven to be non-toxic.* If no response can be observed in the test organism, the experiment has not answered *any* questions; the experiment has failed. Or phrased in another way, absolute safety cannot be proven. This applies not only to industrial chemicals, but also to drugs, food additives, and chemicals made by plants or animals. *It applies to all chemical and physical agents.*

Toxicity testing is not normally done on human subjects. To obtain information for our own species, other tools must be used. One of these is the epidemiological approach we discussed earlier. Another is extrapolation of animal data. This is a difficult process for several reasons. One major limitation in the translation from laboratory animals to humans is the species specificity mentioned earlier. Another limitation is the non-uniformity of the human population as compared to the intentionally bred uniformity among laboratory animals. Major assumptions and large safety margins are therefore a normal part of the extrapolation process.

Given these problems in applying animal data to humans, the question is what experimental toxicity testing means to our own species. At least part of the answer is that there are not only substantial differences but also similarities between species. Some toxic effects are almost universal throughout the animal kingdom. For example, the thiophosphate pesticides, malathion and its close relatives, act as cholinesterase inhibitors on almost all animal species, insects and mammals. The carcinogenic effect of radiation, albeit very weak, is the same in all species, and excessive exposure to ultraviolet light is harmful to almost all living things.

Another part of the answer is a large database on the toxic effects of many chemicals on many different species of animals. When chemical X is found very toxic to several animal species, it would be foolish to assume that it is non-toxic to humans. Thus, at least some generalizations are possible. Nevertheless, there are still vast numbers of chemicals that have yet not been tested, and there are many chemicals that do show substantial differences from species to species.

The post-mortem on all animals at the end of a toxicity study is a very important part of the protocol. It provides information on precisely which organ is affected, or specifically which type of cancer occurred. This is work for skilled pathologists. They must determine if the animals did indeed get sick or died as the result of the chemical and not from other causes. This means that even before the test can start, a veterinarian must examine all the test animals to make sure they are healthy. The test animals must be fed, kept clean, and observed. Longer term experiments require that food and water intake, weight gain or loss and other details, such as behaviour, be recorded for each animal. Chemical analysis of blood and urine during the duration of the test adds further costs. Consider the necessity for repeating the tests with different species and it will be clear that the use of laboratory animals for toxicity testing is extremely expensive.

This cost factor places the unpractical demand for universal safety in perspective and it is a stimulating factor in the search for alternative methods of testing. Particularly in industrial health and safety we often hear demands for "safer products." There are frequent suggestions that certain chemicals or formulations be replaced by "safer materials." Similar demands are made for food additives and agricultural chemicals. However, the numbers of chemicals and the costs of the tests place a severe strain on the capacity for comprehensive toxicological testing. Nevertheless, existing data banks are growing steadily. For industrial chemicals several collections of detailed *material safety data sheets* or MSDSs, required in many jurisdictions, are now available from a variety of sources.

The claim in a recent report that "the vast majority of chemicals in use have never been tested for toxicological effects" is true.[24] However, the implication that untested chemicals should not be used at all, or that all chemicals in use should be tested forthwith, is not realistic in the light of the numbers, the costs, and the risks and benefits involved. Moreover, our continuing exposure to thousands of naturally occurring chemicals, particularly in our food, has demonstrated that a more pragmatic approach can provide at least a partial solution to the testing bottleneck.

How safe is safe?

An individual's exposure to a toxic agent can be at a high level, a large dose, for a short time in the case of an acute exposure. A longer period of exposure, at a lower dose level, gives a chronic exposure. For a health effect to occur, there must be exposure. Exposure means not just being in the neighbourhood where a toxic spill occurred, it means that the toxic agent has actually entered the body. Living in a house next to a storage site for PCBs means nothing in terms of exposure, unless the containers are leaking or there is a fire.

It follows then that to be safe from poisonous materials, all we have to do is to avoid exposure. If exposure is zero, no toxic chemicals enter the body, there is no toxic action and no beneficial action. This appears to simplify our problems a great deal. Never mind the chemicals, as long as we can maintain zero exposure. Most of the time that is an over-simplification, but this relationship between toxicity and exposure is of very fundamental significance. The relationship between exposure and toxicity leads to an estimate of hazard and can be expressed by way of a simple equation:

$$H = T \times E$$

where H is the hazard, T the toxicity of a given dose to a given species, and E is the exposure. Exposure in turn is an indication of the dose to the individual.

If the exposure is zero, there is no hazard. Conversely, if toxicity is zero the hazard is also zero. Since we can never *prove* toxicity to be zero, hazardous chemical or physical or biological agents are almost exclusively managed by controlling exposure. This carries over into the question of safety. Improving the level of safety means minimizing hazard by minimizing exposure.

The toxicity of a chemical can never be changed. Toxicity depends on how the chemical interacts with the chemical components of the living cell. In the above equation T is an inherent property of the chemical. Therefore the idea that a chemical could be "made less toxic" is a contradiction in terms. Changing toxic properties means changing chemical properties. Once we change the chemical properties we have another chemical. This brings us back to the starting point: the only way to control the hazard associated with a chemical is to control exposure.

The regular occurrence of environmental spills is the best illustration of the difficulties of exposure control. Containers can break or rust, and storage facilities can be affected by fires, earthquakes, or floods. Many chemicals occur as contaminants in food, water and air. Numerous circumstances affect exposure. Some of these are due to "human error," and others are beyond human control.

To strive for low exposures to toxic agents is just common sense. In industry we apply the ALARA principle, exposure should be kept As Low As Reasonably Achievable. Since analytical chemistry has shown us that zero is never zero the demand for absolute safety is unrealistic. Also, hormesis shows that the pursuit of zero exposure is not only futile in practical terms, it also denies us the benefits of the "peck of dirt." The question is not about zero pollution. It is about the minimum *acceptable* level of contamination and that immediately leads to the question: acceptable to whom? Having opened that can of worms, we now have no choice but to examine it in more detail.

Risk and benefit

Risk has to do with the probability, the chance, that a certain event is going to happen. It also has to do with the consequences of that event *if* it happens. The relationship between probability and consequences is the basis of formal methods of measuring and calculating risk. Examples from outside the field of toxicology are well known. The automobile insurance industry uses its many claims records to estimate the probability of accidents happening, in a specific geographical area, for example. Averaging the magnitude of the claims gives an estimate of the consequences, and this together with the probability numbers gives an expression of the risk of car travel in that area.

Similarly, the life insurance industry has data to predict life expectancy, which is the reverse of the probability of death. The probability of a person's death combined with the size of that person's insurance policy allows for an estimate of risk. A 90 year old has a high probability of dying, but if that person's life is insured for one thousand dollars, the risk to the insurance company is small. A 50 year old may have a much better life expectancy, but with a policy worth a million dollars could present a much larger risk to the insurer because of the consequences in the case of death. The 90-year-old client does not pay a large premium and the company does not make much profit from his policy. The 50 year old must pay a hefty premium and the company can make a substantial profit on the million-dollar policy, but only if that client lives! Replacing profit with *benefit* gives us a simplistic but useful example of the relationship between risk and benefit.

This relationship between probability and consequences also applies to toxic agents. The probability of a large scale exposure would be estimated from factors such as the number of people present and the potential for a storage tank to rupture or the rate at which a container leaks. The consequences would be estimated from toxicity measurements, costs of work stoppage, and medical and health costs. If the spill occurred outdoors, additional factors would be population densities, wind direction, or proximity to surface waters. Together these various factors determine the risk associated with toxic agents. Just as in the insurance industry, there is also a benefit to the use of these toxic agents. The notion of benefit from the use of toxic chemicals is causing a great deal of difficulty in today's society.

Very few people question the use of diagnostic X-rays or the use of cancer treatment with radiation. We see these applications of a toxic physical agent as beneficial because we accept the risk, real or perceived, as a trade-off for the benefit. When the same radiation comes from a nuclear reactor perceptions change. In the case of chemicals the distinction has almost completely disappeared. The tendency to take the achievements of modern science and technology for granted is now coupled with an unwillingness to accept any negative consequences. A direct result is the "demand-for-zero-risk" syndrome that has pervaded most industrialized nations.

There is a direct link between risks and costs. The business world has made cost-benefit analysis a familiar subject. Since protection from risk usually means an expenditure of money, the risk-benefit relationship can easily become a cost-benefit relationship. This means that a debate about risk and cost can rapidly move to the political agenda. The key questions about cost, benefit, and acceptability are "benefit to whom?" and "acceptable to whom?" The subjectivity of the possible answers and the belief in the common factoid of zero risk, then leads to the demand "to reduce risk at any cost." The risk factor is the preoccupation of those that are, or are perceived to be, at risk. The cost factor affects those that carry the costs, usually manufacturers, corporations, and various levels of government.

Corporations are in business to make profits. Some of them spend vast sums on research and development and on toxicity testing. That toxicity testing can never prove absolute safety. For the same reasons, the absolute safety of mechanical devices cannot be proven. The zero risk factoid has started a flood of law suits against the manufacturers of the materials, that is the chemicals used for the manufacture of heart valves, synthetic blood vessel material and other implants. The judiciary has allowed substantial damage awards to the victims of defective products, often ignoring the statistics that expressed the risk in numbers. The result, summarized recently in the business press, is that many manufacturers simply withdraw their products from the market.[25] They cannot meet the impossible demand for absolute safety, and they cannot afford the costs if they don't. Similar to the demand for zero exposure and the loss of the benefits of low level exposure, the demand for zero risk from medical devices is about to eliminate the associated benefit.

Particularly controversial are estimates of public risk from such low level exposures as radon in homes or trace pesticides in food or water. These estimates are nearly always based on the "zero threshold hypothesis" and the linear dose-response curve. Our examination of hormesis shows the flaws in this approach. Risks to public health calculated on the *hypothetical* linear dose-response curve are in conflict with the *experimental* hormesis dose-response curve. Regulators prefer the hypothesis and therefore the risk-benefit equation is usually discussed on the basis of hypothetically derived calculations. However, there are a few pragmatic perspectives for judging the risks and the benefits.

The last part per billion

Both perspectives deal with the relationship between the size of the costs and the extent of the risk reduction. The reduction of an environmental risk can often be achieved in stages. Consider for example, removal of a toxic chemical from an effluent stream.

Suppose an almost complete removal, which reduces the risk by 99.9 percent, can be achieved at a cost of X dollars. This almost-complete removal still leaves traces of the contaminant in the effluent and raises the question of removal of these last traces. The cost of removing these last traces is often a multiple of X, so a second stage of remedial action will be vastly more expensive. The cost of completely removing the last part per billion becomes astronomical. The costs of removing the last very small risk are extremely high and out of proportion with the original situation. The last part per billion has become a trans-science problem.

The last part per billion is by far the most expensive part of a risk reduction program, and this means that we have to choose our priorities. Is the large expenditure needed to reduce a small risk to zero the best way to spend scarce resources, or are there other priorities? The wrong answer is not only costly, it can also result in important lost opportunities. Moreover, when we achieve zero today, it will no longer be zero tomorrow when we have a new and better instrument.

For example, inhalation of asbestos fibres in an *occupational* exposure situation can lead to asbestosis or sometimes to lung cancer. Suppose a monitoring program is set up in a school for the protection of the young. The program starts with measurements of airborne fibres, and these are found below the detection limit. Next, an inventory of asbestos-containing building materials in the school is prepared. Since this inventory reveals the presence of asbestos in several places, procedures for renovations and repairs are adopted to protect repair and maintenance workers. Encapsulation of some materials in a boiler room completes the activities. This program has reduced the risk for the school children to a fraction of the original. Remember that the original risk was already a low, *non-occupational* risk.

If additional funds would now remain in the health budget for that school, should that sum be spent on a complete removal of all asbestos to reduce a very small risk to zero? Or is there a better way, a larger benefit, for spending that money? What would be more effective in terms of respiratory protection for the school children, the removal of the last traces of asbestos or, for example, a vigorous anti-smoking campaign? If we also consider that airborne asbestos concentrations often increase after removal operations, it would clearly be the wrong choice to try and remove the last part per million.[26]

Synergistic action

While an organism defends itself against the invasion of a toxic agent, it is under stress. As such, it would be vulnerable to the action of a second toxic agent. Usually the effects of the two agents could just be added together. Sometimes there is a more complicated interaction between the organism and the two toxic agents. That interaction is called synergism, or *synergistic action.*

In toxicology, synergism means that the combined toxic action of two agents is greater than would be expected from the toxicities of the two agents if they had just been added together. A simplified comparison would be to say that a simple additive effect of two agents could be expressed by $1 + 1 = 2$, while a synergistic effect would look more like $1 + 1 = 3$, or perhaps more. The agents in question are nearly always chemicals, but there are cases where one is a physical agent. The toxic action of some chemicals can be substantially increased by the action of sunlight on human skin.

Synergistic action is well known for several prescription drugs. Some prescriptions include advice to the patient to avoid exposure to sunlight, and an almost universal warning is not to consume alcohol with prescription medication. Of recent concern is the potential for unknown synergistic action among the multiple medical prescriptions that are often taken by the elderly.

For industrial chemicals the picture is not so clear. Although cigarette smoking substantially increases the cancer risk from asbestos inhalation, not many examples of synergistic action of industrial chemicals are known. Synergism of multiple chemicals is difficult to measure experimentally, and since the number of possible combinations is almost unlimited, the process would be very time-consuming and incredibly expensive. Synergistic action is a part of experimental toxicology where much research remains to be done.

The philosopher's pet

The relationship between humans and their environment has been described by philosophers since the time of Aristotle and Thomas Aquinas, and later René Descartes and Immanuel Kant. The emergence of ecology as a branch of biological science, some sixty years ago, re-awakened this philosophical concern about the impact of human activity on the environment. An anthology edited by VanDeVeer and Pierce presents a compact survey of environmental ethics.[27] Our discussion does not allow for a detailed study of environmental ethics, but the ethicists do use the language of the environment, and, in turn, affect that language.

Ethnobiology has shown how humans learned to distinguish edible from inedible plants, using taste and experience as their tools.[28] They also used one other tool. Part of the experience factor was their observation of animals. The reasoning was, if an animal can eat a plant without ill effects, then it should be safe for humans to eat too. Although we now know that this generalization does not always hold, early humans used animals for testing their food. Of course, the animal was not restrained in any way, or made to suffer. Nevertheless, the animal did perform an early version of a toxicity test.

As agriculture developed, animals were used for helping to do mechanical work, and they also became part of the food supply. From there to the canary in the mines was a small step. Coal mines often contain pockets of gas that contain methane or carbon monoxide or both, the results of anaerobic biodegradation. Canaries are quite sensitive to carbon monoxide poisoning, in terms of the dose per body weight. The death of the canary indicated a threat of carbon monoxide and gave the miners a chance to get out of the danger area. That was another early animal test.

Animal testing

Animal testing in laboratory-based research is found most frequently in toxicology, medicine, and psychology. The objectives of the testing in these settings are quite different. When the use of chemicals in industry proliferated rapidly, concerns about their safety followed quickly. With the canary as a precedent, other animals were found to be suitable for testing the toxicity of a chemical. The tests vary from acute lethality, the infamous LD_{50} test, to protocols for chronic toxicity, reproductive toxicity, and toxic effects on behaviour. Toxicity testing involves many different species of animals. The most common ones are the rodents, guinea pigs, rats, and mice. Dogs, cats, and monkeys of various species are also used.

In medical research, the use of animals is closely related to toxicity testing. For the development of a new drug, from a newly discovered plant for example, toxicity is the first criterion to decide if development should continue. If, in a first approximation, toxic effects are considered acceptable, a second stage of the research deals with the beneficial effects that the drug could have in disease. This requires that the test animal be sick, so that the researchers can see if the drug works. Therefore laboratory animals must often first be treated to become sick.

The classical example of this situation was the work on insulin. Frederick Banting and his co-workers used surgery on the pancreas of dogs to make them diabetic. Then they used insulin to replace the function of the pancreas. Michael Bliss, the author of the insulin story, wrote that "The insulin story is dramatic, wonderful evidence of how the use of animals in medical research can play an indispensable role in lifesaving medical advances."[29]

Animals are also used to develop mechanical devices like pacemakers, mechanical heart valves, and heart-lung machines. The animals may also undergo surgery so that researchers learn more about surgical techniques, how cells react, how "repairs" can be made.

An area that deserves special mention deals with the tests used in the cosmetics industry. Besides general toxicity, an important criterion for many cosmetics is their potential for irritating eyes and skin. Rabbits are most frequently used for this test. Chemicals under study are placed in one of the rabbit's eyes, the other eye is the control. The rabbit eye test, also called the Draize test, is one of the most controversial toxicity tests. For the rabbit skin test the chemical is placed on a shaved area of skin. A standardized scoring protocol of the visible effects then gives an indication of the potential to irritate the human subject. Obviously, the Draize test is a highly unpleasant experience for the rabbits.

Another category of experiments with animals aims to learn more about the complex chemistry of brain cells. Such experiments have contributed much to present-day understanding of brain function and, perhaps more important, brain malfunction. This is where we have learned about the functions of different parts of the brain, the kind of knowledge that relates to conditions like Alzheimer's and Parkinson's disease. Experiments of this kind are used in psychology and medicine. In these laboratories we can find animals, particularly primates, in cages, partly or completely immobilized, visibly unhappy, with electrodes in their heads and "wired-up" to recording equipment. It is easy to turn this view into a dramatic media presentation, similar to the white baby seals on the ice off Newfoundland. The emotional appeal is irresistible.

Very promising in medicine is the search for therapy at the molecular level, *gene therapy*. We have known about some two thousand hereditary, or genetically controlled, diseases for many years. Sickle cell anemia, cystic fibrosis, and some forms of diabetes, are among the best known examples. More recently there is increasing recognition of the important role of genetics in human disease. Beyond the direct hereditary diseases there are many other forms of genetic control that can affect human disease. Discoveries of "genetic markers" and "defective genes" regularly produce headlines in the literature. Gene therapy aims at treating diseases that are based on genetic characteristics. Some examples are malignant melanoma, hemophilia B, familial hypercholesterolemia, and several forms of cancer. The first human genetic engineering experiment, the transfer of gene-marked immune cells into patients with advanced cancer, started in 1989. Techniques for transferring marked or modified DNA into the mammalian body were developed in animal models in the early 1980s.

The term "human genetic engineering" does not mean the "creation" of humans with abnormal characteristics, either good or bad. It is not "science fiction come true." Gene therapy is addressing important aspects of human disease. It is a field of research full of controversy, struggling with serious questions of ethics, and conceivably the most important development in the fight against human disease. The techniques are first developed on animal models.[30]

Pros and cons

A major disadvantage of animal testing is that toxicity is species specific. This means that our extrapolations from animals to humans cannot be accurate and can cause serious errors in interpretation. The thalidomide tragedy of the early 1960s is one of the best known examples.

A further serious difficulty is the cost. The LD_{50} test for acute lethality gives some indication of what is involved. A common protocol for this test requires five male and five female rats for each of four dose levels. The same numbers apply to acute dermal toxicity and an acute inhalation test. This means that three acute toxicity tests on one chemical need 120 test animals. These animals have to be examined for good health, they must be fed, housed, maintained, and observed. At the end of the test the surviving animals are sacrificed and examined by a pathologist. Similarly, for a ninety-one day chronic toxicity test one hundred and twenty animals are needed and in that case the post-mortem pathology requires the examination of some forty kinds of tissue for each animal. During these ninety-one days there are blood chemistry tests and urine analyses, weighings, and measurements of food intake, excretions, and other observations. If there were easy alternatives, the company and its practicing toxicologist would be the first to use them. The enormous cost problems of animal testing have been discussed in the scientific press since 1986.[31,32]

Also part of the cost issue is a commonly held opinion that "most laboratory research with animals is not necessary." The funding squeeze is also relevant to this objection. Research funds in general are becoming very scarce. Most of the granting agencies have standing ethics committees. The merits and the ethical aspects of a proposed research project are closely scrutinized. Under those circumstances "unnecessary experiments" are luxuries few can afford.

Two issues are frequently part of the debates on the use of animals. One deals with the LD_{50} test and the other with cosmetics. A principal criticism of the LD_{50} test is that the information it provides is so limited that it is just about useless. The test is frequently labelled as unnecessary cruelty to the animals. This argument leads us to a conflict in our discussions about all the "deadly toxic chemicals" that are released into the environment. We will use the chlorinated pesticides and the chlorinated dioxins as examples.

The major weapon for condemning and controlling these chemicals is their acute toxicity, and the weapon was fashioned with the help of the LD_{50} test. It goes like this. In *Silent Spring* we find many comparisons of chemicals based on their toxicity. "Endrin is the most toxic of all the chlorinated hydrocarbons," writes Carson. "It is fifteen times as poisonous as DDT to mammals, thirty times as poisonous to fish and about 300 times as poisonous to some birds."[33] Carson does not reference the sources of these numbers, but it seems safe to assume they are based on LD_{50} tests of acute lethality. As we examine the literature on the dioxins, we find repeated mention of the incredibly high toxicity of one member of the dioxin family, called TCDD. Almost the entire dioxin story and the web of controversy around it is based on the acute toxicity of TCDD, measured as the LD_{50}.

This is the conflict: if the LD_{50} test is of little value we must dismiss the evidence based on it. We must then dismiss Carson's comparisons, and those of similar essays on pesticide toxicity. We must also dismiss most of the dioxin story as a "hype" based on a test with little validity. Alternatively, if we wish to continue our belief in Carson's comparative toxicity of pesticides and we wish to retain the dioxin saga as it has been written, we cannot dismiss the validity of the LD_{50} test. As the popular saying goes: "you can't have it both ways."

For some reason, possibly the publicizing of the Draize test and similar test methods, protesters have singled out the cosmetics industry as being "ripe for attack." Several public figures, actresses and the like, threw the weight of their celebrity behind various forms of boycotts.[34] The cosmetics trade began to see the equivalent of the health food store. Very carefully worded labels on "green cosmetics" tell the customer that the manufacturer of the product did not use animal testing. The label usually does not state explicitly that the ingredients of the product *have not been tested* on animals. A detailed report in *Chemical and Engineering News* gives insight into what is happening.[35] The cosmetics trade involves a very large number of companies that formulate shampoos, skin lotions, and cleansers. They use ingredients supplied by a much smaller number of manufacturing chemical firms, and an even smaller number of so-called perfumery houses. The animal rights protests had two effects. Both the manufacturers and the formulators have found that a more critical look at their animal testing programs led to substantial financial savings. The controversy is also "driving many companies to reformulate existing products and ingredients" rather than develop new ones.[36]

The report quoted a company source as: "Even if a company says a final formula is not tested on animals, there's a (good) chance that somewhere down the line the ingredients were animal tested by one of their suppliers or someone else." The formulator did not lie, but the manufacturer did comply with the law. Because in most jurisdictions, legislation requires that the ingredients used for making cosmetics have satisfied certain criteria of toxicity. Those criteria make the use of animals mandatory. The label did not lie, it merely omitted some details. It misled the customer who expected to buy a product that was not tested with animals. The ethical stance against animal testing led to questionable ethics concerning the label on the "ethical cosmetics."

Alternatives

There are viable alternatives to some testing procedures that use live animals. Since toxicity is based on the interaction of chemicals within a living organism, often within specific cells in that organism, toxicity cannot be measured with an instrument.

Fortunately for the rabbits, one of the best *in vitro* alternatives is a replacement for the Draize test. That method involves the chorioallantoic membrane (CAM), the thin, delicate membrane that becomes visible when part of the shell of a fertilized chicken egg is removed. The procedure is called the CAM test. When a test chemical is applied to the membrane, a trained observer can see the evidence of inflammation or irritation. Scoring the observations according to a standard protocol gives a number for irritancy potential. Initially, this alternative method of testing had to be validated by comparison with the original Draize test before it was found acceptable as a replacement.

Cells taken from human skin can be cultured in a nutrient medium. When a chemical that irritates or inflames the skin is added to the culture, changes in the cell structure can be made visible with certain dyes. An instrument can measure the intensity of the colour changes, and give a numerical value that relates to the test chemical's potential to irritate human skin. This method is also a good replacement for the rabbit skin test. Many other interactions between chemicals and living organisms can be observed at the cellular level. Specific groups of cells can often be obtained from animals or from human subjects without harming them.

Probably the most expensive area of animal testing is in cancer research. The testing of a single chemical can cost, in 1989 US dollars, between 500,000 and one million, take two or three years to complete, and take the lives of thousands of laboratory animals.[37] Therefore the Ames mutagenicity test, developed by Bruce Ames about thirty years ago, was a significant step forward in the practice of toxicology. The test employs a simple bacterial culture and can show the mutagenic potential of a chemical. Since almost all mutagens are also carcinogenic, it became possible to test large numbers of chemicals at a fraction of the cost and time requirements of animal tests.[38]

The test is not a replacement of animal testing, but an important supplement to it. It has significant limitations. It detects only certain types of mutations and it responds only partially to chemicals that must first be altered by metabolic reactions in the animal before they can express their carcinogenic potential. Some metabolic changes of the chemical can be detected because of the addition of oxidizing enzymes to the cultures. These enzymes are obtained from a homogenized extract of rat liver. The Ames test does use the fresh livers of laboratory rats; it is not a completely animal free test.[39]

We encountered earlier the mathematical approach to the calculation of quantitative structure activity relationships or QSARs.[40] Theoretical chemistry has developed the basic principles that can explain similarities and differences between closely related molecules. For example, it is not a complex problem to explain why water is a liquid and the closely related hydrogen sulfide is not. The same principles explain why methanol (methyl hydrate or wood alcohol), ethanol (the stuff in wine or beer) and ethylene glycol (antifreeze) are all soluble in water. The problem becomes more complex if we want to explain why methanol and antifreeze are deadly toxic, while ethanol is much less toxic, and the closely related glycerol is essential for life. Such a model must, besides the architecture of the molecule, also consider how the molecule interacts with enzymes and other chemicals in the living cell. These problems are extremely complex and progress in this area is difficult. Explaining and predicting properties like solubility, boiling points and similar properties is an early beginning of the use of QSARs. Predicting toxicity, particularly the many different kinds of toxicity, is still in the future.

A few general observations characterize the current situation with respect to alternative methods.[41] The relationship between the dose of a chemical and the response of a whole organism is an extremely complex interplay of many interdependent processes. It takes considerable time for a new testing technology to find acceptance. It is unlikely that theoretical and *in vitro* methods will completely replace animal tests. New methods have to compete with existing procedures that boast a nearly sixty-year record of application. The relatively few *in vitro* tests that are now available can be extremely useful as supplementary technology that saves vast amounts of money and the lives of many laboratory animals.

Whelan has pointed out that the only real alternative to animal testing, short of using human subjects, is epidemiology.[42] She explained: "Epidemiologists have excelled in identifying cancer-causing substances like tobacco, asbestos, vinyl chloride, diethylstilbestrol, and others, but only after the fact. Epidemiology does not allow us to predict in advance the cancer risk of a new chemical to humans." She emphasized how the long latency period for cancer bedevils epidemiology, since after the long lead time, information of exposures and work habits becomes less accurate, because the data have to "rely on faded memories and lost records." Epidemiologists, at least in the case of cancer, must wait for decades before they get the information.

The development of surgical procedures and mechanical devices is benefitting significantly from graphical computer simulation. Ultimately, a trial on a live animal must confirm the result. Which species of animal that should be, whether the human species should supply laboratory "animals" is a trans-science question that needs to find an answer somewhere else. For the time being, there remain areas of research where there is presently just no replacement for animals.

CHAPTER 6

Cancer everywhere?

Health information, prepared for people who handle chemicals on the job, can contain some disagreeable details. The detailed descriptions of how corrosive chemicals destroy human skin, or the precise action of chemicals that ruin lung tissue, are not enjoyable reading. However unpleasant, these feelings seem to fade in the light of the word "cancer." Chemicals that can cause cancer are feared with an intense fear. They are considered a class apart, the ultimate threat of the industrial world.

Does everything cause cancer? According to Michael Jacobson the answer is: "No. Some things cause mutations and birth defects."[1] There is a general perception that cancer is increasing to the point where many are using the word epidemic. One influence on the shaping of that perception follows from a report in a Dutch newspaper. For five common diseases, rheumatism, cardiovascular disease, cancer, respiratory diseases, and diabetes, the reporter compared percentages of the rates of disease incidence with the percentages of media reporting on each disease.[2] The comparison showed that the media select on the basis of "dread-value" rather than on the basis of disease prevalence. Thus, while the number of cancer patients (presumably in the Netherlands) represented about 11 percent of the total of these five categories, the reporting on cancer filled about 35 percent of the medical items in media reports. Conversely, the 41 percent of total disease cases made up by rheumatism sufferers scarcely merited 10 percent of the media's attention. The terminal nature and premature death aspect of cancer represent a dread that is more newsworthy than the years of chronic pain that rheumatism patients have to contend with. The result is a distorted perception of cancer prevalence. Yet the questions remain. Why so much cancer? Does pollution cause cancer?

The double helix

Cancer, birth defects, and mutations are words that are associated with a communication system that is vital to all living organisms. Not communication in the sense of one human communicating with another, but biological communication from one generation to another. A chicken egg contains information that ensures that, when the egg hatches, what comes out is another chicken. Specifically, if a white chicken laid the egg, what comes out will not only have feathers, but most probably white feathers. What we are dealing with is *genetic information.* Genetic information is the plan, the blueprint for the design of living organisms. An integral part of that blueprint is cancer.

The genetic blueprint

The genetic message is transmitted from one generation to the next. The chicken and egg example show this in the case of *generations of individuals.* This information is also transmitted from one *generation of cells* to the next. This part needs a closer look.

The cells in a living organism continuously undergo cell division, the formation of two cells from one. If the organism is still growing, like young children or an unborn fetus, cell division in nearly all types of tissue is rapid, and there is a net increase in the number of cells. In adults the total number of cells stays more or less constant. Cell division in adults has to provide only for the replacement of worn-out cells and for repair of damage. The rate of cell division varies and depends on the type of tissue. In gum tissue or in the lining of the intestines cell division is still rapid. In others, like bone, the process is much slower. As new cells are formed, old cells die and are discarded. The chemicals these old cells were made of are re-used for making new cells and some waste chemicals are excreted.

Cell division starts with a series of *chemical* changes in the cell. The complicated process is described in plain language in James Watson's *The Double Helix.*[3] For those with a little chemistry background good descriptions are also found in several chemistry textbooks, particularly those aimed at students in the health sciences.[4] Those biochemical details may be interesting for some, but they are not essential for our discussion.

In all living cells, small molecules, building blocks for larger structures, are built or broken down as needed. Some building blocks, called *mononucleotides,* are used for making very large molecules of deoxyribonucleic acid (DNA) and ribonucleic acid (RNA). They are called nucleic acids because these materials are mainly found in the nucleus of the cell. The DNA molecules associate in paired strands that are coiled around each other in the form of a helix. They are coded for information purposes, hence terms like the *genetic code,* and the *double helix.*

There are four different kinds of mononucleotides. The *sequence* in which these small subunits are placed along the strands of DNA works like a code. The function of this code is to guide the synthesis of proteins. Proteins, in turn, are made of simpler building blocks called amino acids, attached to each other in a specific and unique sequence. Thus, the *sequence of units* along a DNA chain is the coded instruction, a blueprint, for determining the *sequence* of amino acids that form a protein molecule.

There are many thousands of mononucleotide subunits on one DNA molecule. Groups of three subunits represent the code that identities the various amino acids. *Genes* are sections of DNA containing several hundreds or thousands of these subgroups of three mononucleotides. *One gene* contains the information needed to make *one specific protein.*

Just prior to cell division the DNA strands in a cell start to cluster and stick together in shorter, thicker structures, the chromosomes. In this form a replication process takes place whereby a second set of chromosomes, identical to the original set, is constructed from the small building blocks in the cell. After cell division, the two daughter cells each have a complete set of chromosomes. Therefore they also have the same genetic information. Although the information varies from one species of organism to the next, the genetic code for all living organisms is written in the same language. It is a language consisting of four different mononucleotides arranged in groups of three. The analogy with a language, consisting of three-letter words written with a four-letter alphabet, readily comes to mind. In precise detail the genetic information is unique for each individual.

In his book *The Dragons of Eden,* Carl Sagan presented the arithmetic to relate amounts of information to the genetic code.[5] One molecule of chromosomal DNA contains approximately five billion (five thousand million or 5×10^9) pairs of nucleotides. In the binary language of the computer, describing five billion letters requires twenty billion bits of binary language. Simple arithmetic allows us to convert the four-letter alphabet to an alphabet of about sixty characters (letters, numerals, and other symbols), which is enough for most human languages. It can then be calculated that the information in one human chromosome is equivalent to about four thousand books of five hundred pages, each with about three hundred words per page. Alternatively, since there are eight bits per byte, the twenty billion bits are equivalent to about 2500 megabytes. That is the information *capacity* of one human chromosome. With twenty-three pairs of chromosomes in each human cell, the language of life is not "lost for words." Why is this relevant? During cell division all this information has to be replicated. Let's look at the problem of copying four thousand books of five hundred pages, within a few hours and without a photocopier.

When hi-fi goes wrong

The information encoded in the chromosomes serves two goals. Every time a cell divides, the information has to be copied or *replicated* so that it can be passed on to the newly formed daughter cells. Within the cells, the instructions in the code must be carried out so that each cell fulfils its specific biological task. Each cell must make the precise proteins it needs for that task. Genetic information ensures that the daughter cells of a red blood cell are indeed red blood cells and do the work of a red blood cell.

Cells must follow specific instructions for converting fats or carbohydrates into energy. That conversion involves many chemical changes that must occur very quickly. Therefore the cells also make proteins that work as catalysts, they regulate the chemical reactions within the cell. These proteins are called enzymes.

Each cell needs to "know" when it has to divide again or when it has to stop dividing. In the early stages of a developing embryo all cells are the same. The information in the chromosomes "instructs" some cells to become liver cells and others to become brain cells. The coded information contains instructions for controlling cell division and *cell differentiation.*

All these functions of a cell are encoded in the DNA of the species. When that information is passed on, errors in the copying process have far reaching consequences. Once an error has occurred, it will be copied with every subsequent cell division, and all further generations of cells will contain that error. If the specific error causes a malfunction, all further generations of cells will malfunction in the same way. Information must be passed on with a very high degree of fidelity, it is a real hi-fi process. To help preserve the integrity of the coded information, the cell also contains several clever mechanisms for repairing damage after it has occurred. The cell has *defense mechanisms* for coping with damage to the code, up to a point.

When the nucleic acid in the cell starts building a copy of itself just prior to cell division, it attracts to itself the individual mononucleotides present in the cell. However, there are many other chemicals that are part of the cell content. There are amino acids, proteins, carbohydrates, fats, and many other chemicals. There are also chemicals that do not really belong in that cell, but came there by accident, in food or drink for example, or in the air that we breathe in.

Many of these foreign chemicals can interfere with the cell division process. They can cause errors in the code, called mutations. A *mutagen* or a *mutagenic agent* is a chemical that has the potential to interfere with the transmission of the genetic message. Thousands of chemicals are known to be mutagenic. Some physical agents, radioactivity, X-rays, ultraviolet light, are also mutagenic agents.

Mutations of the code can have different results. A mutation may simply lead to a minor change in the organism. If that change is favourable for the survival of that organism, it will pass the mutated gene on to its offspring. If the mutation is unfavourable for survival, the organism will die. These many random mutations, studied by Gregor Mendel in the mid-1800s, have been the basis of the natural selection that led to Darwin's *Origin.* Without mutations the abundant variety of living organisms would not have evolved. Without mutations we would not be here.

If a mutation leads to damage of the code, there are two possibilities. When the damage is severe, the resulting daughter cells may not function at all. They die and are discarded. When the damage is slight, the daughter cells may be viable, but with an impaired biological function. Mutated liver cells may be unable to oxidize certain food chemicals. Other mutated cells may no longer synthesize an enzyme their predecessors used to make.

If the control mechanism for cell division is damaged because of a mutation, the cell and its daughter cells may continue to divide in an uncontrolled way. The result is a rapid, uncontrolled growth that usually consumes nutrients at the cost of surrounding tissue: a tumour is growing. When a chemical has the *ability* to cause the formation of a cancerous tumour, it is called a carcinogen. That does not automatically mean that it *will* cause a tumour. Carcinogens are nearly always mutagens, but not all mutagens are carcinogens.

A mutagen can also affect the cells in a developing fetus. Again, if the error is severe, the fetus may not be viable and die, resulting in a miscarriage or spontaneous abortion. If the error allows the fetus to live, birth defects may result. Although birth defects come in many different kinds, physical malformations and malfunctions are probably the best known. The well-known tranquilizer thalidomide, introduced in the early 1960s, is a mutagen that affects the cell differentiation process in a developing human fetus. The result is offspring with deformed or absent limbs, a condition medically known as *phocomelia*. When a mutagenic agent has the *ability* to cause a defect in the developing fetus, it is called a *teratogen* or a *teratogenic agent*. Incidentally, thalidomide is also effective in the treatment of leprosy. For this reason it is extensively used in South American countries where this disease is common. Lack of education and lack of distribution control has caused the continued occurrence of thalidomide babies in these countries as recently as 1994.

Thus, a mutagenic agent, chemical or physical, can cause either cancer or developmental effects in the newborn. There is one more detail. Cells in a mammalian body are of two kinds, the germ cells, egg and sperm, and all the others, the somatic cells. So far we have looked only at somatic cells. When a germ cell undergoes mutation, the damage can easily prevent conception. However, if damage is slight, a viable embryo can result. In that case the mutated code in the germ cells of the parents becomes part of all the cells of the embryo, including *its own germ cells*. When the offspring matures, it could pass on the mutated information, an *inheritable defect*, to future generations. Again, if the mutation is favourable for the survival of the offspring, it will also pass on that advantageous mutation.

An inheritable defect can be caused only if the mutagenic agent affects the germ cells. Many environmental and health information sources fail to make this distinction. One environmental science text treats "mutations" as a distinct health effect and by implication suggests that it is something different from cancer or teratogenicity.[6] Other texts are also unclear about the relationships between mutagenicity, carcinogenicity and teratogenicity and the issue of inheritable defects.[7] The resulting confusion has created the popular factoid that many chemicals cause inheritable defects. However, in the rare cases where a fetus indeed survives with genetically damaged germ cells, the resulting mutations are usually not new. They are nearly always of the kind that is already part of the human gene pool.

The above scenarios represent, albeit much simplified, the possible interactions of a mutagenic agent with the genetic code. The mutagenic agent is either a chemical or a physical agent such as radiation. The agents involved in this mechanism are called *genotoxic* carcinogens.[8] There are other events that can lead to the formation of malignancies or birth defects that do not involve the genetic code directly.

Radiation and a variety of chemicals can affect *control mechanisms* within the organism. For example, some control mechanisms operate by means of enzymes that are induced or "switched on" at the proper time. Whereas a mutagen could prevent such enzymes from being formed, by causing an error in the DNA, other chemicals, which are not mutagens, may interact *directly* with the enzyme and disable it. In this way cancer or birth defects are possible without mutation of the genetic code itself. Although not necessarily genotoxic, the chemicals involved are still acting directly in the form in which they enter the organism. They are still *primary carcinogens*.[9]

Many other chemicals can express their ability to cause cancer only after they have been biochemically changed, or metabolized, within the organism. These chemicals are called *procarcinogens* or proximate carcinogens.

Agents of cancer

Simply listening to friends and reading the local newspaper will leave little doubt that the number of cancer cases has increased dramatically in much of the industrialized world during the past two decades. The media produce a steady flow of numbers, revised numbers, and predicted numbers, produced by various agencies. The term epidemic is becoming a buzzword; in many minds environmental degradation is to blame.

The important details in any numbers game relevant to health problems are the difference between *numbers and rates* and the question of age. Simply saying that the number of cancer cases is increasing has no meaning. As the size of a population increases, the number of cases *per thousand* population is the only meaningful number. In other words, we must look at cancer *rates* rather than at numbers. Furthermore, since age is an important factor in cancer, the rates must be specified for various age groups. In the statistician's language, the rates must be age-corrected.

Two other ongoing arguments affect the cancer scene. One is the perceived connection between cancer and environmental pollution. The other is the argument about the assumption that carcinogens are, from a toxicological viewpoint, a class by themselves. The characteristic feature of a carcinogenic agent, so say the proponents of this idea, is that for a carcinogen there does not exist a so-called "safe dose."

After our brief look at the biological mechanisms, it is time to examine these issues in more detail. We will start with the causative agents, then we will look at the "safe dose."

Carcinogens, mutagens and teratogens

Samuel Epstein, an American cancer researcher with a long list of impressive credentials, made a much quoted statement in 1976 when he asserted that:

> Most human cancer is environmental in origin, and therefore should be preventable.[10]

In a curious way, this statement is both right and wrong. Epstein never explained what he meant by "environmental in origin." From the context of his writing it is obvious that he meant the industrially polluted environment.

The first part of Epstein's thesis is almost right. Almost, but not quite. It is not most, but *all* human cancer that is environmental in origin. Epstein did not invent this somewhat dramatic statement, he just used it. Another cancer specialist made this unfortunate statement quite a few years earlier. Epstein merely embellished the idea by taking the word environment to mean *the pollution in the environment.*

John Higginson, an Irish physician, became the first director of the International Agency for Research on Cancer (IARC), when it was founded in 1966 by the World

Health Organization. By that time Higginson had been in cancer research for many years. In the 1950s he compared the incidence of certain cancers in blacks in the USA with those in Africa. As a result of that research he stated: "About two-thirds of all cancers have an environmental cause and would therefore be preventable." The word was out and it was copied a thousandfold.

In a lengthy interview with *Science* some thirty years later, Higginson explained the misinterpretation of his statements.[11] Whereas he meant to include in "environment" everything from smoking to eating to lifestyle, his unfortunate words from the 1950s were simply taken to mean "pollution." From there on "cancer and the environment" became synonymous with "cancer and pollution."

The entire debate hinges on the all-inclusive meaning of the word environment. For most ecologists the environment is the sum total of the physical surroundings in which the human species lives, together with all the other life on the planet. This includes the indoors as well as the outdoors. It includes food, water and air and all the things that occur on the surface of the planet. This was clearly the meaning of "environment" that Higginson had in mind.

What was probably not so well known in the 1950s is the extent to which this environment is full of many naturally occurring mutagenic agents that have the potential to cause cancer. There are thousands of plants, including many normal components of the human diet, that contain naturally occurring carcinogenic chemicals. Many fungi make chemicals that are potent carcinogens. Naturally occurring radioactivity is plentiful in the environment. All living organisms are made of carbon, and carbon comes in different varieties, including radioactive carbon-14. Nature, that is the environment, is filled with mutagenic agents. To suggest that cancer is preventable is like saying "the environment is preventable."

The scope of our discussions does not allow for the inclusion of the long lists or extensive tabulations of specific chemicals that have been shown to be mutagenic. Very detailed surveys of this kind were presented by Edith Efron in 1984, one listing of industrial chemicals and another for the huge number of naturally occurring carcinogens.[12] That listing is not obsolete, after a further decade of research it will merely be incomplete. We will therefore restrict our discussion to a few general points about mutagenicity.

Bruce Ames is a biochemist at the University of California. Many years ago he invented and perfected a relatively simple method for testing mutagenic activity.[13] The procedure, now known as the Ames mutagenicity test, makes use of bacteria. It is fast and relatively cheap. This test has made it possible to test large numbers of chemicals. Many known carcinogens test positive in the Ames test; they are mutagens. Conversely, if the Ames test shows a chemical to be a mutagen, it is almost certainly a carcinogen or a teratogen.

During the past decade Ames has reviewed the literature of many naturally occurring chemicals in ordinary foods that are powerful mutagens.[14] By implication, most are also carcinogens and many have been shown to be carcinogenic in animals. Many references to the original studies show that foods such as parsnips, black pepper, and edible mushrooms, to name a few, contain potent carcinogens. Ames also proposed an explanation of how the chemistry of fat metabolism can give rise to cancer.

Close scrutiny of the literature reviewed by Ames makes it clear that every single human meal, however carefully and "organically" grown, contains many carcinogens. Many of these natural carcinogens occur in food not in trace amounts, like picograms or parts per billion. They occur in the human diet in *gram quantities*. Since humans have consumed these foods for centuries, the conclusion that carcinogenic chemicals are an intrinsic part of the natural environment is inescapable. This raises serious questions about the no-safe-dose hypothesis. But first we have to look at other potential culprits.

X-rays and radioactivity

The nuclei of some atoms are not quite stable. For reasons we can't go into here, atoms of these elements can easily fall apart, losing a few bits and pieces in the process. The bits and pieces falling off can be nuclei of helium atoms, called alpha-particles, or they can be electrons, called beta-particles. They can also be high intensity X-rays, which in this context are called gamma-"particles." Particles are in quotation marks because they are not really particles. All three forms of debris are ejected from the disintegrating nucleus with considerable force. This means they represent energy. The ejected particles travel at great speed; collectively they are called *radiation*. The element or nucleus responsible for the event is said to be *radioactive*. The release of alpha and beta-particles is called, naturally, particle radiation. The gamma-radiation is of the same kind as visible light and ultraviolet light, but is associated with much more energy.

The atoms that undergo this kind of decay are unstable because the nuclei of their atoms contain "extra" neutrons. Many elements exist in two or more versions. These different versions of one element are called *isotopes*. Some isotopes are stable, others can easily fall apart, they are said to *decay*, and are radioactive. The only difference between two isotopes of the same element is their *physical behaviour* in terms of stability and radiation. Their *chemical behaviour* is the same. For example, the element carbon can exist as carbon-12 and carbon-13, written as 12-C and 13-C, both of which are stable. There is also the carbon-14 isotope, 14-C, which is unstable and radioactive. Chemically all three kinds of carbon are the same, all three become incorporated in organic molecules, both inside and outside the body. Since the unstable isotopes decay, there is usually much less of them around in the environment. Their *natural abundance* is very low.

Some unstable isotopes occur naturally, others can be made in the laboratory. After a nucleus has decayed and the radiation has dissipated, a particle remains that is the nucleus of a different element. For example, when a nucleus of uranium-238 decays, several bits and pieces are released as radiation, and what remains is a nucleus of thorium-234, which is also unstable. The decay of thorium leads to further unstable isotopes until at the end of a short series of decays a stable isotope of lead is formed. These radioactivity decay series of the heavy elements are issues in nuclear chemistry and nuclear physics.

We will encounter the radioactivity question again later in connection with energy issues. At this point it is important to look at radiation energy at the molecular level. This energy can interact with molecules in living cells. When water interacts with the energy of microwaves, the water absorbs the energy and becomes warm. When the molecules of a dye interact with the energy of daylight, they absorb one colour of the light and reflect another colour. If the dye molecule is constructed so that it is easily damaged, energy associated with daylight can make these molecules fall apart and lose their colour. A coloured fabric fades if the dye is *photodegradable.*

When sunlight interacts with some chemicals in human skin, enough energy can be absorbed to cause increased synthesis of pigment in the skin, which then changes colour. If the skin absorbs a lot of energy at once, a painful burn can be the result. Skin that contains a lot of natural pigment can tolerate more sunlight than a light skin with little pigment.

If the energy of the radiation is high enough, it can dislodge an electron from a molecule and create a positively charged particle that is called an *ion.* That kind of radiation is called *ionizing radiation.* Relative to other forms of radiation, ionizing radiation starts approximately at the ultraviolet part of the spectrum and continues all the way to X-rays and gamma-rays. When ionizing radiation interacts with some molecules within a cell, that cell can become damaged in such a way that it becomes a cancer cell. Ionizing radiation has the *potential* to start the formation of a cancer and is therefore called a carcinogen. It is a very weak carcinogen and it is only known to cause cancer at very high dose levels. The perception that low level radiation causes cancer follows only from calculations based on the zero-threshold hypothesis, not from observation of cancer.

Health warnings for ionizing radiation are well known. They vary from the warning signs in and around nuclear reactors to the badges worn by X-ray technicians in hospitals. These warnings deal primarily with *occupational exposures.* Warnings about the ultraviolet (UV) levels of sunlight have become a feature in weather reports on some radio and television programs. The no-safe-dose-for-carcinogens hypothesis that we discussed briefly in the previous chapter, was inspired by the earlier hypothesis of a zero threshold for radiation. That hypothesis, based on the assumption of a straight-line dose-response curve, states that there can be no "safe level" of exposure to ionizing radiation. That would mean that one single chest X-ray could lead to trouble.

The zero-threshold hypothesis for ionizing radiation runs almost parallel to the no-safe-dose-for-carcinogens argument. These hypotheses are being argued in the literature and we will look at that dispute shortly. In the meantime, what are the realities of radiation?

Radioactive isotopes are an integral part of the molecules of life. Bananas and broccoli are healthy foods, in part because they are rich in minerals. One of these essential minerals is potassium. The daily requirement for an adult human is about four grams, which is mostly supplied from meat, fruit and vegetables. Human milk contains about 45 mg of potassium per 100 mL. However, potassium is one of the many elements that exists in different isotopic forms. Of all the potassium consumed in the diet, about 0.01 percent is naturally occurring, radioactive potassium-40. If bananas and broccoli and human milk are rich in potassium, they are also rich in potassium-40. Could it be that there is no "safe level" of eating broccoli or bananas? What about nursing the baby?

Except water and small amounts of minerals, all chemicals in a living organism contain carbon, this is why the chemistry of carbon was named organic chemistry. This means all living organisms also contain carbon-14, the natural, radioactive isotope of carbon, used so conveniently by archaeologists for measuring the age of their finds. If there is no safe dose for exposure to radioactive carbon-14, sleeping next to another human being is a health hazard. And how do we keep on living in this body that contains carbon-14 and potassium-40 and a few other radioactive odds 'n' sods? Clearly, there must be something wrong with this "no-safe-dose" idea. It just does not fit.

Electric and magnetic fields

Although this discussion of cancer and related issues cannot possibly be complete, there is one controversial issue that we must not exclude. That is the question of electric and magnetic fields. This topic has been on the scientific research agenda since the mid-1960s, but it was only in 1979 that the story entered the public consciousness. In that year a report was published in the *American Journal of Epidemiology*.[15] The authors, an epidemiologist and a physicist, found a higher than expected incidence of leukaemia among children living close to electrical power lines, transformer substations and similar concentrations of electrical conductors. True to the rules of epidemiology, the authors considered a number of confounding factors such as street congestion, social class, and diet.

Up to about that point in history, the association between cancer and radiation was strictly about *ionizing radiation*. The Wertheimer–Leeper study suggested that the less energetic, *non-ionizing radiation* could also contribute to cancer. The report had an immediate and alarming effect. Public concern about the safety of an essential service, electrical energy, was expressed in several popular magazines. Many researchers climbed aboard the bandwagon.

If electric fields were involved, what better place to look than among electrical workers. Soon several reports appeared of studies on electrical workers, expanding the concern from leukaemia to brain tumours.[16,17] Both leukaemia and brain tumours appeared to occur more frequently in this occupation. Toasters, electric toothbrushes, and electric blankets became instruments of fear.

Criticism was inevitable, and we will summarize some points briefly. Wertheimer and Leeper did not actually measure electric field strengths, they calculated them. This raised serious questions about the actual levels of exposure. Researchers that had studied electrical workers admitted that these people are occupationally also exposed to many chemicals, including PCBs, solvents, fluxes, and metal fumes.[18] Others ended their reports with the suggestion that the topic deserves further study. A serious flaw in the original Wertheimer/Leeper study appears to have escaped most critics. These authors, in considering various confounding factors, did not consider that the grounds around power lines and transformer stations are often kept free of weeds by the liberal use of weedkillers.

Other epidemiological studies contain similar flaws. Ruey, Lin *et al.* while discussing the possible errors in their study, refer to "occupations categorized as non-exposed to chemicals such as mechanics, electrical maintenance and metals production." Why would mechanics and maintenance workers, who are frequently exposed to solvents, fluxes, PCBs and many other chemicals, be classified as "non-exposed to chemicals"?

In an industrial hygiene study on the measurement technology of electric and magnetic fields, the authors referred to nearly forty studies on the epidemiology of the problem and found that there were virtually no reports on animal studies.[19] A 1988 survey of some 65 of these epidemiological studies found that the results were inconclusive.[20] A common weakness of these studies is that exposures are often not measured but expressed in terms of proximity to a source, such as transmission lines.

A more recent news article mentioned biological effects seen in cell cultures. The problem with such culture experiments is that the normal neural and hormonal controls are absent. The author, emphasizing the complexity of the problem and the resulting contradictions, quoted from two sources. A 1990 US EPA report stated that low level electromagnetic fields "are a possible but not proven cause of cancer in humans." Two years later a White House-commissioned panel reported that "adverse health effects are unlikely and major research expansion in this area is not warranted."[21]

The possible interactions between electromagnetic fields and human genetic material were discussed by two groups of researchers recently. Both groups reported that there were no effects of electromagnetic radiation on the genes of a human leukaemic cell culture. Both groups of researchers questioned the validity of previous reports of measurable responses.[22,23]

A major problem with respect to the EMF issue is the scarcity of animal data. Apart from inconclusive reports of leukaemia and brain tumours in *occupationally exposed* workers, researchers don't really know what specific disease patterns to look for. Animal experiments with clearly observable responses would help to clarify this aspect. Conversely, it is highly probable that, if we make the dose levels high enough, sooner or later there will be some adverse health effect. The real danger then lies in another linear extrapolation to the hypothetical zero-threshold point. In the meantime, there is public pressure for answers and for "doing something." Unfortunately, decisions based on insufficient information have a high probability of being wrong.

A one-shot deal?

In 1962, Rachel Carson introduced us to Dr. Otto Warburg, a German biochemist.[24] Warburg's theory on the mechanism of carcinogenesis led him to believe that there is *no safe dose* for a carcinogen. The phrase "no safe dose" for this dreaded disease became a powerful tool for Carson in raising the alarm.

A few years later, in *The Politics of Cancer*, Epstein combined the views of Higginson and Warburg.[25] When he restated his ideas about cancer and the environment, and elaborated on the no-safe-dose concept, we read "There is no known method for measuring or predicting a safe level of exposure to any carcinogen below which cancer will not result." Unlike Warburg, Epstein "explains" the safe level by adding: "That is, there is no basis for the threshold hypothesis which claims that exposure to relatively low levels of carcinogens is safe and therefore justifiable."

We have seen that the zero-threshold or no-safe-dose hypothesis is still used, particularly in risk assessment and for the making of regulations. Although at first glance the argument appears reasonable, there are several flaws.

Epstein's statement is wrong on two counts. Even if it is true that a safe level cannot be measured, it is improper logic to conclude that therefore it does not exist. There have been many things that could not be measured in earlier years and which have since been found to exist. More importantly, both Epstein and Warburg failed to explain what they mean by "safe." Presumably the idea is that the smallest possible amount of a carcinogen, even a single molecule, could cause one single cell to misbehave and start forming a cancer. Our earlier discussion of the dose-response concept has shown the flaw in that reasoning with respect to carcinogenic chemicals. Instead of using a hypothetical linear dose-response curve, let us have a brief look at the events.

A chemical must clear many hurdles as it travels through the body. It must pass intestinal walls or the membranes of the respiratory tract. It must survive many chemical interactions on its way to the target site. A chemical carcinogen meets many opportunities to be metabolized, that is chemically changed, so that it can be excreted. Ottoboni describes the numerical odds against a chemical carcinogen reaching a target cell with her example of the carcinogenic benzo[α]pyrene, also known as 3.4-benzpyrene, an inevitable ingredient of the barbecued steak.[26] Only a very small number of the molecules of the carcinogen make it to the target; many carcinogens are completely metabolized and excreted. Numerically, the zero threshold makes no sense for a chemical.

Matter, including animal tissue, contains at the molecular level a remarkable amount of empty space. When energetic radiation passes through tissue, most of it will exit unchanged. There has to be substantially "more than zero" for one single quantum to hit one molecule of DNA that *could* then cause a cell to misbehave. But a damaged DNA molecule can be repaired by a number of repair mechanisms within the cell. Even if the damaged DNA is not repaired, a misbehaving cell also faces many obstacles before it actually starts growing into a malignancy. It has to evade the immune defenses of the organism. Numerically, the zero threshold makes no sense for radiation either.

A further argument against the zero-threshold concept follows from the observed carcinogenicity of many chemicals that are vital for life. The situation was eloquently described by Efron. Examples of such natural essential chemicals are several of the steroid hormones, cholesterol, urea, and many essential trace metals. The notion that these chemicals would *not* be carcinogenic in the small amounts essential for life was apparently based on the belief that "nature wouldn't do that to us."[27] This faith in the goodness of nature contradicts the zero-threshold hypothesis.

The experimental opposition to the zero-threshold hypothesis is the *observation* of hormesis. Earlier we placed hormesis in the context of general toxicity as expressed in the dose-response curve. Here we need to look at hormesis specifically in the context of carcinogenic chemicals and radiation. In addition to the examples of growth stimulation by chlorinated biphenyls, Stebbing also listed studies on the beneficial effects of carbon tetrachloride, an animal carcinogen and, in regulatory language, a suspected human carcinogen. Arsenic compounds, which are known human carcinogens, have been shown to stimulate the growth of chickens, calves, and pigs.[28] Some chlorinated dioxins and furans, widely considered as carcinogens, were patented as anti-tumour agents a few years ago.[29] The evidence is even more abundant in the case of ionizing radiation.

Both Smyth and Stebbing give several examples of studies showing the beneficial effects of low level radiation on a variety of laboratory animals, including mammals, fish, and invertebrates.[30] More recent reports confirm these early results. A particularly relevant study involved the survivors of the atomic bomb attacks on Hiroshima and Nagasaki in 1945. The Japanese survivors that received low doses of radiation were compared with the population of Japan as a whole. The survivors had lower general mortality rates and lower cancer mortality. Also, the infant mortality among their offspring was significantly below Japan's national average.[31] In an equally comprehensive survey, Luckey lists a multitude of studies, both on animals and on humans confirming the beneficial effects of low level radiation. These benefits are not restricted to enhanced growth, they include improved reproductive capacity, improved immune responses, lower cancer rates, and longer lifespan. Luckey's survey contains over 1000 references to the literature on these studies.[32]

Notwithstanding this experimental evidence, a large part of the scientific community, including those involved in risk assessment, holds on to the zero threshold for carcinogens and radiation. The issue will probably remain unresolved for some time. On one side of the argument stands a *hypothesis* concerning the nature of the dose-response curve. On the other side stands a large amount of accumulated *experimental evidence.*

Prospect and retrospect

Mutations and natural selection have shaped life on this planet. The human species is one of many that evolved in an environment full of chemical and physical mutagenic agents. For the past several decades scientists have made a credible start identifying and enumerating these mutagens. We have a huge amount of information on animal tests, bacterial tests, and epidemiology relating to cancer. What do we do with all this information?

Testing "rodent carcinogens"

A common method for testing chemicals for their carcinogenic potential is with the use of laboratory animals, particularly rats. In order to obtain answers in reasonable time, carcinogenicity testing on rodents usually involves the use of fairly high doses of the chemical concerned. In other toxicity tests the starting point is usually a smaller dose, which has to be increased to exceed the no observable effect level (NOEL). The higher doses used in carcinogenicity tests, which coincidentally save on the costs of a very expensive process, also cause a problem. The amounts that are used give no information on long-term exposure to very low doses, which is specifically important in the case of cancer.

One of the alternatives to laboratory animals is the use of bacteria in the Ames mutagenicity test. This test is used to measure the *mutagenic potential* of a chemical, which is *related* to its carcinogenic potential, but it is not the same. Despite the widespread use of this test, carcinogenicity testing with laboratory rats continues. Ames' review studies on carcinogens in food led him to criticize the use of these rodent tests for the assessment of cancer risk. Ames' contention is that a basic flaw in the testing method leads to "too many rodent carcinogens." That basic flaw is the use of the high doses in animal carcinogenicity tests. Ames believes that other toxic damage to the animals at these high doses causes the cancers seen in the test animals. In other words, the cancer is a secondary effect. Ames is not alone in his criticism, but others disagree with him. They bring the hypothetical "zero-threshold" or "no-safe-dose" hypothesis into the dispute.

Another part of the counter-argument is that humans presumably have evolved with defenses against "natural carcinogens" which are thought not to apply to industrial (synthetic) chemicals. Ames and Swirsky-Gold have explained in detail why this "selective evolution" thesis is wrong.[33] Their main arguments are that:

1. Biochemical defenses that have evolved in animals, including humans, are general rather than specific, like shedding of cells and the induction of general, non-specific oxidizing enzymes.
2. Many natural toxins, such as the aflatoxins, were present during our evolution, but we have not adapted to them; they do cause cancer.
3. Very few of the plants we eat today were present in the old hunter-gatherer's diets. Humans did not have the evolutionary time to evolve into a "toxic harmony" with nature. Evolution by natural selection takes much longer than the few hundred years since we started to consume the carcinogens in potatoes, broccoli, and coffee.

The journal *Science* contains continuing news updates on the issue.[34]

In this dispute about natural versus industrial, two arguments are conspicuous by their absence. Ames' thesis of the "general defense systems" against natural carcinogens could easily be replaced by the observation that at these low concentrations hormesis makes the need for defense unnecessary. On the other side, the notion that many industrial chemicals are "unnatural" and therefore more dangerous quickly disappears in the light of recent discoveries of thousands of naturally occurring organochlorines.[35] The issue is important to the environmental debate for the answers to a few basic questions. What is the role of industrial pollution in cancer? How important is it to identify "rodent carcinogens?"

A carcinogen is an agent that can be shown to cause cancer in laboratory rats or guinea pigs. Such experiments show that a carcinogen has *the potential* to cause cancer. Whether it does cause cancer depends on several factors. One of the most important of these is the exposure of the organism to the carcinogen, the dose.

The time span between the period of exposure and the actual appearance of disease is called the latency period. For most cancers latency periods are many years. In view of these long latency periods, the specific exposure that caused the cancer is very difficult to pinpoint. If an individual gets cancer at age 60, is it because of a lifelong addiction to broccoli (and potassium-40) or is it due to a five-year exposure to saccharin during a weight control effort that took place ten years ago? Is a cancer at age 70 due to a life-long indulgence in barbecued steak or is it caused by living close to a chemical factory for ten years? How reliable is our information about any occupational exposures in the earlier life of the cancer patient?

It seems to make sense to consider seriously the proposal by Whelan to make a clear distinction between carcinogens and cancer-causing agents.[36] There are several agents that have been shown, without a doubt, to have caused cancer in humans. Cigarette smoke is one, high level ionizing radiation another. In the occupational arena we could add several industrial chemicals to the list. Cigarette smoke contains chemicals that have caused cancer, they are true cancer-causing chemicals. Perhaps the cancer debate would make some progress if we changed the terminology a little. A clear distinction between *carcinogens* and *cancer causing chemicals* could well help solve the argument on rodent carcinogens. If we concentrated on chemicals that are truly cancer causing we could perhaps describe safety as the absence of real, well-identified hazards. It would help setting priorities for the use of scarce resources.

Anticarcinogens

A series of reviews published in 1986, which included Ames' work on carcinogens in food, led to two important observations.[37] The first is that cancer is very much an age-related disease, a natural part of the aging process. Cumulative cancer risk is directly related to age, both in short-lived and long-lived species. About 30 percent of rodents, rats and mice, have cancer by the end of their two to three year life span. The same percentage is found in humans by the end of their 85 year life span.

A second observation is that the human diet contains vast *numbers* and vast *amounts* of naturally occurring carcinogenic chemicals. The exposure to these naturally occurring chemicals is obviously much larger than the exposure to industrially derived carcinogens. After all, we eat every day. Naturally growing plants biosynthesize large amounts of toxic chemicals. It has been suggested that these toxic plant chemicals are the plant's defense against attack by bacteria, fungi, and insects. They are nature's own *pesticides*.

Potatoes contain several chemicals toxic to both insects and humans. Cross breeding of potato strains have produced one cultivar that not only had a high resistance to insects, it also had to be withdrawn from consumption because of unacceptably high toxicity to humans.[38] The occurrence of these natural pesticides is also implicit in the practices of natural gardening, a topic we will consider in more detail later.

The variety of these plant-derived chemicals is very great. They have been the subject of chemical studies for many years and new ones are still being discovered. Plants in the human diet are no exception. These natural pesticides are often as toxic to humans as they are to insects. Many have been tested and found to be mutagenic. Why should naturally occurring pesticides that are proven carcinogens be any worse or better than synthetic pesticides, most of which have *not* been proven to cause cancer.

This, of course, is only one half of the story. We have already mentioned the defense mechanisms we have against disease, including cancer. Cellular repair mechanisms, the body's immune system, and the shedding of cells from internal and external surfaces are all part of our defenses. Particularly the lining of stomach and intestines is almost continuously exposed to dietary carcinogens, so it is important that cells that are about to turn malignant are removed and replaced regularly. Another important weapon in our defense arsenal is the taste of food. Virtually all animals reject food with a bitter taste. Indigenous South Americans, over many years, selected wild potato varieties which are low in toxic chemicals, on the basis of taste. This aspect of *ethnobiology* is described in detail by Johns.[39]

A potentially important defense against cancer is emerging from the reviews by Ames and others. Plants used in human food also contain a variety of chemicals that counteract some of the mechanisms by which carcinogens cause cancer. A variety of *antioxidants* have been identified that can intercept or inhibit the action of dietary carcinogens. These chemicals occur in a variety of food plants and appear to hold promise for a defense against cancer. Ames called them *anticarcinogens.* These chemicals include some of the common vitamins like C and E. Ames' idea of anticarcinogens is not new. As early as 1952 researchers at the University of Chicago found antimutagenic activity in naturally occurring chemicals.[40]

Much further research on these anticarcinogens is needed to confirm the preliminary observations and to convince the justly critical scientific community. In the meantime epidemiological studies on cancer appear to offer some indirect support for the food–lifestyle theory. The incidence of stomach cancer in Japan is very high and much lower in the USA. However, Japanese immigrants that move to the USA become subject to the lower American rates for stomach cancer. These same immigrants also suffer the rates for colon cancer and breast cancer, which are much higher in North America than in Japan. Since both population groups live in highly industrialized environments, it has been suggested that diet and lifestyle are more important than the effects of industrialization.

If the presence of anticarcinogens in food is important, the reverse must also be true. The absence of anticarcinogens in the diet would then be a cancer risk. Confirmation of these preliminary observations needs solid experimentation and reliable epidemiology. In terms of returns on the investment in human health, this direction of cancer research should have a much higher priority than studying the effects of traces of industrial chemicals in the context of the zero-threshold hypothesis.

An ounce of prevention

Ames repeatedly makes the point that cancer is a degenerative disease, and that there is a biochemical relationship with other diseases that affect the elderly. Other researchers support that view.[41] Although this is certainly not the only issue, there seems little question that the probability of getting cancer increases rapidly as we get older.

After the cancer-virus theories of the 1940s and 1950s the Higginson and Epstein statements shifted the question of cause towards the industrial developments of the past decades. Now we have added the lifestyle approach.

In cancer research the line between frontier science and textbook science is easily lost. The field is moving so fast that a textbook is almost obsolete by the time it is published. Consequently, many results of a very preliminary kind are released prematurely. Stories on the health benefits of certain foods or the avoidance of others are daily headlines. Suggestions that cancer could be prevented by eating health food are almost irresistible. For others, health food has become a defense against the effects of industrial pollution. There is no question that diet and lifestyle play leading roles in the cancer problem. They are not the only factors.

Viruses do play a role in cancer. Cancer of the cervix in humans is attributable almost entirely to human Papillomavirus. Human T-cell leukaemia is caused by a virus and is particularly prevalent among the Japanese. Hepatitis B virus is mainly a problem in developing countries.[42,43]

A genetically transmitted disease, *xeroderma pigmentosum,* interferes with DNA repair in the cell.[44] The patients are hypersensitive to the carcinogenic effect of ultraviolet light. The number of scientific reports on the genetic aspects of various kinds of cancer, particularly breast cancer, are rapidly increasing.[45,46,47] The increasing recognition of the genetic aspects of many diseases, including cancer, was discussed in more detail by Levine and Suzuki.[48] Where such genetic factors operate, cancer is obviously not a degenerative disease that affects only the elderly. Genetic predisposition towards cancer explains, albeit only in part, the occurrence of the disease in younger persons.

There is another factor that possibly plays a role in predisposition. As a result of advancing medical technology, infant mortality has decreased, but an increasing number of babies are born under difficult circumstances. They survive the traumatic adventure of birth only with extensive help of ever-more sophisticated technological methods. We have little information about the potential handicaps these children carry through their lives. Do prematurely born babies have the same defenses against disease, any disease, as a baby carried to full term? Do cancers in young people have to do with unidentified handicaps acquired at birth?

A major difficulty with cancer is that the problem is outgrowing the resources available for the fight. We can not follow up every lead, pursue every idea. However, the ordering of the priorities in this area is influenced by political decision making, which in turn is influenced by lobbying. A few well-known cases illustrate the point.

Asbestos will once more serve as an example. It is not as closely associated with lung cancer as many people think. Nevertheless, when children are involved, different criteria are used. Therefore asbestos containing building materials, installed freely in earlier years, have been systematically removed from schools and other public buildings all across North America. These removals are driven by regulations based on zero-threshold models, cost millions, and health benefits cannot be observed.

In another example, studies reported in the literature have attempted to link breast cancer with organochlorines, chemicals mostly of industrial origin.[49] The fatty tissue in the breasts of cancer patients contained pesticides, PCBs and other organochlorines. Some of these chemicals are indeed rodent carcinogens and all are fat-soluble and enter via the fat in the diet. None of the studies explains if these patients got cancer because of the chemicals, or whether the chemicals were just there because of dietary fat. In other words, there is no indication that the link is indeed a *causal relationship*. The issue of breast cancer and organochlorines is discussed in more detail in a later chapter. At this point we merely raise the question if a search for a causal link between diet and breast cancer would be a better investment of resources.

In suitable climates micro-organisms grow readily on stored food. One fungus, *Aspergillus flavus*, is common on stored grains and peanuts. It biosynthesizes a group of toxic chemicals, the aflatoxins, that are dangerous liver carcinogens. Rapid post-harvest drying of crops and controlled storage conditions reduce mold formation and minimize aflatoxin exposure. In societies not equipped with the technology to apply such methods, the much higher risk of aflatoxin exposure was confirmed by a survey, by geographical region, of occurrence of liver cancer caused by aflatoxin contamination.[50] This is a clear example of how the absence of technology puts large populations at the risk of premature cancer.

We can now summarize the main causes that have been demonstrated to contribute to cancer: viruses, our genetic inheritance, and the degenerative effects of advancing age. In small groups of humans certain cancers have been related to *occupational* higher level exposures to industrial carcinogens. But our exploration of the literature has failed to turn up any evidence that cancer in the general population has a causal relationship to any form of pollution.

In a review that mainly addresses the lifestyle question, Henderson *et al.* write about "the primary prevention of cancer."[51] However, epidemiologists often say "the number of cancers in the study was more than expected." Risk assessors talk about the "acceptable" number of cancers or the number "above a natural background." This means that some or other background cancer rate is *accepted as natural*. Therefore, is cancer preventable? A negative answer to this question is nevertheless an answer: the important issue about cancer is not if we get it, and how, but when.

Our discussion on cancer did not address issues in risk assessment and regulation. These issues require some background in statistics, but ironically, once the numbers have been collected and processed, we often take the statistics into the territory of trans-science. When that happens we frequently find that:

> In our present culture, arcane and even trivial mathematical formulations often gain persuasiveness beyond their real meaning.[52]

CHAPTER 7

Energy and the laws of nature

In the environmental language there is good energy, such as solar energy, and bad energy, from nuclear reactors. The perception is that solar energy, available without limits, is good because it does not pollute. Nuclear energy is bad because of radioactive waste. Wind energy is good because it is free, but coal-derived energy is bad because it gives acid rain and greenhouse gases. Energy is an important word in the environmental language.

Foray into physics

The *Concise Oxford Dictionary* defines energy as "the ability of matter or radiation to do work" and mentions the difference between kinetic energy and potential energy. The simplicity is misleading. The *McGraw-Hill Dictionary of Scientific and Technical Terms* contains thirty-eight entries under "energy."[1] Energy means more in a scientific context than it does in everyday English. In the environmental language, we must place the emphasis on the scientific sense.

The many faces of energy

First we need a few illustrations of the idea of "matter doing work" or "radiation doing work." Waterfalls have been used for many centuries to do work. When the water of a river falls to a lower level, and onto boards mounted on the perimeter of a wheel, the wheel will turn. This turning wheel can be made to grind wheat, saw wood, or drive an electric generator. The falling water is "matter" that, because of its movement, can do work. At the top of the waterfall the water has *potential energy*, the potential to do work. When the water falls, with a certain speed, it has *kinetic energy*, or energy of motion. When the paddle wheel turns, the kinetic energy of the water is transferred to the wheel. Now the moving wheel has kinetic energy.

The kinetic energy of the wheel can be transformed into yet another form, depending on the machinery it drives. The moving stones of a flour mill will generate heat because of friction. Some kinetic energy of the wheel has become heat energy while the wheat was ground to flour. If the flow of water stops, friction brings the wheel to a stop. A wheel with vanes or sails can be driven by the wind. Windmills of all shapes and sizes have been used throughout the ages. If we use a fuel-driven engine to drive the mill stones, the energy comes from the *combustion of fuel*. Now the fuel is "matter with the ability to do work."

Potential energy, kinetic energy, and heat and electric energy are part of the terminology in the textbooks dealing with energy.[2] Such textbooks also introduce the idea of the *energy equivalence* of different fuels and they contain the arithmetic needed to calculate these. Other calculations can show how much energy is lost as heat in an energy conversion process. The difference between the energy input and the output, as heat loss, then leads to the concept of energy efficiency.

The need to "feed" machines with fuel has its parallel in biology. Energy, the ability to do work, is needed by all living organisms. Animals must convert their food into heat energy to keep warm. They use electrical energy to send nerve impulses from brain to muscles, and mechanical energy to run. Plants must also stay at operating temperature. They use the energy associated with sunlight for synthesizing complex organic molecules such as carbohydrates. There is no living organism that does not convert energy from one form into another.

Humans first used simple amplifiers to use their own mechanical energy more efficiently. They used sticks as levers and wheels and cords as pulleys. Both amplified the mechanical energy of muscles. Mechanical energy needed for chewing food was partially replaced by heat energy for cooking. Animals were made to pull sleds and wagons. All these applications are *conversions* of one form of energy into another. Food is needed to generate heat, sunlight is used to synthesize chemicals, water at a high point is allowed to fall to a low point. Every time energy is *used* it is *converted* from one form into another. This means matter and radiation are *storage forms of energy*. When matter, coal or food, is used to do work, it becomes fuel.

Oils are burned in a machine to produce mechanical or electrical energy. Food is fuel that is also "burned" to give heat and electrical and mechanical energy. "Burning up food" is used as an analogy. Although energy is extracted from food at body temperature instead of actually burning, the analogy is valid in the chemical sense. Fuel for the machine and food for animals both consist of chemicals. If the animal is a herbivore, it eats mainly carbohydrates. If it is a carnivore, it eats mainly proteins and fats. Chemicals, in the energy budget, contain chemical energy. Chemicals *are* stored energy. Some chemicals store more energy per molecule than others. Carbohydrates and fats and petroleum are high energy molecules, water and carbon dioxide are low energy molecules.

The primary form of energy is associated with radiation from the sun. Light from the sun is stored, by green plants, as chemical energy, which is then converted in many different ways. All energy essential for living organisms comes from the sun. This is not a scientific version of sun worshipping. It is the vital condition for making life on our planet possible.

From a purely biological viewpoint the only energy needed by all living organisms is food and a suitable ambient temperature. Energy consumption beyond that basic level is not essential for life, but it provides comfort. To use heat energy from a cooking fire is more comfortable than eating raw meat. A machine that can lift heavy objects allows us to build more comfortable shelter. A machine used for transport is more comfortable than walking. Early humans used little energy. They consumed chemical energy by eating and some additional energy by burning wood to keep warm. They lived on a very low energy level. Even with the use of falling water and a paddle wheel, in small settlements near a river, their impact on the environment was insignificant as long as their numbers stayed low.

As soon as population numbers went up, conditions changed. Wood was the earliest energy resource. It was used as fuel for cooking and to keep warm. Then it was needed to build protection from the weather. These primitive "technologies" allowed the population to increase further. The amounts of wood that were consumed became more significant. Deforestation around the Eastern Mediterranean caused soil erosion, desertification, and migration to new, forested areas in Northern Europe. When the "end of the forest" was reached on the big island that became Great Britain, the alternative fuel was coal. The Industrial Revolution started with the need for coal as an energy source. Once underway, industrial development quickly became a reason in itself for the consumption of energy, and the need for energy began a steep upward path. Early energy sources, wind and falling water, were no longer enough. Extensive use of coal was soon followed by a vigorous exploitation of petroleum in the early twentieth century. Basically nothing had changed. Both coal and oil are derived from biological material. It is biological material in a fossilized state, but it still contains the chemical energy derived, albeit long ago, from the sun. These *fossil fuels* are now the main energy source for the industrialized world.

Energy is still derived from falling water where possible. The paddle wheel has become an electrical generator and "waterfalls" were created by building dams. However, there is not nearly enough falling water. Soon after the Second World War it became possible to tap yet another source of energy. Huge amounts of energy are locked up in the nuclei of atoms. This energy can be converted into heat and therefore into electricity. Nuclear energy became the promise of the future. As recently as 1974 a vice president of Atomic Energy of Canada said: "Nuclear power promises a world in which 10 billion people can live in harmony and affluence. Abundant energy would enable the world to feed and support an expected world population of 10 billion to 20 billion at present North American standards."[3] Somehow, it didn't turn out that way.

This is the law

All energy conversions have an important limitation in common. Every machine or device used to convert energy from one form into another *appears to lose* some of the energy. The output is always less than the input. The conversion of energy from one form to another never results in a net energy gain. No machine or device can ever produce more energy than is supplied to it. The perpetual motion machine, seriously pursued in earlier times, does not exist.

In our example of the falling water the paddle wheel ground to a halt when the water supply stopped, because of the friction around its own axle and in the stones of the flour mill it was driving. Friction generates heat. The wheel lost its kinetic energy. That energy was not truly lost. It was converted into heat energy. In mechanical devices friction inevitably causes some energy conversion into heat. The heat warms up the surrounding air, and the movement of that air dissipates the heat energy into the surroundings. Numerous careful measurements have shown that, although some of the energy appears to be lost, there is never a real net loss. The total amount of energy remains the same. However, the energy lost as heat is often not recoverable.

First, experiments confirmed that this preservation of energy is a general rule. Then, complex theoretical considerations showed this rule to be a universal law, the *law of the conservation of energy*. Thermodynamics is a branch of physics concerned with heat and energy conversions. The law of the conservation of energy is the *first law of thermodynamics*. All known events in the universe obey this law. At present there is no reason to doubt the absolute universality of the first law. In other words, it is a law *that cannot be broken*.

The production of heat during all these conversion processes causes both theoretical and practical problems. Theoretically, energy loss, due to heat, seems to be a contradiction to the first law which says energy cannot be lost. Practically, unwanted heat cannot be allowed to build up and must be dissipated. It reduces the efficiency of the process.

The heat loss problem is explained by another basic principle of physics. Heat always "flows" from a warm place or object to a colder one. Heat generated in an energy conversion process makes the surroundings a little warmer; it is dissipated into the environment. The term energy loss is used because this heat *cannot be recovered* and used "to do work."

Many industrial processes involving energy conversion have to deal with the practical problems of managing excess heat with cooling towers and condensing devices. They often discharge warm water into the adjacent surface waters or steam into the atmosphere. The continuous loss of heat energy has far reaching consequences.

You can't have your cake and eat it

Heat dissipated into the surroundings is not a loss of energy. It is merely energy in a *less useful form*. Gasoline, a mixture of chemicals, has a certain energy content. That energy content can be measured experimentally and written in numbers. When gasoline burns in an engine, part of the energy becomes kinetic energy. It makes engine parts move and it makes the automobile move. A great deal of the energy from the gasoline becomes heat that is transferred to surrounding air via the radiator. The warm air dissipated to the atmosphere can no longer be used to do work.

Sometimes it is possible to use this "surplus" heat energy one more time. Especially in Western Europe, progress has been made with heating apartment buildings or greenhouses with the hot cooling water from electrical generating stations. This is obviously better than throwing the hot water out as waste, but eventually heat is lost to the environment. The apartments, once heated, will eventually transfer the heat energy to the surrounding air. Transportation of energy by pipeline or by power lines also causes heat loss. In this case the lost heat energy cannot be used a second time.

Even with a partial re-use of the heat energy, every process involving energy conversion results in *a net decrease in usable energy*. This is also a law of nature that cannot be violated. The conversion of energy into another form to derive a benefit, heat or work, leads to a net loss of usable energy. It is called the *second law of thermodynamics*. This continuous loss of heat energy to the surroundings means that all processes involving energy conversion must eventually come to a stop unless energy is added from outside the system. Once the sun stops sending energy this way, all life on earth will come to a halt. It is the second law that prohibits the perpetual motion machine.

The mathematical language used to describe these physical processes makes use of another idea that plays an important role in environment problems. The second law uses the word *entropy*. An example readily explains the entropy concept. When wood is used as fuel and interacts with oxygen in the air, major chemical changes take place. Wood consists of large complex molecules, lignin and cellulose. Many thousands of atoms are neatly attached to each other in an orderly fashion. The long, string-like molecules of cellulose lie side by side, "glued" together by the equally large molecules of lignin. Seen on a molecular scale, wood is a highly ordered arrangement of many atoms of carbon and hydrogen and oxygen. When the chemical energy in this arrangement is converted into heat as the wood burns, these millions of atoms become a much less ordered system. Particles of carbon called soot, molecules of carbon dioxide, water, and carbon monoxide blow away in all directions and assume a state of *disorder*.

As chemical energy was converted into heat, the level of disorder increased. That degree of disorder is called entropy. The second law of thermodynamics says that when energy is converted from one form into a less usable form, entropy increases. When carbon dioxide and water from burning the wood are taken up by the tissue of a green plant, the plant can put these small molecules together in the highly ordered arrangement of wood. But it can only do that if there is *an addition of energy* in the form of sunlight. A decrease in entropy costs energy. The only way we can get around the second law is by spending energy.

The laws of thermodynamics are universal. In a perfectly insulated, closed system energy cannot be lost. Earth is not such a perfect system. It gains heat energy from the sun, and loses energy by radiating heat to outer space. Many energy conversions in between cause a continuous increase in total entropy of the system. Disorder continues to increase.

The laws of thermodynamics are of crucial significance to environmental science. Energy can be used wisely, some of it can be conserved temporarily, but eventually it will be lost. Even "renewable" energy such as sunlight and wind are not a solution to the industrial energy crunch. Their conversion into other forms still obeys the second law, and entropy continues to increase. We place renewable in quotation marks because solar energy is only renewable on a human time scale. Eventually the sun's energy will run out. After turning into a nova, the sun will become a cold star. This is not an urgent problem.

The decrease in order caused by the burning of wood illustrates a further piece of the thermodynamics story. The soot and the smoke are examples of pollution. The carbon dioxide may affect the way the atmosphere behaves. This means that entropy has another name. It is called waste. The use of energy results in waste. Burning coal gives similar wastes. Nuclear energy gives spent fuel as waste. Wind energy produces worn out windmills and hydro dams flood habitats and dissolve metals from the minerals in the soil.

The consequences of the laws of thermodynamics touch on all issues in the natural environment. A formal treatment of thermodynamics is, of course, outside the scope of our discussion. A translation into common English will suffice for several down-to-earth versions of the laws of nature. James Hadley Chase, a writer of detective stories from a few decades ago, took the second law as the title for one of his novels: *There is Always a Price Tag*. Other colloquial expressions state the second law as: "you can't have your cake and eat it too" or "it's no good crying over spilt milk." Barry Commoner stated his fourth law of ecology as "There is no such thing as a free lunch," and said his version was borrowed from economics.[4] This fourth law of ecology is yet another version of the second law of thermodynamics.

Resources

Energy available on earth comes from the sun or from leftovers from "the big bang." Solar energy is available as radiation, mainly in the ultraviolet, visible, and infrared part of the spectrum. The sun plays a role in the movement of the ocean tides, and solar energy is also stored as falling water and as chemical energy in wood, coal, and oil. Leftovers from the complex events that took place during the formation of stars and galaxies are the atoms of nearly all the elements except the hydrogen and helium from which they were formed. The formation of these elements required the vast amounts of energy that were involved in the supernova type explosions of old stars. Some of that energy can be released from the nuclei of those atoms. The core of our planet also contains a lot of heat left over from the formation of our solar system.

Three forms of this stored energy are most easily tapped: falling water, fossil fuels, and nuclear fission. Most of the industrial world uses a mixture of these three, most frequently in the form of heat or electrical energy. Other sources such as wind, solar radiation, and geothermal energy are less readily converted into electricity. All have advantages and disadvantages, but the viewpoints of user, generator, consumer, and environmentalist often differ. The advantages and disadvantages are often interchangeable. Favouring one energy option over another is nearly always wrong, because of thermodynamics. A brief overview of the various sources will illustrate the point.

Conventional sources

The advantage of hydro-energy is that it uses a natural event, a waterfall. This apparently has no negative impact on the environment. Unfortunately, the amount of naturally falling water is not nearly enough. Therefore hydro must be expanded by technical methods such as building dams in rivers. However, flooding land drowns vegetation, which on rapid decomposition forms a compost that releases nutrients into the water. These nutrients in turn promote the growth of aquatic plants. Our discussions on water quality will show how algae benefit most from these nutrients. Decomposition of plant material also produces methane, an effective greenhouse gas. Of course, the flooded land is no longer available for wildlife, agriculture, or industry.

The reservoirs created by big dams tend to fill up with silt. As the movement of river water is slowed down, silt carried down by the river settles out in the reservoir. This problem has already been observed in the case of Egypt's Aswan dam. It is also a concern in the case of the huge dam presently under construction on China's Yangtze River.[5] Chinese engineers have expressed fear that the reservoir of this dam could fill up with silt in about twenty years. Nevertheless, plans for further hydroelectric expansion are unstoppable in energy-hungry Asia. Another environmental concern with hydro dams is that the flooding can convert the metals in some minerals to more soluble forms, particularly if acidification enters the picture. This way hydro dams can cause the release of unwanted metals into the water.

Hydro-energy is also not completely free from safety concerns. In 1928 a dam broke in Santa Paula (California) with a death toll of 450, and in 1959 a dam failure in Malpasset (France) cost 412 lives. In Belluno (Italy) a dam failure in 1963 resulted in 2000 fatalities. Some dams in the USA could cause several hundred thousand deaths in case of structural failure.

The large scale use of coal started in Europe with the Industrial Revolution. Initially coal was used to make steam and the steel industry used the coke. Coal soon moved from the steam engine to the electric power generator. The principal advantage of coal is its abundance. A recent estimate by the Coal Association of Canada shows coal as more than 90 percent of the world's fossil fuel reserves, with natural gas at about 5 percent and petroleum at 4 percent.[6] In many parts of the world coal could last as a power source for many years. However, it is not inexhaustible and there are problems.

The most well-known problem with burning coal is the formation of carbon dioxide. Carbon dioxide is normally present in the earth's atmosphere in a concentration of about 0.03 percent (300 ppm). This concentration has been increasing steadily for the past few centuries. Carbon dioxide absorbs part of the heat that the earth's surface radiates back to space. It acts as the glass in a greenhouse. It affects the average temperature of the planet and is therefore called a *greenhouse gas*. We will look at greenhouse gases in a later chapter, for the time being it is sufficient to keep an eye on a possible causal connection between the combustion of fuel and the climate of the planet.

Closely related to the use of coal, but on a much smaller scale, is the combustion of peat. Peat is decomposed plant material that is geologically speaking much younger than coal. It is found in vast quantities in Northern Europe and in sub-arctic Canada. The advantage of peat is that it contains virtually no sulphur, an impurity in coal that has a major environmental impact. This does not mean that peat is free of problems.

Other prehistoric plant materials became fossilized into a liquid or gaseous form: petroleum and natural gas. While the use of coal goes back to the mid-1600s, it took another two hundred years before oil wells were drilled in the state of Pennsylvania. Known reserves of crude petroleum and natural gas are very much less than those of coal. Oil and gas are more easily transported and used because of their liquid or gaseous form. Initially most crude petroleum was useless because of its high sulphur content. An

American immigrant of German origin, Hermann Frasch, invented a process for removing sulphur compounds from the crude "sour" oil in 1887. Frasch later applied his petroleum experience to the recovery of sulphur from underground deposits, a technique now known as the Frasch process.[7] Soon the internal combustion engine led to automobiles and diesel trucks. In most industrialized countries mass transportation of goods and people now depends on burning fossil fuels. Although many countries in Western Europe have sophisticated electrical rail systems, most of the electrical power is generated by the combustion of natural gas or petroleum derivatives. Only France has made substantial commitments to nuclear energy.

In 1971 Commoner presented figures that illustrate how transportation changed in a short time.[8] Between 1947 and 1970 there was a nearly 4 percent *annual reduction* in rail horsepower and a nearly 8 percent *annual increase* in automotive horsepower. These figures become even more significant when we consider that on a per load basis road transport consumes roughly four times as much energy as rail transport. Our turning away from the environmentally more friendly rail transport started early and is still continuing.

An important aspect of gas and oil is their use for non-fuel purposes. Thousands of materials are made from petroleum derived chemicals. Automobile tires, upholstery, textiles, building materials are derived from petroleum. Entire industries, like cosmetics and pharmaceuticals, depend on petrochemicals. The *petrochemical industry* has become the very basis of modern industrialized societies. Although there are fewer automobiles in many less developed countries, the consumption of oil via petrochemicals is spreading rapidly. Petrochemical factories of all kinds are appearing in India, China, and all around the Pacific Rim.

This rapidly increasing consumption focuses our attention on the supply of oil and gas. Pessimists estimate the end of petroleum in about twenty years; others are more optimistically calling for another fifty years. Much depends on whether rates of consumption are decreasing or increasing, and whether new supplies are found. The gigantic natural gas bubble that feeds much of North-Western Europe is a typical example. Reserves there are estimated to last ten years for wells presently in use, twenty years if known reserves can be exploited by drilling new wells. The drilling of more gas wells is opposed by environmental action groups.[9] Different and uncertain time spans apply to other areas in the world. All are subject to the same certainty: oil and gas will run out much sooner than coal. Once they have been burned, they cannot be used again.

By far the most controversial form of energy conversion is nuclear fission. By the end of the Second World War physicists had learned to unlock the energy stored in the nucleus of atoms. Initially that energy was released as the destructive force that ended the war, then nuclear energy became useful for generating electricity. The fuel, uranium, was abundantly available. The "affluence for billions" was yet another triumph of modern technology. Not so, say the opponents.

Although the fuel is abundant, there are problems in getting it. Although uranium represents very low level radioactivity, there are health risks associated with the mining. Also, mines produce tailings. In the case of uranium these tailings are also a source of low level radioactivity. That radioactivity can blow around in the wind or it can be leached out by rain and surface waters. Tailings from uranium mining can be a problem and must be properly managed.

Breaking a heavy atomic nucleus in smaller pieces releases energy that is readily converted to heat. That heat is used the same way as in coal-fired plants: to generate steam, which drives a turbine, which drives a generator, which produces electricity. As with coal, a vast amount of that heat is "lost" during the conversion steps. Cooling water transfers this "waste heat" to air or water and eventually into the environment. A common feature of a nuclear reactor is *heat pollution*.[10]

The fragments of the broken atomic nucleus are, in turn, the nuclei of smaller atoms. If the uranium-236 atom is split in two, the pieces that are formed could be an atom of xenon (a gas) and an atom of strontium for example. Most of the nuclei formed this way are unstable, that is, they are radioactive and give off radiation. These radioactive elements cannot readily be used to generate more energy. They have become waste, nuclear waste. The spent fuel rods of a nuclear reactor are a waste disposal problem. A problem not so much because of space, but because of the perceived danger of radioactivity.

Some radioactive isotopes in the spent fuel take a long time to decay to a point where most radioactivity is lost. The time needed for decay is measured in seconds for some isotopes, in thousands of years for others. For practical reasons decay time is measured as the element's *half-life*. The half-life, indicated in technical language as $T_{1/2}$, is the time needed for an amount of that element to lose half of its radioactivity. Uranium-238, which is most of the naturally occurring uranium, has a half-life of 4.5 billion years, that means it takes that long for a given lump of uranium-238 to lose half of its activity. It will take another 4.5 billion years for the remaining uranium to lose half of *its* radioactivity. The original sample has lost 75 percent of its activity after nine billion years. Iodine-129 and strontium-90, both common products of the fission of uranium-234, have half lives of seventeen million years and twenty-eight years respectively. Another iodine isotope, iodine-131 has a half-life of eight days.

For practical purposes, a radioactive material loses almost all of its activity after about seven half-lives. For strontium-90 that would be some 200 years. For iodine-129 it would be over a hundred million years. Therefore, in human terms, nuclear waste lasts forever. It cannot be disposed of in the sense that disposed means "it's gone." The flippant suggestion that "it comes out of the ground, so just put it back where it came from" does not hold. Uranium is part of the earth's crust in a very disperse form. Although mining produces this element in a much more concentrated form, it is not of major concern in terms of the waste and its management. The spent fuel rods do not just contain uranium, but also the radioactive fission products. They must be stored in a "safe" place where they can be *managed*. This again raises the question about the meaning of "safe." Much of Weinberg's essay on trans-science is about the nuclear safety issue.[11]

The other issue of nuclear energy is the safety of the process itself. Compared to other sources of energy the nuclear industry has an admirable safety record. Notwithstanding that record, reactor incidents continue to feed the fear of nuclear power. This is unfortunate. It makes rational discussion of energy options difficult. We will return to the safety issue shortly.

Environmental impact

According to the laws of nature, waste is inevitable in any energy conversion process. Therefore energy issues are environmental issues. The various different energy sources have their own characteristic problems. These problems generally fall within the closely related areas of human health and environmental health.

All fossil fuels have the common problem of producing carbon dioxide during combustion, a gas that may have a global influence on climate. This also applies to combustion of any biological material or biomass. Coal contains many impurities. Among these are sulphur compounds. During combustion, the sulphur is oxidized to sulphur dioxide. When sulphur dioxide dissolves in water, it forms sulphurous acid. Part of the sulphur dioxide is oxidized to sulphur trioxide which forms sulphuric acid in water. The amounts of acid coming down in rain or snow are large enough to have noticeable environmental effects.

Acid forming gases can be removed from the smoke of coal-fired generating plants. Bubbling the combustion gases through alkaline materials, such as lime, neutralizes the acids. Depending on the sulphur content of the coal, a 1000 MW (megawatt) plant could discharge several tons of sulphur dioxide daily. This would need several tons of neutralizing material. Once used, this material cannot be regenerated; it becomes waste. The leaching of soluble chemicals from this waste into soil and ground water complicates the disposal problem. The quantities of lime needed and the subsequent waste disposal problem present enormous difficulties: the logistics of neutralization technology are inhibiting its large scale use. Another factor that must be included in the scrubbing argument is that the scrubbers also consume energy. About 5 percent of the energy generated must be used to operate the scrubbing equipment.[12] Attempts to remove sulphur compounds from coal prior to combustion are only partially successful. We will consider the effect of acid rain on water quality in another chapter.

Hydrologically generated electricity has problems in addition to the disadvantages of flooding of habitat. Expansion of the hydro option is limited by availability of suitable rivers and space for flooding. The well-being of animals and plants and biodiversity are mostly at issue in controversies over new hydroelectric projects.

Ironically, the abundance of coal is also a disadvantage. In many areas large coal reserves are close to the surface, so that strip mining is economically advantageous. Strip mining improves safety for the miners, but the scars on the land make the environmental impact very visible. Usually the opponents have little confidence in the promises of land remediation.

Coal and petroleum also produce air pollution other than carbon dioxide. Coal is normally contaminated with a huge variety of other minerals, including uranium and heavy metals like lead, mercury, cadmium, and chromium. These metals become part of the ash, most of which falls to the bottom of the burner. A part of the ash, the *fly ash,* is carried up in the smoke. Most of the fly ash is trapped by electrostatic precipitators or other control devices. It can be used for cement manufacturing, but that is only a partial solution for its disposal. The efficiency of fly ash capture has improved over the years so that presently less than half a percent of the particles escapes. If that sounds impressive, let us do some arithmetic. A modest 400-megawatt generator burns about 220 tons of coal and produces 20 tons of fly ash per hour. A 99.5 percent capture efficiency means that "only" 100 kilograms escapes from the stack per hour.[13] That is 2400 kilograms per day. That is 876 tons of ash per year. As an example, Saskatchewan coal contains about 25 ppm of uranium and thorium, that is 5 kg in the 220 tons of coal burned per hour. Most of that 5 kg will stay in the ash, and in 20,000 kg of ash that represents 250 ppm. Thus, 250 ppm in the 876 tonnes of fly ash that are spread annually over the countryside is 200 kg of thorium and uranium. As we saw earlier, China wants to build some 200 of these "medium-sized" power plants by the year 2000.

Attempts have been made to use blends of peat with either coal or with natural gas. The advantage of such mixtures is not so much the additional fuel source, but the reduction in sulphur dioxide emission. Mixing coal and peat in a 1:1 ratio would cut acid emission in half as compared to coal only. However, the presence of appreciable amounts of naturally occurring organochlorines in peat could lead to the formation of chlorinated dioxins.[14] We will discuss this issue in more detail later. From the energy viewpoint it is strange that a probable link between peat combustion and dioxin formation has not received any attention in the energy literature.

Petroleum products have their own problems. Gasoline does not burn very efficiently in an internal combustion engine. High temperatures are needed to control the emission of carbon monoxide and unburned, volatile components of the fuel. These higher temperatures also contribute to the formation of the oxides of nitrogen that play a key role in the formation of smog. A lesser known problem is caused by fuel evaporation. Gasoline losses by evaporation from stationary vehicles and during refuelling can amount to many tons per day in the world's larger cities. The resulting air pollution is of concern on both sides of the Atlantic.[15,16,17]

Diesel engines present a different pollution problem. Incomplete combustion of diesel fuel produces the well-known diesel "fumes", the black smoke coming from the exhaust of large, diesel-powered transport trucks with improperly tuned engines. The black colour is caused by particles of carbon. Adsorbed on the surface of these particles are chemicals called polycyclic aromatic hydrocarbons or PAHs. They are known human carcinogens.[18,19] Another aspect that applies to all combustion processes, particularly biomass and automotive fuels, is the formation of carbon monoxide. Much of this is oxidized to carbon dioxide by micro-organisms in soils. The amount of carbon monoxide taken up by the soil is only slightly less than the amount resulting from fossil fuel combustion.[20]

Then there is the nuclear reactor. The association with atomic bombs created the fear of nuclear energy. Relatively recent incidents at Three Mile Island and Chernobyl reinforced that fear.[21,22] Public fear of the nuclear option has drastically slowed down the construction of new nuclear generating capacity. Wagner and Ketchum, in their book *Living with Radiation,* describe in detail how this public fear of radiation, specifically aimed at the nuclear industry, is fed by "the Politics of Fear."[23] In addition to politics, education also has to do with that fear.

A textbook on environmental science lists, among several objections to nuclear energy, questionable reactor safety and health effects of low level radiation. Reactor safety was discussed almost entirely in terms of the Three Mile Island incident in 1979, and the Chernobyl incident in 1986. The health effects of low level radiation were not discussed in this book; they were considered negative by implication. The large body of experimental evidence for the benefits of low level radiation received no mention in

Figure 7.1 Preventable air pollution in the city – carcinogenic PAHs in diesel exhaust

this discussion; the concept of radiation hormesis was not mentioned.[24]

The safety details of conventional forms of energy, as presented by Yaffe in 1979, can serve as a comparison for the safety issue. This author compared, in terms of occupational hazards, the *real* death toll of the various energy industries with the *projected* death toll of the nuclear industry. He also offered figures on the substantial environmental damage caused by fossil fuel combustion, even without considering the carbon dioxide problem. In this straight comparison of safety and environmental impact, the nuclear energy option came out on top.[25]

However, the issue with nuclear power plants is not the safety of the uranium miners or the plant operators. The problem is not with the hundreds of nuclear plants that have operated safely for decades. The problem with nuclear power plants is with the *perceived* effects of a "what-if" scenario. "If an accident occurs, how many cancer cases will result from the released radioactivity."

Supposedly the reactor design at Chernobyl was old and inefficient and such an accident could not possibly happen in the much better plants in the USA or in France. But something happened at Three Mile Island.

Yaffe's figures on the radioactivity released by coal-burning power plants are now mostly obsolete, because they were based on less efficient fly ash control.[26] However, the calculations we just did tell the story clearly. Even with dust control in place, a coal-burning power plant releases radioactivity into the atmosphere in quantities that far exceed the amounts released at Three Mile Island. Although the total amount of radioactivity released in that accident was quite high, nearly all of it was in the form of radioactive *gases*, xenon-133 and krypton-85, which rapidly dispersed into the atmosphere. Human exposure was limited to a very small amount of radioactive iodine. One study of the incident called itself "a disaster survey of a non-disaster."[27] Wagner and Ketchum, in discussing health problems associated with indoor radon, note that people living in and around Harrisburg at the time of the accident received more radioactivity every day from the radon in their houses than they did from the reactor incident.[28]

These authors also give details of the Chernobyl incident. In that accident released radioactivity has been *estimated* to cause an increase in thyroid cancers and leukaemia of about 5 percent in the town of Levlev.[29] Among the 45,000 people in the nearby town of Pripyat exposure levels *could* give rise to about 156 deaths from cancer. At least 9000 of the people in Pripyat will die of cancer due to all other causes. That is the "background." Against that background the increase of 156 due to the accident will be too small to be detectable. This is little comfort for those 156, but it means that *we will never know* if there was indeed an increase of 156. Similarly, among the 60 million population of the former Soviet Union there will be an *estimated* 2900 cancer deaths due to the Chernobyl accident. At the same time, 15 million of the 60 million will die from cancer from all other causes. We should note that all these estimates were based on the zero-threshold hypothesis. The theory of radiation hormesis, which is supported by experimental evidence, predicts a *reduction* in cancer incidence in most of Europe as a result of the Chernobyl fallout.[30]

The disposal of spent nuclear fuel has become a very controversial issue. Canada wants to bury it deep in the rock of the Canadian Shield. The USA would like to store it in Yucca Mountain in Nevada, a plan that has become a political football.[31] Most European countries, except Sweden and Finland, reprocess their spent nuclear fuel. Both Sweden and Finland contemplate burial in deep rock, similar to the plans in Canada and in the USA.[32] Canada participates in an international consortium studying deep rock burial at the Stripa mine in Northern Sweden.[33] Many other countries, including the UK, have not announced plans for the management of the wastes from their fuel reprocessing operations. While the nuclear reactors themselves are building up one of the best safety records in comparison with any other industry, opposition to the nuclear option in general, and particularly to the waste disposal plans, continues to bedevil this energy option.

An interesting insight in perceptions comes from a recent discussion of nuclear waste disposal in the journal *Environmental Ethics*. The author introduces the interesting concepts of the various *dilemmas* she sees in the nuclear issue. The *legacy dilemma* refers to the legacy of radioactive waste left for future generations. The *locus dilemma* deals with the selection of disposal sites and the Not In My Back Yard (NIMBY) syndrome. The *consent dilemma* deals with the ethics of "buying" consent in small, underdeveloped communities by promising money for libraries and swimming pools.[34]

The author did not mention the ethical dilemmas of coal-derived energy. Almost the entire article could be rewritten, just substituting coal for nuclear, and little else would need to be changed. The legacy would now be the radioactivity and heavy metals spread over the countryside and the carbon dioxide released into the atmosphere. The locus would include a very much larger surface area, the whole planet in the case of carbon dioxide. The consent dilemma would have to deal with the ethics of the fact that most people don't even know that they have not been asked for their consent. Specifically in terms of the ethics of the energy dilemma, we must compare the mainly local risks of nuclear energy with the global risks of fossil fuel combustion.

Options and alternatives

There are many problems that beset conventional energy sources. Therefore many discussions on energy eventually move toward conservation and alternatives. Of the alternatives, only passive solar *heating* uses the sun's radiation directly. Biomass combustion is partially used directly for heating, particularly in regions where wood is an important energy source. Geothermal energy is often used directly as heat energy. Other alternative energy sources involve conversion into electrical energy. Two common errors frequently slip into these discussions of alternatives.

The first is caused by the inappropriate use of the term "clean energy." Countless media reports describe the electric automobile as clean, while no questions are asked about where the electricity comes from. California's legislation requiring the availability of electric cars in the near future will simply export a significant pollution source from the streets of Los Angeles to the smoke stacks in the countryside.[35] Whether the energy used came from dirty coal or from "clean" nuclear fuel is, in the minds of many, not a question. The benefits for the city dweller are the removal of the typical big city smog, and the easier point source control of pollution at the power plant.

A second error in the alternative energy picture is the notion that "renewable" energy has no impact on the environment, that it is essentially "pollution free."[36] This is in conflict with the second law of thermodynamics. Close scrutiny of any energy conversion will show that there is waste. Waste means environmental impact. However, it is obvious that some wastes are worse than others. Therefore we must examine the most promising alternatives.

Extending the biomass option

Gasification of coal, the conversion of coal into something like natural gas, is an old technology that was used for many years as natural gas is used now. It was inefficient because a substantial part of the energy content of the coal was lost in the process. It was also dirty technology, acceptable only before the term "air pollution" had been invented. The conversion of coal into liquid fuel was driven by potential fuel shortages near the end of the Second World War. Technologies developed separately in Germany and in the USA were expensive, but liquefication of coal has become commercially feasible. South Africa, a country without oil or natural gas that spent decades in political isolation, has developed these technologies and made them work. Coal liquefication is not an energy alternative, it is merely a more *convenient*, and somewhat cleaner form of coal utilization.

A possible extension of petroleum supplies is the development of tar sand and oil shale. Tar sand is a sand formation containing a syrupy, semi-liquid material called bitumen. Tar sand formations are found in Canada, the USA and the former Soviet Union. The bitumen can be manipulated to produce liquid fuels and petrochemicals. Oil shale is a sedimentary rock, more widespread in Canada, China, the USA, Australia, and some Middle East countries. It contains a thick, oily, organic material resembling crude petroleum. The extraction of bitumen and shale oil could extend the world's liquid fossil fuel supply by several decades. Also decades of carbon dioxide production, of course. At present, economic conditions favour the use of conventional petroleum.

Another way to extend the supply of liquid fuel is the use of alcohol. Two kinds of alcohol are suitable, methanol and ethanol. Methanol is available from wood, other vegetation, coal, natural gas, and petroleum. When wood is heated in the absence of oxygen a mixture of liquids containing water and methanol is produced. The latter used to be called wood alcohol. Methanol from wood is expensive and inefficient although improvements are still being made.

Ethanol, the same alcohol we find in alcoholic beverages, is the other alcohol that can be made from biomass.[37] Some yeasts can biodegrade starches and sugars. They use the carbon atoms to build more yeast cells. Carbon dioxide and alcohol are their waste products. This fermentation takes place in a water medium. Since most yeasts die when the alcohol concentration exceeds 12 percent, the alcohol must be separated from the water by distillation. This requires energy input. Fermenting surplus sugar and corn to produce alcohol appears attractive for stretching the fuel supply. There is a wrinkle in the plan, though.

Using typical industrial agricultural methods, it takes more than *two gallons* of automotive fuel, diesel or gasoline, to plant, cultivate, fertilize, harvest and transport the corn needed for making one gallon of alcohol. Since it takes about five gallons of crude oil to make two gallons of automotive fuel, producing one gallon of alcohol costs about five gallons of crude petroleum.[38] Clearly, making alcohol for use in automotive fuel has a *negative impact* on the energy balance sheet, unless....

Brazil has used alcohol as a mixture with gasoline called gasohol, for more than a decade. Sugar and sugar cane are the raw materials, cultivated for the specific purpose of making alcohol. The solution to the energy deficit of the process is simple. Brazilians use only a fraction of the energy in their agricultural processes as compared to industrialized farmers. Tilling the plantations with oxen, fertilizing with the animals' own manure, and transporting materials in animal drawn carts, allows Brazilians to get more than two gallons of alcohol for an investment of only one gallon of fossil fuel. Questions remain about the amount of available land. One author suggested the conversion of cropland to "fuel farms."[39] The World Bank questioned whether enough agricultural land should be sacrificed to large scale ethanol production, citing the importance of croplands for food production.[40] A big stumbling block to the North American and European nations would be the impossible "reversal" to labour-intensive, manual agriculture needed to achieve a net positive energy output. Yet, a discussion of the methanol-from-biomass as used in Hawaii suggests that biofuels can be valuable supplements to a nation's conventional energy resources.[41]

The issue of automotive fuel from biomass has become a political hot potato in Western Europe. Changes in international agricultural agreements affect the amounts of agricultural land that are available for growing crops specifically for the making of "biofuel." The question of energy input and output has just been added to the European biofuel debate.[42]

Alcohol does improve a gasoline's burning inside the internal combustion engine. When mixed with gasoline, ethanol has an effective octane rating of about 135. Adding 10 or 20 percent ethanol could significantly improve the behaviour of a gasoline engine. Conversely, in terms of fuel consumption per mile, ethanol must lose. Yet there is merit in the alcohol story. Sugar, sugar cane, and corn are not the only sources for making alcohol.

Municipal waste contains vast amounts of fermentable materials. Several fungi other than beer and wine yeasts can produce alcohol. The mold *Trichoderma viride* can break down cellulose such as cotton and paper into simple sugars and then convert these into alcohol. This versatile biodegrader, known by Vietnam veterans as "jungle rot," can also break down several pesticides.[43] Fermentation of waste cellulose could produce useful quantities of alcohol and at the same time bring relief to the waste disposal problems of large cities.

The manufacture of methanol from coal gas was discussed by Gray and Alson.[44] There seems little question that methanol could perform well in internal combustion engines. Unfortunately, as these authors point out, it does not perform so well in the political and economic arenas. The advantages of both alcohols would be the elimination of the multiple pollutants released by coal and liquid petroleum. The alcohols burn more cleanly and besides the inevitable carbon dioxide there is little else by way of emissions. The issue of automotive fuels is an increasingly complex problem.[45]

Alternatives and "free energy"

All these "biomass" alternatives have the formation of carbon dioxide in common. As international pressure for the reduction of carbon dioxide emissions keeps mounting, a less environmental damaging alternative *must* be found.[46] Virtually the only alternatives that can supply energy without the drawback of global climate change are the nuclear option, wind power, and solar energy. The latter two are still far from able to supply the quantities of energy needed for a modern, industrialized society. Nuclear energy brings another challenge in addition to the commonly perceived disadvantages. The international cooperation needed for policing such things as reactor safety and spent fuel management is going to be a significant challenge in international politics.[47]

The most popular alternatives to conventional energy sources are those that are "free," sun and wind. A significant advantage of these alternatives is the avoidance of air pollution. Of course, they are not free, they do obey the law. Solar energy and wind energy are only free in the sense that the supply is, on a human time-scale, unlimited. They are not free in the thermodynamic sense in that they too convert energy into a less usable form. Their use leads to an increase in entropy. In a simple industrial context, energy is used to make things; making things means making waste, whether or not they are made with "free energy."

Considerable advances have been made in the development of both solar and wind energy. Pressure from the environmental movement, increasing costs of conventional sources, and political uncertainty about oil supplies are the contributing factors in this development. Projections into the middle of the next century, although still heavily counting on fossil fuels, do include significant increases in the utilization of the solar and wind options.[48]

The most important aspect of solar energy is the development of the photovoltaic cell, a device that converts light energy into electrical energy. It is already in use in solar-powered calculators and watches. There are realistic expectations that photovoltaic installations will become a significant contribution to the electrical energy supply.[49]

A few percent of the sun's radiant energy that hits the earth is converted into air movement in the form of wind. Wind has been used since humans started making tools. Today, wind energy is used to drive electrical generators. The design of these devices has improved, and reliability and output are within reach of large scale commercial development. In North-Western USA and in Europe some commercial wind farms are operating successfully. Placement of these wind farms is important; they can generate a considerable noise problem for nearby residential areas. Nevertheless, wind derived electrical energy will soon make its entry into the energy mix.

Economic hurdles must be overcome before solar and wind energy can make significant contributions. The first of these is the actual cost per unit of electrical energy produced. This is still much higher for solar and wind energy than it is for hydro or coal or nuclear fission. Since taxation already plays a large role in the economics of these conventional supplies, many jurisdictions have accepted the idea that alternate energy supplies can be economically stimulated by tax incentives. The other problem for solar and wind energy is their intermittent nature. Solar energy is not available at night, for example. This means that storage facilities for electrical energy must be available more economically than at present. The survey by Starr *et al.* gives an overview of the status of the world's energy requirements and the question of availability from a 1992 viewpoint.[50]

The price disadvantage of alternative energy sources could in part be corrected by adjusting the artificially low price for conventional energy. Present prices for energy are based on the costs of production, market demand, and profit considerations. The environmental cost is as yet not included on the energy balance sheet. The cost of nuclear energy does not include the eventual cost of decommissioning an entire plant, nor does it include the billions spent over several decades on researching the radioactive waste management issue. Similarly, the cost of coal energy does not include the price that may have to be paid, eventually, for the *potential* damage from global climate change. While politicians charm their electorates by promising lower energy costs, the environmental argument should be for *much higher* energy costs. Unfortunately, it is obvious who would win such a political contest.

Two extreme opposites on the technical scale deserve our attention briefly. They are the high-tech nuclear fusion and the wood stove. Technically speaking, fusion energy is generated by the joining of four atoms of hydrogen to form the nucleus of a helium atom. The reaction takes place at a very high temperature and emits small amounts of high energy radiation. This fusion process is what happens in the sun. It produces the temperatures of millions of degrees celsius in the mass of the sun and energy is radiated into space. At that scale the amounts of energy and radiation are truly immense. This reaction can be replicated on a small scale. The fusion of two deuterium nuclei has more or less the same energy effects as the fusion of four hydrogen nuclei, or the fusion of one tritium nucleus with one hydrogen nucleus. Fuel for the process is abundantly available. Naturally occurring deuterium is easily isolated from seawater and tritium can readily be made from abundantly available lithium.

The problem is the containment of the process. The temperatures needed are much higher than any known material on earth can withstand. The high temperature flux can be contained within a strong magnetic field. Such a magnetic "flask" must be created by strong electromagnets. The energy input for the electromagnets still exceeds the output of the process. Research during several decades has not yet reached the break-even point. Thomas Erber of the Illinois Institute of Technology commented, during a conference on the controversial cold fusion project, that fusion research so far has been "inefficient, dangerous, complex and very expensive."[51] How expensive? About $750 million a year, presumably a figure that relates to the USA. No doubt, commercial energy from nuclear fusion is still far in the future.

At the complete opposite end of the scale is the wood burning fireplace. Many larger homes in Western Europe and North America have them. They are a desirable status symbol, provide a romantic atmosphere in the winter, and are often a poor way of heating a house. Because of the cost of conventional energy and the availability of cheap wood in some areas, more sophisticated wood burning devices have been developed. Wood burning is a fashionable partial replacement of oil or gas. On a large scale, wood burning is not a viable alternative energy source because of limited fuel availability and air pollution problems. We have already discussed the early problem of deforestation in many parts of the world as a result of wood burning. This happened only a few thousand years ago in what are now industrialized countries. In many poorer countries this deforestation is still continuing at an alarming rate. The resulting further deforestation of large areas of the planet may well have an effect on the world's climate. Also, wood burning produces smoke.

Complete combustion, of any fuel, produces only carbon dioxide and water, provided the fuel contains no sulphur or chlorine compounds. Incomplete combustion produces smoke. Smoke is a mixture of combustion gases and particulate matter. The gases are carbon dioxide, carbon monoxide, and many volatile organic chemicals, VOCs in incinerator parlance. The particulate matter is mainly carbon. Regardless of the nature of the fuel, the organics in the smoke always contain smaller or larger amounts of PAHs, the polycyclic aromatic hydrocarbons.[52] Just as with diesel fuel, the carcinogenic PAHs are adsorbed on the carbon particles and carried away by the wind. Wood smoke is also a significant source of the controversial chlorinated dioxins and furans, an aspect of wood combustion we will address in a following chapter. Wood burning on a large scale is a significant source of air pollution, and the smoke has a negative impact on public health. Burning wood is not an environmentally friendly form of energy.

A sustainable future?

Discussions about sustainability normally include a variety of resource issues, but energy is by far the most important. The solutions for virtually all resource problems presented by the cornucopians are based on the availability of "abundant and cheap energy."[53] The cornucopian view tends to bank heavily on nuclear energy, both fission and fusion. The agnosophobia factor in nuclear fission and the technical obstacles to nuclear fusion are seen through strongly rose-coloured cornucopian spectacles. While these nuclear problems are being solved, so say the cornucopians, economics will take care of the infinite availability of fossil fuels. Fossil fuels are creating some pollution but after all, scientists obviously cannot agree on global warming anyway. Cornucopians rarely address the question of developing nations aspiring to Western levels of affluence.

Another approach to the supply problem involves estimating projections of future energy needs. One such estimate suggests that by the middle of the next century, global energy needs would be about four times present levels.[54] This would place a significant strain on the availability of relatively low cost resources like coal, oil, and gas. At the same time international pressure to reduce carbon dioxide emission can be expected to increase. The problem with such estimates is that they are frequently overtaken by political events, so that they become obsolete within a few years. For example, a 1983 "status report" on the supply and consumption of coal in the USA starts with the observation that "the structure of the coal industry is driven by the oil embargo and world oil prices." The oil embargo has disappeared, oil prices have come down, and the projections made in that status report are no longer valid.[55] However, the view of fossil fuel consumption on a longer time-scale does deserve our attention. That view presents the "epoch of fossil fuel exploitation" in the perspective of human history. The "epoch" is shown as a tiny peak on a time-scale that starts about five thousand years ago and projects an equal five thousand years into the future.[56] The implication of this dramatic picture is that within a few generations humanity will see a gradual decrease in fossil fuel consumption. The 1973 date of this survey means that global warming and greenhouse gases were not yet part of that picture.

Virtually all literature on energy and energy resources starts with the basic assumption that conservation must stretch existing reserves while alternative resources are being developed. This allows the industrialized nations to continue their *present mode of operation.* The new energy sources, particularly solar and wind are usually discussed in terms of 1990s consumption levels. However, neither solar nor wind energy development is anywhere near supplying a large part of the energy mix. Together with the carbon dioxide issue this really leaves very few options. The question of sustainability will force us to have a close second look at nuclear energy.

CHAPTER 8

Air for breathing

In this corner of the galaxy earth is unique because of the composition of its atmosphere. The main ingredients are slightly less than 80 percent nitrogen and a little over 20 percent oxygen. Several minor components, water vapour, carbon dioxide, methane, argon, helium, and a few other trace gases complete the 100 percent. The oxygen makes life possible on this planet. How the earth's atmosphere came to be this way is part of the story of the evolution of life itself. That story is told in many earth science textbooks. Popularized versions can be found in various atlases. Lovelock's Gaia hypothesis describes the formation of the earth's atmosphere in easily understandable terms.[1]

Some organisms, the biodegraders, need oxygen to metabolize, or biodegrade, their food. Oxygen combines with the carbon and hydrogen in the food to form carbon dioxide (CO_2) and water, which are then excreted as waste. Other organisms, the photosynthesizers, use energy from the sun to recombine the CO_2 and water into the complex architecture of carbohydrates. A third group of organisms, the anaerobic bacteria, break down their food in the absence of oxygen. Their major waste product is methane. In the atmosphere, chemical interaction with oxygen oxidizes part of the methane so that its concentration remains low. These conversions from one chemical into another, brought about by living organisms, keep the composition of the atmosphere in a state of *dynamic equilibrium.* Things have been this way for many thousands of years, until recently.

Oxygen using organisms cannot separate one gas from another. When an animal breathes, it inhales not only oxygen, but also nitrogen, carbon dioxide, and any other gas or vapour that is in the air. Particulate materials, dust and soot are partly removed by filter mechanisms in the respiratory system. Similarly, when fish take oxygen from the water, they also take up other chemicals dissolved in the water. The exchange of these gases, inhaled air and exhaled waste gases, needs a large surface area. The lungs of mammals and birds and the gills of fish provide these large surface areas. We will let the precise details of these gas exchange processes stay in the biology books. We just need to be aware of them.

Air for breathing

Human activities have affected the composition of the atmosphere during the past few hundred years. Some of these changes are global, others are localized, even limited to indoor air. The nature and magnitude of these changes can be measured with precision. The effects of the changes on the life of the planet are less certain and subject to much argument. Some of these effects may be beneficial, others could be detrimental to life. The nature of the problem was stated, rather dramatically, in the closing statement of a 1988 Conference on the Changing Atmosphere: "Humanity is conducting an unintentional, uncontrolled, globally pervasive experiment whose ultimate consequences are second only to nuclear war. It is imperative to act now." The problem is, we don't really know what these "ultimate consequences" are. So what kind of action is "imperative?"

The great outdoors

Evidence from many studies shows that the atmospheric CO_2 concentration has increased from about 270 ppm in the year 1600 to about 350 ppm today.[2] Geophysical records show that there have been other, earlier changes in the atmosphere. It could be that the present increase is *coincidental* with human activities. However, a graph of the increase against time shows that this would stretch coincidence beyond credibility. The trend and magnitude of these changes now raise questions about the ability of the system to maintain its equilibrium. The possibility that these anthropogenic changes to the atmosphere are permanent and have a deleterious effect must be seriously considered.

The good old days

The composition of the atmosphere is part of the larger system that ensures the stability of the entire biosphere. The movements of air and water, the circulation of materials and other details of the biosphere are discussed in ecology and geography textbooks. A few of these books make for easy reading in plain language. Nisbet's *Leaving Eden* concentrates on general physical geography and also lists materials for further reading.[3] Readers with some chemistry background will appreciate the book by Nigel Bunce.[4] In *Global Environmental Issues*, Kemp uses the approach of the climatologist.[5] These detailed discussions provide good background reading, but they go beyond the scope of our discussions. Therefore we will concentrate on a few specific points that deal with the buzzwords.

Air pollution is usually thought of as chemicals, present as gases, vapours and particulates, that "are not supposed to be there." The notion of what is supposed to be there, and what is not, is complex and somewhat subjective. Standard ecology texts contain details of the carbon cycle, the movement of carbon and carbon containing molecules between land, air, and water. Many chemicals other than carbon also travel between these compartments. Their movements are not quite so generally known.

Living organisms also need sulphur and iodine as essential nutrients. Biomass and minerals containing these elements are continuously carried to the sea by rivers. Without a mechanism to return them to the land, life on land would end when sources of sulphur and iodine were depleted. The path back from the oceans to the land is via the atmosphere. The sulphur and iodine stories are told in plain language by Lovelock.[6]

Dimethyl sulphide is a gas made by marine micro-organisms from dissolved sulphur compounds. It is readily oxidized in air to sulphur dioxide and then to sulphuric acid; both contribute to acid rain. At lower altitudes sulphuric acid droplets may play a role in cloud formation. Eventually, precipitation returns part of the sulphur to the soil for the benefit of plants and land animals. This global sulphur transport means that acid rain also occurs naturally.

A similar situation applies to iodine. Some marine plants make methyl iodide, a toxic gas and a carcinogen. When precipitation returns it to land it is changed into other iodine chemicals that are essential nutrients for land organisms. An exception to this cyclic transport of minerals is phosphorus. The only mechanism for returning phosphorus from the oceans to the land is in the form of bird droppings. Isaac Asimov, slightly tongue-in-cheek, labels phosphorus as one of the land's non-renewable resources.[7]

Acid rain and toxic iodine compounds thus have a legitimate presence in the atmosphere, provided the quantities are controlled as part of a global equilibrium. Volcanic eruptions can also eject vast quantities of sulphur and chlorine into the atmosphere, and the accompanying ash particles can affect the global climate. Ash particles can obstruct sunlight reaching the earth. Another theory is that the formation of a stratospheric sulphuric acid haze, which reflects sunlight back into space, contributes to a cooling effect.

The cooling effect from a single volcanic eruption can last years, demonstrating that natural events can significantly affect the atmosphere. Eventually the equilibrium restores itself. Human activities can also disturb the equilibrium. While natural disturbances are intermittent and infrequent, human activity is sustained and increasing. That is why questions about acid rain and ozone and carbon dioxide have become more urgent.

Problems, causes, and consequences

Small tribes of prehistoric humans could not have lasting effects on the atmosphere. Six billion people using the sophisticated tools of modern technology can. In fact, only about one billion have done most of the damage so far. This damage is of two kinds. One is global and it can affect the entire atmosphere and therefore it is of interest to all living organisms on the planet. The other kind of damage is restricted to smaller areas and affects smaller populations.

Combustion of fossil fuels produces several gases. Sulphur containing coal produces sulphuric and sulphurous acids. All fossil fuels and all biomass produce carbon dioxide during combustion. Combustion needs the oxygen in the air, but the remaining nitrogen cannot help being involved also. At high temperatures some of the nitrogen combines with oxygen to form mainly nitric oxide, another acidic chemical. We will first look at the carbon dioxide, or in chemistry shorthand, CO_2.

Naturally occurring carbon dioxide and water vapour have an important function. The sun's radiation transfers *heat* energy to the surface of the planet during the day. The resulting warm surface radiates much of this energy back into space, but in a different form: infrared light, invisible to the human eye. This "low energy radiation" would quickly carry all heat energy away into outer space if there was nothing to delay it. What delays this loss of heat is the mixture of water vapour, carbon dioxide, methane, nitrous oxide, and ozone in the atmosphere. These gases absorb the infrared energy, trap it within the atmosphere, and reflect it back to earth. This blanket of gases acts like the glass of a greenhouse, it lets the heat in, but it does not let all of it out. These gases in the earth's atmosphere are called greenhouse gases, and the energy trapping mechanism is called the greenhouse effect. A very efficient greenhouse effect makes Venus too hot, the absence of an atmosphere and a greenhouse effect make Mars too cold, and a moderate greenhouse effect makes earth just right. Climatologists have dubbed this the Goldilocks phenomenon.[8]

The windows of the greenhouse control the temperature inside. If the glass is made thicker, or the windows doubly glazed, the greenhouse gets much warmer. Some plants like that, others may die. By analogy, the *magnitude* of the earth's greenhouse effect is determined by the amount of greenhouse gases in the atmosphere. When our solar system was formed, the sun's energy output was about 30 percent less than it is today. All water would have been frozen and life could not have started. It has been proposed that instead a much higher concentration of CO_2 in the atmosphere, perhaps a hundred times today's level, provided a much stronger greenhouse effect. This kept the temperature up, the water liquid, and allowed the evolution of life. Since then, the CO_2 concentration has decreased and has been almost constant for centuries.

Industrialization brought the large scale combustion of organic materials, and the resulting carbon dioxide. Since about 1800, carbon dioxide has increased by about 26 percent.[9] Other greenhouse gases have also increased. Methane concentration has doubled, and nitrous oxide has increased by 8 percent. The synthetic chlorofluorocarbons (CFCs) are efficient greenhouse gases that entered the atmosphere from many industrial applications. All these additional greenhouse gases in the atmosphere are believed to have the effect of double glazing the greenhouse. As a result, the planet could experience an *enhanced* greenhouse effect. The CFCs have additional effects on the upper atmosphere.

There is little doubt about the amounts and the measurements. There is, however, widespread controversy over the possible consequences of these increased concentrations. It is very difficult to calculate the potential enhancement of the greenhouse effect by atmospheric pollution. Only a few decades ago some biologists issued urgent warnings about the *cooling* effect of atmospheric pollution.[10] The problem is not very suitable for experimentation. Therefore computer modeling is a common approach. Even with access to the biggest available computers and the use of sophisticated models, it could take many more years before reliable estimates become available. At present, there is a considerable amount of consensus within the scientific community that global warming is indeed underway, although the magnitude of the effect is still unknown. Dissenting opinions have it that increased CO_2 concentrations could actually *decrease* global temperatures.[11]

By *assuming* a given rate of warming, consequences of the effect can be *estimated*. The polar ice caps would shrink, and the additional water would raise sea levels, affecting low lying coastal areas. Climate changes would destroy agriculture in some areas and enhance it elsewhere. Life on land and in the oceans would change in terms of species diversity. Some organisms would thrive, others would become extinct as the result of temperature changes or the flooding of habitats. Humans could probably adapt, but the economic costs and the cost in human lives could be very high.

On the other hand, Cure and Acock reviewed hundreds of studies showing that increased CO_2 levels would greatly benefit the planet's plant life.[12] Many researchers believe that this "regreening of the planet" is already observable.[13]

All these predicted scenarios result from *the assumed magnitude* of the *enhancement* of the greenhouse effect, and are clearly in the realm of Weinberg's trans-science. However, a detailed review of the 1992 state of the problem explains that we are slowly but surely coming closer to an understanding of this very complex system.[14] Also, in view of the finite nature of fossil fuel reserves it makes sense to curtail the further production of CO_2 by conservation and consideration of other energy resources. Leo Yaffe's "health hazards of *not* going nuclear" may soon be viewed in a new light.[15] Furthermore, CO_2 is not the only problem.

When acidic chemicals, formed from the oxidation of sulphur and nitrogen, dissolve in water, they can lead to acid precipitation. Taken together, coal-fired power plants and automobiles put more acidic chemicals into the lower atmosphere than natural compensating mechanisms can handle. The accumulating acidity in rivers and lakes causes observable damage to vegetation and animal life.

It has been pointed out that the quantities of *natural* acid producing substances are so enormous that the impact of the human contribution is questionable. Naturally occurring sulphur compounds from volcanic eruptions and from marine micro-organisms are estimated to make up from 35 to 85 percent of the total sulphur dioxide in precipitation. The other important contributor to acid rain, nitric acid from nitrogen dioxide, is readily formed when oxygen and nitrogen react in an electrical discharge. Calculations suggest that lightning could produce enough natural nitric acid to impart considerable acidity to precipitation over the entire planet.[16]

Nevertheless, some consequences of acid precipitation are sufficiently localized and visible to demand explanation. The damage to vegetation in many areas and decreases in some freshwater fish populations cannot readily be explained otherwise.The die-back of heather in Western Germany is an example of such vegetation damage. We will have a further look at the effects of acidity on surface waters in the following chapter.

At an altitude of about 25 km, the thin atmosphere contains a small amount of ozone, chemically a very reactive gas. Even in this very dilute concentration, ozone absorbs a major portion of the ultraviolet light from the sun, the so-called UV-B portion. Most of the longer wavelength UV-A part is not absorbed and reaches the surface where it is important for several natural processes. Short wave ultraviolet light, UV-B, carries more energy than UV-A and can interact with molecules of biological importance. Therefore UV-B *can* have adverse health effects.

Ozone molecules are readily formed from oxygen under influence of ultraviolet light. They also fall apart easily to form molecular oxygen. When ozone interacts with other molecules, the oxygen atoms of the ozone become incorporated into these other chemicals or they form ordinary oxygen gas. For example, nitrous oxide, made by bacteria, is oxidized in the upper atmosphere to nitric oxide, which readily destroys ozone molecules. A dynamic equilibrium between formation and destruction keeps the ozone concentration approximately constant.

About twenty years ago it was discovered that the stratospheric concentration of ozone is declining. Experiments confirmed that ozone interacts with the synthetic chlorofluorocarbons, the CFCs. Many studies, both in laboratories as well as via satellite and other high-tech tools, paint a clear and comprehensive picture of the stratospheric chemistry of the CFCs. While the enhancement of the greenhouse effect is still uncertain, the ozone events are supported by well-understood, well-described chemistry. Two researchers who blew the ozone whistle at the time, Rowland and Molina, recently reviewed the twenty year history of the problem.[17]

CFCs are synthesized by the interaction of natural gas components, like methane, with chlorine and fluorine. They have intriguing properties. CFCs do not burn, they are gases, can easily be condensed to liquids, and are chemically quite unreactive. These properties make the CFCs ideally suited for several industrial uses. They are good solvents, especially in the electronics industry. They are ideal drivers in aerosol products for the consumer market. Their stability and compressibility make CFCs excellent refrigerants for domestic and commercial refrigeration and air conditioning equipment. Refrigeration and air conditioning are the largest components of CFC use. The lesser known *bromo-chloro-fluoro-carbons* are important as fire extinguishing agents under the trade name Halon(s). Halons are very similar to CFCs in their behaviour toward stratospheric ozone.

It takes a long time for CFCs to travel the 25 km altitude to the ozone layer. This is why today's tests measure the effect of CFCs released several years ago. There is a time lag between cause and effect. Also, one molecule of a CFC can destroy many ozone molecules. There is a large amplifying factor in the process so that relatively small amounts of CFCs have a large effect on ozone destruction.

Although there is little doubt about the chemistry and the measurements of the ozone problem, a good part of the problem is in the realm of trans-science. When ozone at high altitude decreases, more ultraviolet-B radiation will reach the surface. Qualitatively, excessive exposure to UV-B is known to affect living organisms. Common forms of damage are cataracts and skin cancer in humans. Damage to plants varies from one species to another.

The controversy, as usual, lies with the quantities rather than with the qualitative nature of the effects. The question is not whether UV-B can cause skin cancer but how much. Do observed increases in skin cancer result from increased UV-B radiation or from increases in leisure time spent at the beach or increases in the amount of skin, per person, that is exposed? The nature of this controversy is illustrated by a note concerning the effect of UV-B on marine algae.[18] While one research group observed reduced growth rates of algae exposed to UV-B *in the laboratory*, other scientists found no effect of increased UV-B on large algal blooms *in the ocean*. This could be explained by the abundance of ultraviolet absorbing chemicals dissolved in the water.[19] Since light intensity decreases rapidly with increasing water depth, field observations are very important in the marine environment. Conversely, effects on plants and animals are easier to observe. Thus Loblolly pine was found to be sensitive to UV-B so that an increase in exposure could significantly affect harvest yields.[20] About two-thirds of the wood supply for the US paper industry is Loblolly pine. As with global warming, the consensus of the scientific community has come down on the side of caution.

Local air pollution problems are caused by the release of pollutants near the planet's surface. These chemicals are changed in the lower atmosphere when they interact with each other or with atmospheric oxygen. Many of these chemical changes are promoted by sunlight.

Nitrogen dioxide, one of the nitrogen oxides formed in combustion processes, readily decomposes to form nitric oxide, and in the presence of oxygen this also leads to the formation of ozone. In a pristine environment that reaction is readily reversed, to regenerate the nitrogen dioxide and oxygen. In this balanced, non-hazardous equilibrium condition the ozone does not build up to hazardous concentrations. In the presence of other air pollutants, however, the nitric oxide reacts more easily with various hydrocarbons, for example from automobile exhaust. Then the low level ozone accumulates. At low altitude, where it no longer protects against UV radiation, ozone is simply a toxic air pollutant that adversely affects both plants and animals.

Interaction of nitric oxide with petroleum hydrocarbons forms a family known as peroxy-acetyl nitrates, PANs in air pollution parlance. PANs are toxic and irritating to eyes and mucous membranes. Many of these airborne chemicals become adsorbed on particles of soot and dust. When this whole cocktail stays together because of lack of wind and temperature inversions, the result is the well-known smog that often covers many large cities. Some books refer to the phenomenon as *photochemical* smog, because the complex chemistry of the whole process is catalyzed by sunlight. Although industrial processes can cause slightly different forms of smog, the main problem is the automobile. The evaporation of gasoline from stationary vehicles and during refuelling operations, mentioned earlier, also contributes to smog formation.

An organic, lead-containing chemical, tetraethyl lead, has been used for decades to improve the performance of automotive fuels. When unburned hydrocarbons first caused smog formation on a large scale, the approach was to "burn" this unused fuel by means of the catalytic converter. This device, as an integral part of the exhaust system, functions like an "after burner." It converts carbon monoxide into carbon dioxide and helps to burn other volatile organic compounds (VOCs) to carbon dioxide and water. Because lead deactivates the catalyst, which contains platinum and rhodium alloys, the catalytic converter is incompatible with leaded fuel. Although the initial reason for lead-free fuel was the chemistry of the catalytic converter, we now know that lead compounds are a serious pollution problem by themselves. Lead containing gasoline was completely phased out in most of North America by 1989. It is still in use in many countries in Europe.

Lead containing chemicals resulting from automobile exhaust and other combustion processes can remain in the atmosphere for some time before precipitation brings them down.[21] Precipitation run-off can transfer airborne lead pollution to a public water supply, so that lead ingestion via drinking water can also be a problem.[22] The replacement for tetraethyl lead, a manganese compound, is now under scrutiny for both its performance and for possibly adverse health effects.

Smoking is often mentioned as the major cause of lung cancer, but some attempts are now underway to study the effect of urban transportation. Low level air pollutants are mostly associated with heavy automobile traffic and with industrial activity. Although they can harm plant and animal life, the main concern is human health. Lave and Seskin examined the relationship between air quality and the frequency of bronchitis, lung cancer, cardiovascular disease, total respiratory disease, and infant mortality.[23] Their results indicate that big city air pollution does adversely affect human health. A more recent survey in California placed temporary health effects in an economic context by calculating the appreciable value of lost production time.[24] Another writer raised the importance of the transportation problem by pointing to the lung cancer rates in rural China which are about a tenth of those in Britain notwithstanding similar smoking patterns.[25]

We briefly looked at the incidents at the Three Mile Island and Chernobyl nuclear power plants. The air pollution aspects of these incidents deserve some additional comments. In case of an accident in such plants, different kinds of radioactive materials can escape. Radioactive gases like krypton, or vapours of "radioactive water" can escape to the atmosphere or into the cooling water. The release of radioactive solids like iodine and strontium occurs only during a major incident where the integrity of containment structures is compromised.

Media reports on such accidents often refer to "radioactive heavy water." However, heavy water, or deuterium oxide, is not radioactive. In normal reactor operation water or heavy water is used as a coolant. This coolant is often contaminated with small amounts of tritium oxide. Tritium is a radioactive isotope of hydrogen and tritium oxide is truly "radioactive water." Depending on weather conditions, these materials can disperse rapidly into the atmosphere or in surface waters. There they are quickly diluted and can usually no longer be measured against natural background radioactivity. This release of radioactivity does not generally pose a health hazard to either plants or animals. Most of the Three Mile Island release was of this kind.

Emission of solids like radioactive iodine or strontium can become *temporarily* airborne. These particles eventually come down as fall-out or with precipitation. That was the case at Chernobyl. Although much of the fall-out was localized, increased radioactivity from that incident was measurable in much of Western Europe. The levels were very low and we have already looked at a summary of the numbers of that incident.[26] The medical and scientific communities are monitoring the possible effects from the Chernobyl incident closely in the immediate area. Considering the abundant literature on radiation hormesis, it is unlikely that any adverse effects will be seen elsewhere in Europe. If anything, the effects of Chernobyl for European areas that received low levels of exposure, may be expected to be beneficial.[27] Incidentally, those interested in the numerical details of low level radiation, and units and conversion factors, can find such information in Luckey's treatise on radiation hormesis.[28]

Remedies and solutions

Air pollution problems are clearly of two kinds. Stratospheric ozone depletion and atmospheric loading of carbon dioxide are global issues. Both result mostly from the use of heating, air conditioning, refrigeration, and transportation in the industrialized nations. In other words, it is closely associated with the consumption of energy. The long-term effects of this kind of pollution are far from certain and many questions remain. Although these *appear* to be scientific questions, it is obvious they cannot be answered by science. The outcome of the "globally pervasive experiment" lies in the domain of trans-science. Whatever the effects turn out to be, they affect the entire world. This means a political rather than a scientific debate. With respect to the CFC-ozone problem that approach seems to have worked to some extent.

A popularized story presents a useful chronology of events in the twenty-year history of the ozone saga.[29] Unfortunately, this book misses the point by placing the principal blame on household aerosols, which were only a minor part of the problem. Perhaps this bias was caused by an early ban, in the USA, of aerosol spray cans in 1978. International talks were held in 1983, followed soon thereafter by an international agreement, the Montreal Protocol, on the phase-out of CFCs. After the aerosol cans, use of CFCs as a blowing agent for foamed plastics was also phased out soon after 1987.

A related political event was the formation of an international stratospheric protection fund in June 1990. Since the industrialized world had enjoyed the use of CFCs for so long, less developed nations questioned the fairness of being denied this privilege. The fund was intended to help all parties to develop less environment damaging alternatives to CFCs.

Such alternatives are now coming on stream, but the solution is only a partial one. If *all* the hydrogen atoms in a hydrocarbon are replaced by fluorine and chlorine, the resulting CFC molecule is resistant to photochemical breakdown in the lower atmosphere. It can reach the stratosphere and interact with ozone. When *not all hydrogens* are replaced, a chloro-fluoro-*hydro*carbon (HCFC) is formed instead of a chlorofluorocarbon (CFC). These HCFCs do break down in the troposphere; they have a *tropospheric sink* and do not readily reach the stratospheric ozone layer. HCFCs are already available for automotive air conditioners and domestic refrigerators.

Compression equipment designed for CFCs cannot be used with the new coolants without modifications. This is no problem for the automotive industry, which designs its products with built-in obsolescence so that the new system can be introduced within a few years. The food industry's refrigeration plants, responsible for some 20 percent of all CFC use, will be very slow to switch because of the gigantic costs of retrofitting. The numbers for such a changeover are formidable.

Optimism over "ozone friendly" replacement coolants must be tempered by reality, however. The HCFCs are "destroyed" in the lower atmosphere in the same way as automotive fuels. They are smog producing pollutants and can contribute to air pollution of the local kind. One of the replacement chemicals, a chloro-fluoro-ethane, has been found to cause benign tumours in rats. This is a serious obstacle for a chemical that was considered one of the better replacements. Also, even if all use of the original CFCs were to be terminated now, which is not the case, it could still be many years before the stratospheric ozone layer can repair itself, provided the trans-science questions have been answered correctly. Stratospheric effects of CFCs may still get worse before they get better.

Wallington *et al.* have reviewed the literature on these CFC replacements in detail. They confirmed that these replacements are easily photodegraded in the lower atmosphere and have a much smaller effect on stratospheric ozone than the original CFCs.[30] That effect is not zero, however, and their use will continue to affect the ozone layer. In terms of global warming, the HCHCs have about 10 percent of the warming effect of the original CFCs. The relatively non-toxic breakdown products have a negligible environmental effect. Although most of the HCFCs that are released decompose in the lower atmosphere, a portion will still reach the ozone layer. Ozone damage by HCFCs has been estimated to be from 5 to about 20 percent of that of the CFCs. Therefore HCFCs will still cause ozone damage, albeit at a greatly reduced rate.[31]

In the meantime, the use of the HCFCs has also been brought into the sphere of international control. An international agreement, signed in Copenhagen in 1992, called these replacements "transitional substances," and there is further political pressure for these transitional HCFCs to be phased out completely. There are two serious wrinkles in the ozone story.

The first of these is the fact that the entire ozone concern is based on observations during a few decades. There are no records that go back for thousands of years like in the case of atmospheric carbon dioxide concentrations. There is no doubt about the actual chemistry of the ozone chlorine interactions. But we don't know if the decrease in atmospheric ozone is a new phenomenon in the history of the earth, or if, on a geological time-scale, this is just one such event of many; a blip in geological history. Why would there have been other, earlier problems with ozone?

The Stratospheric Ozone Review Group (SORG) is a study group set up by the British government in 1985. In a 1993 report this group stated that it is not yet possible to say how much of the Northern hemispheric ozone decline is due to the eruption of Mount Pinatubo in 1990 and how much is caused by (anthropogenic) chlorine.[32] Pinatubo's contribution to the problem was probable because of vast amounts of sulphuric acid blown into the stratosphere by that explosion. However, volcanos can make other contributions. They produce large quantities of hydrogen chloride (3 million tons/year) and hydrogen fluoride (11 million tons/year). These gases are ejected with enough force so that a portion can reach the stratosphere. In addition, about three-quarters of the world's active volcanos also produce CFCs.

Nature also contributes to the formation of the HCFCs, the "transitional substances" in Copenhagen language. Chemicals like chloroform, bromoform, methyl chloroform, ethylene tetrafluoride, and tetrachloroethylene are produced in huge quantities by marine algae. For chloromethane alone the annual formation from such natural sources is estimated at about five million tons per year, compared to the human industrial emissions of about 26,000 tons per year.[33] Contributions of that magnitude from natural sources do raise questions about both the geological history of the ozone layer and the efforts to regulate the chemicals involved.

The other global problem, carbon dioxide, is of a different kind. Carbon dioxide is the inevitable product of combustion of all biomass. The term "cleaner burning fuels" addresses only the visible part of air pollution. Cleaner burning still means carbon dioxide. Electricity is widely touted as the "replacement" for automotive fuel. Fortunes are involved in the development of the electric automobile. California has forced this issue by bringing in state legislation that will drastically control air emissions and requires the availability of electrical vehicles by the late 1990s.[34] Prototypes are already on the roads both in Europe and in North America. Science writers and technical journalists are hailing the electric car as a solution to air pollution.

The previous chapter showed that alternative energy sources are still in a state of early development. For the foreseeable future the energy for the electric car is the power of coal, hydro and nuclear. In the political anti-nuclear climate California will simply be exporting air pollution to outside the big cities. Air pollution from conventional cars will be replaced by increased air pollution from coal-burning power plants. This is, of course, no argument against the development of the electric car. Initially, some gains will be made by eliminating evaporative fuel losses and smog. Also, point source control of a power plant is easier than controlling the dispersed pollution from automobiles. Sometime in the future alternative sources of electrical energy will become available. When that happens, the electric car must be ready.

CFCs and the electric car have produced a replacement mentality that signals a covert problem. The cornucopian view is that the *consumption of energy* can and will be kept at present levels, possibly increased, and no sacrifices are asked from anyone.[35] The political pressure, as applied in California, confirms this attitude. Replacement without sacrifice is not only the prerogative of the cornucopians: most catastrophists also like to keep their vehicles. We have found here another example of the environmental industries becoming more important than the environment.

Remedial efforts for localized air pollution show similarities with the global issues in the sense that solutions are sought in the political arena. Fossil fuels are also the main difficulty here. Since alternatives are not yet readily available, solutions tend to be more simplistic and are mainly concerned with conservation. A tax on the amount of CO_2 produced, or hydrocarbons consumed, could be used as an attempt forcefully to encourage conservation. Legislators at all levels are already dreaming of this new source of revenue called the carbon tax. The effect of such a tax would ripple through the entire economy of the developed world. It would have serious economic repercussions for middle and low income people as well as middle and low income countries.

In the meantime, the reversal of North America's "Live Better Electrically" has had little result. The efficiency of electricity consumption has increased substantially due to improved home insulation and other energy saving devices. Community workshops by environmental organizations have also made contributions in this area. However, the number of electricity consuming devices in our homes has also increased dramatically. Thus, North Americans are still the biggest energy users in the world and so far there is little indication that the increased efficiency has any effect on a global scale. Some consequences of the attempts at energy savings are noticeable elsewhere, however.

The safe indoors

The level of outside air pollution to which an individual is exposed is strongly influenced by weather conditions. At times when air pollution is particularly serious, people with respiratory problems and the elderly are often advised to stay indoors. By implication it is "safer" to be indoors when the outdoor air quality is bad. Compared to our ancestors who cooked food inside a cave over a smoky fire, our modern dwellings offer a considerably better respiratory environment. Nevertheless, indoor air quality has suffered considerable deterioration during the past few decades. Indoor air quality problems in domestic dwellings and general occupancy buildings are common. Manufacturing companies face air quality problems on the shop floor. The "sick building syndrome" has become part of the office vocabulary.

Tell your doctor where you work

Besides the common pollution in outdoor air, there are some other agents that can affect the health of people whose work keeps them outside. Most of these are dusts of various kinds. Sandblasting operations, earth moving equipment and road repair all involve potential exposure to silica and other dusts. Farmers can be exposed to grain dust and airborne agricultural chemicals and cotton farmers sometimes suffer from farmers' lung.

Many people are affected by the indoor air at work. It has been known for a long time that air quality inside a building is affected by indoor industrial activity. As early as the year 1700 Bernardini Ramazzini, an Italian physician, wrote *De Morbis Artificum*, a treatise on health effects of chemical and physical agents in the workplace. Occupational medicine became a specialty in medical schools. Still later, legislation for protecting workers' health appeared in most industrialized countries. Toxicology grew an additional branch: industrial toxicology.

Technical expertise for monitoring workplaces and interpreting such measurements came together under the label "industrial hygiene." Professional industrial hygienists, with their backgrounds in biology and chemistry, have become an indispensable part of the industrial workforce. They play an important role in monitoring workers' exposure to solvents, paints, metal dusts, noise, radiation, and many other *potential* hazards. These practices have been in existence for many years, but the past few decades have seen a considerable expansion. Detailed texts on industrial hygiene are available for further reading on that topic.[36]

The energy crisis of the early 1970s turned out to be not much of a crisis. When the threats of OPEC evaporated, many earlier consumption practices were resumed, with one exception. We had learned that saving energy meant saving money. We had seen the huge amounts of heat energy that were lost from indoor heating. Since warm air just leaked out of buildings through cracks and holes, all that was needed was to plug the holes. Weather stripping, sealers, and improved window construction delivered lower heating costs and increasing energy efficiency. Then a new problem crept in through one remaining crack.

People inside a building produce many types of waste. They breathe out CO_2 and water, release volatile components of perfume and aftershave. Old photocopiers contained liquids that evaporated and new photocopiers and laser printers produce ozone. New carpets and building materials release volatile chemicals from glues and stain- and fireproofing additives. In older buildings some of the pigments used in interior paint may contain lead compounds. Redecorating can readily produce airborne, lead-containing dust. More recently attention focussed on mercury compounds. Phenyl mercuric acetate is a common preservative in latex paints, and potentially hazardous exposures are possible during painting operations.[37] Cleaning agents and building deodorants, toilet cleaners, glues and correction fluids, can also be added to the list.[38,39] The whole cocktail becomes part of the ambient air.

When, in cold weather, leaks and cracks and half-open windows allowed expensive, warm air to escape, cold outside air replaced it. The volatile wastes that had accumulated also escaped. When energy conservation sealed the cracks and the windows were closed, the complex mixture stayed inside the "tight" building. The resulting indoor air became unpleasant, oppressive, and stuffy. The tight building led to the "sick building syndrome."

Sometimes there are physical symptoms, watery eyes or a sore throat. On rare occasions, in an occupational setting, there are actual symptoms of poisoning. In many office buildings there are often only vague complaints of just feeling unwell, or a headache that goes away at home. This "disease without symptoms" has become a major controversy.[40] Many cases of this disease without symptoms occur in environments where no air contamination can be measured. That readily results in a variety of confrontations between employers and employees, and one common scapegoat is that "it's all in your mind."

Fortunately, modern biochemistry and psychology have shown that stress does have biochemical implications. In a society of single parent families, job insecurity, and family violence, stress is a common problem. That stress, combined with the unpleasant odours and stuffy atmosphere of a tight building readily brings on the "disease without symptoms." Paradoxically, some people seek relief from stress by deliberately exposing themselves to unusual odours and vapours. Then it is called "aromatherapy."

The measurement of a few "diagnostic" chemicals in indoor air often helps to solve indoor air quality problems. Indoor carbon dioxide is only produced by human respiration and smoking. At low levels it is non-toxic and indoor concentrations up to 2000 ppm can be tolerated without ill effects. However, levels significantly above 500 ppm may be *indicative* of poor air circulation that causes the build-up of other contaminants.

Formaldehyde is another diagnostic chemical that gives information on air circulation. It is formed in small quantities by frying, baking, and smoking, and it can be released in significant amounts from new carpets and furniture upholstery. Adhesives in new plywood and particle board can also release formaldehyde. The urea formaldehyde foam insulation (UFFI) of the early 1980s, although now banned in many jurisdictions, still causes an occasional problem. Formaldehyde is a mutagen in the Ames test and can cause nasal cancer in rats. However, a human health hazard is not confirmed by excessive cancer rates among embalmers, mirror makers, and other occupational users of formaldehyde. The British Medical Research Council's Environmental Epidemiology Unit (Southampton) looked at 7716 men with occupational exposure to formaldehyde. The study found no excess nasal cancers or lung cancers.[41]

When CO_2 and formaldehyde levels point to poor air circulation, improved ventilation or adjustments to air conditioning systems often solve the problem. Air-to-air heat exchangers can recover the heat from exhausted air to conserve energy. Other remedial measures include elimination of tobacco smoke and better exhaust systems for cafeterias to eliminate food odours.

For industrial establishments the limits of allowable contamination are determined by legislation in most jurisdictions. Not surprisingly, the setting of such limits is a continuing source of controversy. Employers, who pay the costs, and employees, who suffer the consequences, both make their respective pet errors. Employers often don't understand that workers' health directly affects productivity and profits, and therefore must carry a price tag. Many employees don't understand that a "safe" workplace does not exist and that risk is never zero.

In industrial settings more specialized remedial action is often needed. This usually involves modifications of work schedules, process changes, substitution of chemicals, and sometimes the use of protective clothing. This is a highly specialized field that should be handled by trained professionals who understand the processes and the health consequences of change. The complex area of industrial hygiene falls well outside the scope of the present discussion. Notwithstanding enormous progress in workplace health protection made in the past decades, problems remain.

Most full-time employees spent seven or eight hours a day at work. That means about one-third of each day in workplace air. Therefore many types of work are associated with specific diseases, such as pneumoconioses in miners, or welding fume fever. Physical agents can also have adverse health effects. White finger disease or Rainaud's syndrome, caused by the use of vibrating tools, noise induced hearing loss, and carpal tunnel syndrome are common work-related health problems. Many of these occupational problems have been studied in detail, and these studies teach us two important lessons. The first one is important for working people. The very first question your new doctor should ask is: "where do you work?" If he doesn't, shop for a new doctor. Failing to ask this question may be fatal at a critical moment in the future. The second question is of general importance in matters that relate health to environment. For a given toxic agent there is a large difference between an occupational exposure and the non-occupational exposure of the general public. In other words, there are large differences between *occupational health hazards* and *public health hazards*. That distinction is frequently lost in the heated arguments about health and environment. Failing to make this distinction can have costly consequences.

Is your home your castle?

Most air quality problems in private homes involve volatile chemicals that are produced by activities like food preparation, cleaning, or painting. This means that most of these situations can be dealt with by improving air circulation and ventilation. The presence of volatile mercury compounds in latex paint requires preventive action in poorly ventilated spaces during renovation. Ventilation does not always solve the problems related to chemicals that contain toxic elements like lead and arsenic. Pigments containing these chemicals are frequently found in the wallpaper and paint in older homes. The dust generated in removing these old paints and wallpapers must be carefully controlled in order to avoid serious health hazards.[42]

Improving the ventilation or air conditioning in a large office building is just "the cost of doing business." For the private home owner ventilation problems are not so easily solved, particularly after a costly energy saving project. After a home has been tightly sealed with weatherstripping, additional glazing and painting, it can be very expensive to improve ventilation with heat exchangers or heat pumps. This unforeseen, additional cost of an energy saving project is frequently overlooked. A simpler solution could be just to open a window when weather permits. A few specific problems deserve a closer look.

An important issue has entered the awareness of the private home owner in recent years: radon infiltration. Radon is a naturally occurring, radioactive gas. It is colourless and odourless. It is part of the series of radioactive decays that starts with uranium-238. Decay of uranium leads, after several steps, to radium, the immediate precursor of radon. Radon has a half-life of about 3 days. It decays further to give other radioactive elements called "radon daughters." Radon is the only element in the whole sequence that is a gas; all the others are solids. Therefore radon's movement in the environment differs from that of all other radioactive isotopes. The radon daughters are all radioactive, solid particles formed from the radon gas. A few of them have much longer half-lives, up to several years. They become part of the dust normally present in ambient air.

Uranium is a leftover from the formation of the planet. It decays very slowly, so vast amounts of it are left. It is almost universally present in the earth's crust, in some places abundant enough for mining, in most places just in trace amounts. However, these trace amounts do add up to large numbers. One square mile of average soil, excavated only five feet deep, would contain about thirty tons of uranium. Because of the continuous formation of radon from the uranium, surface soil constantly releases radon gas.

Because it decays rapidly it has a short half-life, and the resulting radon daughters and any remaining radon are rapidly dispersed in outdoor air. Indoors the gas readily enters through basements or crawl spaces. Poor air circulation, the result of energy conservation, allows radon to accumulate indoors. A heating system that involves on-site fuel combustion, such as natural gas, oil, or wood, can create a negative pressure inside a tight building, which further promotes radon infiltration. Radon can also be emitted from building materials that contain uranium as part of the sand or gravel used for making the bricks, mortar, and concrete. Regulatory agencies in many countries have accepted the idea that these indoor radon accumulations are a public health hazard. The gas itself and most of the decay products are radioactive and are potential carcinogens. A US Environmental Protection Agency (EPA) report stating that radon in homes could be responsible for some 20,000 lung cancer cases annually in that country catapulted the radon issue into the public's awareness.[43]

This awareness and considerable publicity led to public pressure on governments, both in North America and in Europe, to take action. In the USA recommendations for remedial action were issued, and legislation was enacted.[44] Guidelines for "recommended action levels" are now in existence in many jurisdictions. It has been estimated that in the USA billions of dollars would be required for implementing a national radon control policy.[45]

The nuclear chemistry of radon and the physics of its movements in air are well understood. Measurements in homes and other buildings can easily be made. Those who wish further details on the various units of measurement will find a useful tabulation in the discussion by Nazaroff and Teichman.[46] Some surveys have yielded information on the geographical variations in indoor radon levels.[47] Considerable controversy came with the interpretation of the numbers and the assessments of the risks to public health. It all started with the initial EPA estimates. These were based on the observations of uranium miners who were exposed to high concentrations of radon. These measurements were extrapolated by means of the hypothetical straight line dose-response curve we encountered earlier.[48] At the low dose end of this straight line, where no measurements were made, the 20,000 hypothetical lung cancer cases appeared. This extrapolation suffers from several flaws.

The uranium miners were heavy smokers but no correction was made for this confounder: of 380 lung cancer cases reported only 25 were non-smokers. The miners were also exposed to vanadium compounds, which are known respiratory irritants and to selenium compounds, which at high doses are carcinogenic. The silica dust these miners breathed in is also a known carcinogen.[49] The epidemiology of the sample population on which the EPA's predictions were based was therefore affected by several confounding factors. Consequently, the extrapolation from these findings to the low dose levels in homes and buildings is also questionable.

Questions also arise about the measurements. The methodology requires that measurements be taken with all windows closed and close to the possible sources of radon infiltration, such as floor drains in basements. This way the recorded exposure of the occupants of a home represents a "worst possible scenario." For its risk estimates the US EPA therefore assumed an indoor radon concentration that is higher than is representative of the living space in a single family dwelling. Then this higher concentration was applied to the entire US population including those living in apartment buildings, mobile homes, and other accommodation that can be considered radon free.[50] Not surprisingly, a further editorial in *Science* criticized the EPA's predictions.[51] In a subsequent issue several letter writers took the editor to task for his errors, others agreed.[52] The control of this hypothetical public health problem would lead to enormous costs. For a closer look at the issue we must carefully separate experiments and observations from the hypothetical approaches.

The many contradictory reports and statements concerning the radon issue rapidly led to calls for implementing the so-called "precautionary principle." The reasoning is that, if only inconclusive information is available, "we should act now" even if we don't know if our actions are necessary, "before it is too late." This approach is generally rejected by many scientists. It is nevertheless often used to force political or regulatory action. Before we invoke the precautionary principle in the case of radon, let us look at some of the literature.

A survey of several published studies found that there is no solid scientific evidence for linking residential radon to lung cancer. Such a statement could still tempt us to invoke the "precautionary principle" were it not for several studies reporting results of an opposite nature.[53] A Scandinavian study came down on the cautious side, indicating that the radon-related risk for non-smokers is probably much smaller than was previously assumed.[54] A study in Florida found *no association* between radon exposure and lung cancer.[55] An inverse relationship, that is a *reduced rate* of lung cancer with increasing exposure to low level radon concentrations was found in Washington State.[56] One of the most detailed epidemiological studies found definite reductions in lung cancer rates with increasing radon exposure. This author discussed in detail how over fifty confounding factors were used in his statistical analysis. The only possible explanation for the results of this study is the failure of the linear no-threshold theory for carcinogenesis from inhaled radon.[57]

A recent conference on the issue indicated that more than twenty epidemiological studies were still in progress (in 1991), and at the end of that four-day symposium "no one was prepared to claim that radon did not cause lung cancer, but differences remained about the magnitude of the risk."[58] Conspicuous by its absence in these discussions was the theory of radiation hormesis.[59] It will take several more studies indicating the beneficial effects of low level radon exposure before that theory becomes generally accepted. In the meantime, the cheapest way to buy some peace of mind for a perceived radon problem is to open a window occasionally and let in a breath of fresh air.

Asbestos

A much publicized contaminant in indoor air is asbestos. This material is not often found in private homes, but it is very common in office buildings, hospitals, and schools. In the asbestos literature these are collectively called "general occupancy buildings." Asbestos has many applications in many areas, from building materials to heavy industry, because of its unique properties.

Asbestos is the collective name for several fibrous silicate minerals. A few commercially valuable varieties are fire resistant, have unique friction properties, and because of the fibrous structure, have appreciable material strength. Asbestos is used where friction is important such as brake linings. Applications in building materials like ceiling tiles, floor coverings, sheeting, and pipe insulation benefit from its fire resistance. The most troublesome application of asbestos is as a fire retardant sprayed on structural building components. Asbestos cannot be metabolized by living organisms. It is completely resistant to biodegradation. We will omit the mineralogical details.

About the late 1920s a causal link was established in asbestos workers between the inhalation of asbestos dust and lung disease. When mineral particles are inhaled and embedded in the deep lung, they become encapsulated in scar tissue. This impairs the elasticity and the capacity of the lungs, and a condition generically known as pneumoconiosis is the result. With coal dust the result is coal miner's lung, silica leads to silicosis, and asbestos results in *asbestosis*. Pneumoconioses are painful and debilitating diseases. The commonly used varieties of asbestos rarely cause lung cancer. There is a synergistic effect with tobacco smoke that greatly increases the cancer risk.

In the mid-1970s Irwin Selikoff, a leading US researcher, published his studies on asbestos workers and included lung cancer in the asbestos picture.[60,61] When a rare, fast growing cancer, called mesothelioma, was found to be associated with a rarely used amosite or "blue" asbestos, the regulatory fat was in the fire. The most serious problem arose because the US health authorities again applied the "no-safe-dose" principle and made direct extrapolations from the occupational exposure to public exposure. Asbestos was legislated, regulated, banned, and became a major controversy.

A potential danger is the use of other mineral fibrous materials as replacements for asbestos. So-called *asbestiform* minerals like wollastonite and sepiolite could occur in "asbestos-free" brake linings. It has been found that these fibres can have the same adverse health effects as asbestos itself. Swedish health authorities have taken action to control these materials.[62] In North America "mineral wool" remains an almost unregulated substitute for asbestos.

The radon and asbestos problems show significant similarities. In both cases the health risk for the general population is estimated by extrapolation from occupational exposure data. For radon occupational exposure data from uranium miners were used, for asbestos the numbers used were the exposure data measured for workers in the asbestos manufacturing industries, and also from miners. In both cases there is no evidence that the agents, radon and asbestos, actually do cause cancer in the general population. A difference between the two is that for radon the indoor concentrations are often higher than outdoors. That is not the case for asbestos. Outdoor air contains substantial amounts of asbestos due to the large scale use of asbestos brake linings and the natural occurrence of asbestos in many soils.

When we apply the precautionary principle to the asbestos controversy, serious errors that could cost vast amounts of money are easily made while the advantages to public health are at best questionable. The removal of asbestos from public buildings, where it had been applied as a fire retardant, is another practice fraught with difficulties. It is particularly risky for the people who do this work. Removal operations disturb the asbestos and can actually increase the airborne concentrations. In some cases the airborne fibre counts are higher after the removal than before.[63]

Asbestos-containing building materials often remain in older buildings as floor coverings, asbestos cement panels and other hidden construction components. As these buildings get older, asbestos may become friable, dislodged and the microscopically small fibres become airborne and can be inhaled by occupants of the building. The US regulations requiring removal of all asbestos before an old building is demolished have become a major liability in the real estate market. These policies have been strongly criticized.[64] One environmental consultant estimated that almost 80 percent of removals was unnecessary and called the situation a problem out of control.[65] The asbestos scene has come to the point where a commentator wrote in *Science*: "The debate on the health hazards of asbestos has become so polarized that researchers from one camp no longer go to the other camp's meetings."[66] Most professionals with experience in the area agree that for a large majority of asbestos installations the most prudent action is to "let it be." In the case of asbestos, the breath of fresh air is just as good indoors as it is outdoors. In terms of setting priorities for the protection of public health, the asbestos problem has not been served well by the precautionary principle. Asbestos as a public health risk in general occupancy buildings has become a trans-science issue that cannot be resolved by science.

Other indoor air quality problems can often be solved with minor inconvenience. Most easily dealt with is tobacco smoke. The causal relationships between tobacco smoking and lung cancer, emphysema, and heart disease are supported by an enormous weight of evidence. Whether the building is a domestic dwelling or a general occupancy building, dealing with tobacco smoke pollution should have a high priority on the list of actions for improving indoor air quality.

Improvement in air circulation is a second step. This serves to dilute and remove volatile contaminants. An important aspect of ventilation improvements is the need for adequate make-up air. Dirty air removed by a fan must be replaced. This is particularly important since the creation of negative pressure in a building can increase the infiltration of outdoor pollutants, e.g., from passing traffic. Obviously, the necessity for adequate ventilation can be in conflict with the zeal for energy conservation.

Outdoor air pollution issues are not usually as easily solved as those indoors. Potential global problems like climate change and ozone depletion need solutions at the global level, which means international politics. Combustion-related air pollution must move us in the direction of alternative energy sources. Since most of these alternatives involve electrical energy, it is clear that global air pollution is closely linked to the energy resource question. Global air pollution problems, including possible climate change, may have to become much worse than present levels to induce a reconsideration of the nuclear option.

CHAPTER 9

Water for drinking

> You and I are about 65 percent water. You may consider this undignified, but that's the way we're built. A one-month-old fetus is 90 percent water, and babies are more watery than grown-ups. Most mothers are aware of this. It is also the reason babies can suck more out of their thumbs than we can.[1]

In a search for evidence of life on other planets, the first question is if there is an atmosphere. The second question is about the presence of water. Without water life as we know it is not possible. The entire chemistry of living organisms uses water as its solvent. When micro-organisms, plants, or animals are dehydrated, life processes stop. Micro-organisms can often be revived by adding water, for the other two forms of life dehydration is fatal.

Although water covers about three-quarters of the surface of the planet, only about one hundredth of 1 percent of all liquid water on the planet is fresh water in creeks, rivers, and lakes. Brackish and salt water make up the other 99.99 percent. Water travels continuously, sometimes as liquid, sometimes as vapour, from the land to the atmosphere to the oceans and back. This hydrologic cycle is not just a geological process, it is also a human process. During these movements water comes into contact with many other chemicals, most of which come from natural processes, some from human activity. Many of these chemicals can dissolve in the water and remain in solution for a shorter or longer time. Really pure water is a rare commodity. Water is a major component of the biosphere. It is also a very precarious resource.[2]

Plenty of water...

Besides serving as the solvent for the chemistry of life, water performs two other tasks for living organisms: getting needed chemicals into the system and getting unwanted chemicals out. When living organisms exchange chemicals with their environment, water is often the carrier. Therefore the solubility in water is a very important property of chemicals. This is true for both wanted and unwanted chemicals. At room temperature it is possible to dissolve about 2 kg of sugar in a litre of water, therefore sugar can readily enter the body as a solution in water. On the other hand, the solubility of many oil-like chemicals in water is very low, often measured in parts per million (ppm) or parts per billion (ppb). Vegetable oils, animal fats, PCBs, dioxins, and chlorinated pesticides fall in this category. They rarely enter living organisms as solutions in water. They can enter another way.

Most of the waste produced by living organisms is removed by water. For all living creatures that waste is biological, biodegradable material that is subsequently used by other organisms as nutrient. Only one species adds chemicals to the waste stream that are not of biological origin.

Precarious resources

Water is in several distinct compartments of the biosphere. Rivers, lakes, and oceans form the surface water compartment, water in the cracks and pores of soil makes up the groundwater compartment, and the atmospheric compartment contains water vapour. Yet another "compartment" is the water that is part of all living organisms. Frank Herbert's science fiction novels about the planet Dune emphasized how important that water compartment can be. On that fictitious planet the most important ritual associated with death was the recovery of precious water from the body of the deceased.

Water in the various compartments is not static. Surface water evaporates to the atmosphere. When it condenses, liquid water or snow returns to the surface, where it replenishes surface water and ground water. Springs and other seepage can transfer ground water to the surface. As was the case with air and atmospheric movements, discussions of these physical processes are in the ecology or physical geography textbooks. We will not deal with these details; instead, we will complement these *physical* descriptions with a brief look at some of the *chemical* events that occur at the same time.

Water boils at about 100°C. The exact temperature depends on the barometric pressure, which in turn depends on weather and altitude. When the vapour of boiling water is guided through a clean glass tube, and the outside of the tube is cooled, the vapours condense, and liquid water can be collected at the other end of the tube: the water has been *distilled.* If we had started with a solution of table salt, the salt would have stayed behind because it cannot form a vapour at this temperature. Distillation is therefore a way of removing dissolved impurities from the water. Not all impurities can be removed by distillation. For example, during the distillation of a fermented beverage, both water and alcohol form vapours and end up in the distillate, albeit in different amounts. Understanding the distillation process helps us to appreciate *pure water.*

Water is constantly in contact with other chemicals. Percolation through different soils allows many water soluble minerals to dissolve. When calcium and magnesium salts dissolve this way, and the water is then used for domestic purposes, the water is called *hard water*, and the dissolved calcium and magnesium salts interfere with the use of soap. The nature and the amounts of dissolved minerals in the water depend strongly on the soil from which the water was obtained. Fluorides, a group of inorganic fluorine compounds, occur in most soils and small amounts can dissolve. In some regions drinking water contains higher levels of fluoride than in others. When water is obtained directly from a well, it can contain dissolved radon gas. When that water is used in a shower, aeration releases the radon into the ambient air. The claim that this radioactive gas presents a health hazard in this way has not been borne out by epidemiological studies.

When rain water moves across the surface towards creeks and rivers, soil minerals dissolve and become part of the surface water compartment. Rivers and streams then carry part of this chemical mixture to the seas and oceans, from where several chemicals return to the land via the atmosphere.

Dissolved minerals and gases are only one kind of impurity that ends up in water through natural processes. Water also contains many *insoluble* materials. Dust from dry surface soils, even if it is insoluble, can readily become suspended in water or encapsulated in snow and ice particles. Some of this dust is naturally occurring radioactive fallout, for example from radon decay products, the radon daughters that came from radon gas from the soil. All these dissolved and suspended chemicals are carried back to earth in precipitation. Really pure water, like we obtained in the distillation experiment, does not exist in nature.

Most of the dissolved and suspended "impurities" are not very toxic. In reasonably clean environments, ground water and surface water are quite suitable for human consumption. In general, however, it may be better to give rain water the time to filter through soils and sand so that this natural filtration process can remove some of the debris of nature. In a not so pristine environment the picture is different.

The few bands of nomads that walked the planet many thousands of years ago used water only for drinking, cooking their food, and washing themselves. The only waste they dropped on the land was phosphorus-rich biological waste. A portion of this waste was washed away by rivers, and the rest was dispersed on land as nutrient for other organisms. When agriculture appeared, more water was used for raising animals and for irrigation. The course of surface waters changed and more biological waste washed to the sea. An important property of human waste is that it often contains pathogenic micro-organisms. Some of these are responsible for the spread of dysentery, cholera, and typhoid fever. Even without industrialization, water supplies can easily be damaged by this form of contamination.

Rapid population increase has placed heavy demands on water resources. Developments in Egypt are a direct result of overusing an irrigation resource that worked well for thousands of years. Human intervention causing the silting of dam reservoirs and soil salinization because of irrigation, has brought that system close to collapse.[3] Similarly, the disastrous droughts in the countries in Northern equatorial Africa, the area known as the Sahel, are a direct result of overuse. In this case the disaster was accelerated by misguided foreign aid from industrialized countries.[4]

Today about three-quarters of all water withdrawn from the earth is used in agriculture. For example, it takes about forty litres of water to produce one can of vegetables.[5] Vast quantities of water are also used in industrial activity. Relevant to

Asimov's story about phosphorus, the increase in human population and the invention of indoor plumbing drastically accelerated the rate at which phosphorus is washed into the sea.[6] But on its way to the sea water is used many times by different people. When rapid industrialization was added to population growth, human impact on water resources took on significant dimensions. In considering that impact, we easily forget the basic "law" of water consumption: "water is always *borrowed* before it makes its way back to the sea."[7]

Water pollution

Industry assigns yet another task to water. Factories use water mostly for boilers, cooling machinery, and for washing equipment and products. The used water now contains a large variety of chemicals, it has become dirty and it is given another name. It is now called *industrial effluent*. Chemicals in effluent can be organic materials that also occur in nature, for example from fish and meat processing plants. Effluents from mining operations contain naturally occurring, mostly inorganic chemicals. In foundries or oil refineries the chemicals in the effluent are mostly synthetic, some of them alien to nature.

Effluent is normally returned to wherever the water was taken from in the first place, the local river or a nearby lake. Often there is an effluent treatment facility where sludge and other solids can settle out. The effluent is then released and joined by natural run-off and precipitation. The assumption has been that dilution and natural purification would deal with the water-borne waste.

Natural purification sounds reassuring. However, a perspective for this natural process follows from the quantities that are involved in the case of a bigger city on the shores of the Great Lakes. When thousands of pounds of chemicals are involved, the diluting and washing away are not so effective anymore.[8] This is why in most industrialized jurisdictions the approach has changed. Rather than looking at the concentration of a given pollutant in the effluent, the emphasis is now on the total loading of the receiving body of water with respect to that pollutant. The policy of "pollute and dilute" has been largely replaced by the "end of pipe quantity" approach.

A factor that played an important role in this change in approach is the huge improvement in analytical chemistry. Only a few decades ago, we could detect chemicals in water only in the parts per thousand range (one part per thousand is 0.1 percent). This is a little like finding one second in a sixteen minute time span. Today, we can measure concentrations of water-borne chemicals in parts per quadrillion, which compares with finding the one second in a span of about thirty-two million years. In other words, however the "disease" has changed, the *diagnosis* has improved a millionfold. Contaminants that now cause public indignation and regulatory panic, were not even detectable only twenty-five years ago. This enormous improvement in detectability has led to a better understanding of the true effects of the old "pollute and dilute" attitude. The end of the pipe became the focus of our attention.

While the end of the pipe is a *point source* that is easy to control, at least in theory, *airborne chemicals* are a different problem. Combustion processes give rise to the oxides of nitrogen and sulphur which form acids when they dissolve in water. The resulting acid precipitation increases the acidity of surface waters. The acidity itself does not normally present a direct health threat for humans, but there are several consequences to consider. Several forms of aquatic life cannot tolerate the acidity. Also, when water becomes more acidic, metals tend to be changed into more soluble forms and naturally occurring metals such as mercury and aluminum are leached out of the soil. This changes the availability of these metals as nutrients and their toxicity to different species.

Increased acidity also facilitates the leaching of lead from the distribution system of municipal water supplies. Lead in solder, and possibly lead pipes in older systems, are the source of this contamination. As the awareness of the toxicity of lead increases, lead containing water lines are being recognized as a public health hazard.[9] As a result, guidelines for allowable lead concentrations are being tightened in many jurisdictions. The problem is a very difficult one. Although replacement of the offending water lines is technically a simple problem, the capital cost would be staggering.

Qualitatively, it is easy to understand that airborne acidity has a negative effect on water quality and aquatic life. A quantitative assessment of the problem is not so easy. For example, a statement that acid precipitation has caused a decline in fish stocks should be substantiated by *relevant* field data. Such data would show that fish stocks were higher at some time before acidity increased, and that the decline is not due to other, unrelated toxic chemicals, or that fish was not lost to over-fishing.

An example of this difference between correlation and causality is found in the lakes in a region in the North-Eastern USA known as the Adirondacks. Many lakes in that area have been strongly acidic in the past, and were neutralized by the burning of logging waste and particularly by a large forest fire in 1903. The present acidification of these lakes may just be a return to their natural state because reforestation in the area has stopped the flow of alkaline ash into the water.[10]

Another aspect of acid precipitation deals with the notion that airborne chemicals can travel over large distances.[11] Yet, although Great Britain has managed to reduce sulphur dioxide emissions by some 30 percent since 1979, the acidity of lakes in Norway and Sweden has not changed.[12] Conversely, other chemicals such as PCBs, probably do travel in the lower atmosphere over very large distances.[13] The airborne transport of volatile chemicals is a complex issue that does not allow for uncritical extrapolations. On the other hand, local effects of acidity are readily observed. Examples are the dying-off of the forests in Western Germany and the displacement of heather by more acid resistant grasses in Eastern parts of the Netherlands and in Western Germany.[14,15]

The path of chemicals into ground water is through the soil above the aquifer. Therefore soil permeability, that is the coarseness of the particles and the size of the pores, is critical. Sandy soils and strongly organic soils like peat are more permeable than clay or rock. Rain water percolates through the various soil layers, somewhat like hot water percolates through a layer of ground coffee. The results are not unlike the coffee-making process. Chemicals *in* the soil as well as *on* the soil have a chance to dissolve and travel down with the water until they become part of the aquifer. The water solubility of the chemical in question, the amount of water falling on the soil, and the permeability of the soil determine how efficiently the chemical travels down to the aquifer.

In regions with extensive agricultural activity, another concern about water quality is with the presence of fertilizers and pesticides. Fertilizers, which are generally more water soluble, can easily cause problems. In rural South Ontario some 40 percent of farm wells were found to be contaminated with both agricultural chemicals and bacteria.[16]

Besides the effects of pesticides and fertilizers, farming can impact on drinking water supplies in other ways. A peculiar problem in the Netherlands is a large surplus of livestock manure that is affecting both air and water quality in that densely populated country. The release of large quantities of ammonia from the manure is a source of air pollution and the seepage of manure components into ground water is causing water pollution problems. Since a substantial part of that country's water supply depends on ground water, there is reason for concern. The problem is of such magnitude that the Dutch government is contemplating limiting the size of some livestock herds.

A controversial issue with respect to the quality of drinking water is a group of industrial chemicals collectively called the organochlorines or chlorinated organics. They have names like PCBs and dioxins, and some people would add pesticides like DDT to the list. These chemicals, which are of industrial origin, are discussed in detail in a later chapter, but they are also relevant to the water issue. Depending on the amounts, PCBs, dioxins and furans, and some pesticides like DDT, dieldrin and aldrin, can affect human and animal health. They are, however, extremely insoluble in water. Released in surface waters, these chemicals tend to be *adsorbed* very strongly on the surface of suspended particles and the amounts actually *dissolved* in the water are incredibly small. So small, that measuring them becomes very difficult, even in a modern, well-equipped laboratory.[17] A reasonable filtration or settling system in the drinking water supply can readily remove the suspended material and with it the adsorbed organochlorine chemicals. The likelihood of ingesting these contaminants via a municipal drinking water supply is negligible.

These same strong adsorption properties also help to reduce the presence of this group of chemicals in ground water. They become adsorbed as the water seeps through various soil types into underground aquifers. In short, PCBs, dioxins, furans, and chlorinated pesticides do not normally pose a health risk via the human water supply.[18]

Finally, acute accidents and spills can seriously affect drinking water supplies at the local level. For example, in the fall of 1992 a dam near the Tara River, in Yugoslavia, was in danger of failure. The dam was built about twenty years ago to hold the waste from lead and zinc mines in the area. Breakage would have caused a major chemical spill into that river. The Tara is part of the Danube system which supplies large numbers of people with drinking water. Also in Europe, industrial activity along the Rhine continuously threatens the water supply of many people in Germany and the Netherlands. A listing of such incidents serves no purpose here. Newspapers and television newscasts provide such information almost continuously. These continuing threats show clearly how many large population groups continuously live with a very precarious drinking water supply.

This inevitably leads to a level of anxiety that affects regulatory aspects. Therefore some experts believe that, although the best scientific data must be used as the basis for risk assessment, the process also includes a value judgement "which must be made on a societal level."[19] Unfortunately, they do not address the dual problem of how society can learn to comprehend the problems, and the scientific community's ability to understand "the societal level."

Nature cleanses

The simple "dilute and disperse" method worked well, up to a point. Natural bodies of water do have an enormous capacity for breaking down complex molecules into simple and harmless substances like carbon dioxide. That capacity is not infinite, however, and it did not keep up with increasing quantities of waste. Soon the "dilute and disperse" became "dilute and pollute" until the point where we had to start to "concentrate and contain." Nevertheless, the "self-purification" of water is very important. It is based on a process called *biodegradation.* Micro-organisms, occurring naturally in the water, break down complex organic molecules and use the carbon, phosphorus and other elements to grow and multiply. It is less well known that many chemicals also break down under the influence of sunlight. This *photodegradation* is as important in the self-purification of surface waters as is biodegradation.[20] Biodegradation has become an important buzzword. It is considered as one of the *good words* in the environmental language. That this is not always true will follow from our discussion on waste management in a later chapter.

It was the *persistence* of early laundry detergents and pesticides like DDT that raised the ire of Commoner and Carson. While at that time detergents were indeed a problem, in the case of pesticides that is difficult to understand. The literature on biodegradability goes back to well before 1962.[21,22,23,24] Even more information has become available in the thirty or so years since then and the persistence problem of synthetic detergents has been almost eliminated.

When micro-organisms cause chemical changes in the molecule of a chemical, two different processes can operate. On land, in the upper layers of the soil and in surface waters, like rivers and lakes, micro-organisms use oxygen to change the molecules they are digesting. On land that oxygen is available from the air; in water it is present in solution. The bacterium or fungus causes the molecule to be *oxidized,* and the smaller bits and pieces formed are called *oxidation products.* The micro-organism has caused an *aerobic biodegradation.* The oxidation products are generally more soluble in water than the original chemical. They are usually easily degraded further, by a different organism or by sunlight. Eventually, complete biodegradation will produce carbon dioxide, water, and possibly some nitrogen containing molecules as final products. In water, the process is perfect for "natural" purification and biodegradability is a highly desirable property for chemicals that are being released into surface waters.

Obviously, sewage and food waste are readily biodegradable. A frequent generalization has it that synthetic chemicals are not biodegradable and are not very soluble in water.[25] This perception is wrong because it is largely based on obsolete information. In 1972 Barry Commoner wrote that although modern synthetic detergents do biodegrade partially, they leave behind a chemical called phenol which "is foreign to aquatic ecosystems."[26] Today, almost twenty-five years later, the literature of plant chemistry shows that phenol itself and many chlorinated phenols occur abundantly in nature, both in terrestrial and in aquatic ecosystems.

While tires and house paint do not readily biodegrade, most of the chemical building blocks from which these consumer products were made do break down. Many synthetic chemicals, including modern household detergents, are soluble in water and are readily degraded by sunlight or bacteria. Somehow the exceptions to this rule have attained enough notoriety so that wrong generalizations are easily accepted. A few synthetic chemicals biodegrade very slowly, so they stay around in the aquatic environment longer. But there is abundant evidence that even so-called *persistent* chemicals will eventually be broken down, by micro-organisms or by sunlight, into harmless substances.[27,28,29] As early as 1971 it was known that DDT biodegrades in lake bottom sediments faster than was previously believed.[30] Even the chlorinated dioxins and furans "rapidly photodegrade" in the natural environment.[31]

Natural biodegradation also had its effect on the conditions in the Great Lakes basin.[32] In this region the amounts of DDT in the milk of nursing mothers has fallen by 90 percent since this chemical was banned, and the concentration of PCBs in Lake Michigan trout has also dropped by about 90 percent since 1973. This report also contains a comment on the risk of eating fish from the Lakes and suggests that "a calculated risk estimate suggests that about 30,000 people will get cancer over 70 years from eating them." It fails to mention how many people will get cancer over *seventy years* in any case. Calculations like these, based on the straight line dose-response curve, clearly show that the "no-safe-dose" hypothesis derived from such curves cannot be valid.

Biodegradation is an important process with respect to water quality. Because of the early environmental literature, biodegradation has received much attention and it has been so widely promoted that it is now generally assumed that biodegradability is not only desirable, it is essential. Biodegradability assures that "nature can purify itself." As long as the waste produced by humanity is biodegradable, all will be well. This is an incomplete picture. There is a limit to biodegradation and there is a price to be paid. The limit is reached when the *amount* of degradable waste disposed of exceeds the *capacity* of the system. This limit is usually ignored by those who espouse the cornucopian view, although it is clearly visible in many parts of the world.[33]

The price that is payable is the oxygen in the water. The solubility of oxygen in water is quite limited, about 10 to 20 milligrams per litre, depending on temperature, and that oxygen is essential for aquatic life. Additional waste dumped in the water demands additional oxygen for biodegradation, hence the term *biological oxygen demand* or BOD.

A simple test can determine how much degradable waste is present in a given water sample. The result of that test is expressed in terms of the biological oxygen demand of the waste. The higher the BOD value, the more degradable pollutants there are in the sample. The BOD test gives no information about pollutants that are not readily degradable, like suspended minerals or dissolved metals. Nevertheless, measurements of the BOD and the actual amount of dissolved oxygen are a good diagnosis of the general health of a river or lake.

This need for oxygen for biodegradation has two important consequences for the self-purification of water. First, the wave motion of the water affects the amount of dissolved oxygen. If biodegradation is incomplete, remaining pollutants and products of partial degradation must be transported further downstream. In the oceans extensive wave motion ensures further oxygen supply and a different microbial population will complete the biodegradation process. Second, fish and water plants also need oxygen to survive. If oxygen is consumed by waste degrading bacteria, not enough may remain for other aquatic organisms.

If a river moves too slowly, wave motion decreases, there is less oxygen, and impurities accumulate. An example is the Volga river in Russia which, due to hydro-electric development, now flows about ten times slower on its way to the Caspian Sea than it did before. This has significantly increased pollution levels.[34] The other contribution to increased pollution is, of course, the increased number of people whose waste the river must carry.

Entirely different conditions apply to ground water. It is very poor in oxygen and does not readily undergo "self-purification" by means of biodegradation. When an underground aquifer becomes polluted, it quickly becomes unsuitable for human use. *Anaerobic biodegradation*, that is biodegradation by bacteria that do not use oxygen, is very much slower. Depending on the degree of pollution and the local geology, it could take hundreds or even thousands of years for a polluted aquifer to recover.

Chemicals that were spilled or spread on the surface can enter an aquifer in the same way as chemicals that are an integral part of the soil. Accidental industrial pollution on land as well as chemicals used in agriculture are a potential threat to groundwater quality. There is, however, an important modifying factor in this percolation process. As the water and the dissolved chemicals percolate through the soil, chemicals can become *adsorbed* on soil particles and are then removed from the water. Chemicals with a very low water solubility are more strongly adsorbed than those that are very soluble. Soils rich in clay or organic material, such as peat, are better adsorbers than soils that consist mainly of sand. Many studies on the subject of soil adsorption address issues such as soil type, humidity, temperature, and the nature of the chemicals involved.[35] Peat, which is abundant in the circumpolar regions of Canada and Northern Europe, is rich in organic matter and has a high adsorption capacity. Therefore peat holds promise for environmental remediation.

The filtering action of the soil, which removes insoluble particles, and the adsorption of chemicals with a low solubility are important mechanisms for ground water purification. These natural mechanisms have severe limitations. As the limited adsorption capacity of the soil is exceeded, ground water becomes contaminated. This endangers wells and purification plants must spend more energy to produce potable water.

... And not a drop to drink?

It follows that water in many areas is contaminated by natural processes, and human activity is adding further problems in many cases. This means that, in order to protect public health, water must be purified before it is suitable for human consumption. There are several approaches to water purification, all with their own difficulties. While removing impurities, purification often creates new problems at the same time.

Treatment and public health

Although water from both surface and groundwater sources can be contaminated with soluble chemicals, this problem is difficult to correct by treatment. The supply of drinking water primarily involves the removal of suspended particles and protection against bacterial infection. Serious contamination with dissolved chemicals can mean looking for another source. For surface waters that is often temporary, but a contaminated groundwater source is usually permanently lost.

Surface water often contains natural micro-organisms that can present a hazard to human health. Although ground water does not normally contain many bacteria, once a well is dug contamination can easily take place. The safest methods for further purification involve killing micro-organisms with strong oxidizing chemicals, usually chlorine, sometimes ozone.

Problems related to sanitation are relatively minor in the industrialized world. In less developed countries, particularly those already affected by marginal water resources, sewage treatment is generally poor. Raw sewage is often dumped into surface waters without any treatment at all. When raw sewage enters a river, the biological oxygen demand increases and the amount of dissolved oxygen remaining in the water goes down. A comparison of average amounts of dissolved oxygen and the wealth status of countries shows a direct relationship. Higher income countries have cleaner rivers.[36] A 1990 survey estimates that some 900 million people are affected by diarrhea, a similar number by roundworm, and about 200 million suffer from schistosomiasis, also known as bilharzia.[37] Relatively simple improvements in the water supply and the disposal of human waste could readily yield improvements of some 30 to 40 percent in these health statistics. A brief outline of the common purification processes will place the supply problems in perspective. The details of the purification process depend on local conditions, but a few basic rules apply across the board.

Suspended solids in the supply can effectively be removed by sedimentation or by filtration. The sedimentation process can be improved by the addition of aluminum salts. These salts become part of the precipitated material and are removed from the water, but some aluminum stays in solution. This is a concern to some medical researchers. An association between aluminum and Alzheimer's disease was suggested about ten years ago. That this association is a *causal* relationship is now being regarded as questionable. Although these concerns have recently been picked up by the media, a scientific study of the effect of aluminum in drinking water does not seem an urgent issue in view of current research on this disease.[38]

Other *suspended* materials, wood fibres for example, do not readily precipitate from the water. Wood fibre does not necessarily present a health hazard, but it clogs aerators, valves and other equipment, and the appearance of the water is unattractive for the consumer. Suspended material that does not readily settle out can be removed by filtration.

To remove *dissolved* solids like salts and minerals from water is difficult, expensive, and requires large amounts of energy. These methods are almost exclusively applied in desert countries for the removal of salt from sea water. Water *desalination plants* use large amounts of electrical energy and are expensive to operate.[39] Desalination methods are rarely applied to water supplies that draw on ground water or surface water. Dissolved solids are not normally removed from public water supplies. Dissolved salts of magnesium and calcium present no health problems but this so-called hard water interferes with the use of soaps. Sometimes local problems may be caused by unusually high concentrations of iron or fluoride.

High fluoride concentrations can have adverse health effects, particularly on young children. The development of healthy teeth needs a certain minimum amount of fluoride in the diet. Pediatricians and dentists often recommend fluoride supplements and some municipalities add fluoride to the public water supply. The difficulty is that too much fluoride affects tooth development in a negative way, leading to discoloured and brittle tooth enamel. Since very young children often swallow the fluoride containing toothpaste, it is easy to overstep the narrow margin between essential and excess amounts of fluoride.[40] For fluoride the difference between hormesis and toxicity is very small.

Bacterial counts in a body of water depend on the amounts of contaminated water discharged and the concentration of nutrients in the water that support microbial growth. Water taken near the outlet of a sewage plant can contain large amounts of such nutrients. The most common method of sterilizing the water is treatment with chlorine. At 0°C a fair amount of chlorine can remain dissolved in the water, but aeration, particularly at the kitchen faucet, and boiling remove much of the chlorine. The consumption of trace amounts of chlorine does not constitute a health hazard, but there are health-related questions about the chlorination process.

Water from many sources contains small amounts of methane, which is the smallest member of the *hydrocarbon* family. Methane is made by anaerobic micro-organisms that live in the absence of oxygen, like under ground or under water. Anaerobic biodegradation occurs stepwise and many organic chemicals are formed as intermediates in the process. The end product is methane.

Methane is also a component of natural gas and it is common in mines, particularly coal mines. When underground formations are disturbed by mining operations, methane can escape. It burns readily in the presence of oxygen and can cause fires and explosions. Methane is common in swamps and stagnant waters. Many sources of ground water and surface water contain smaller or larger amounts of dissolved methane resulting from anaerobic biodegradation.

Treatment of the water with chlorine easily incorporates chlorine atoms in the methane molecule, leading to formation of methyl chloride, methylene dichloride, and chloroform.[41] The first two are very volatile and are easily removed by aeration. The traces of chloroform that remain in the water supply of many cities and towns have become a source of controversy. Since chlorine, together with fluorine, bromine and iodine, is one of the *halogens*, derivatives of methane made by introduction of halogen atoms are called *halomethanes*. When other hydrocarbons such as ethane are involved the more general term *halocarbons* is sometimes used.

Chloroform is an animal carcinogen; it can induce cancer in laboratory animals. Whether or not the trace quantities of chloroform in drinking water make any contribution to human cancer depends on the method of calculating the risk. The assumption of the "no-safe-dose" hypothesis will overestimate the risk considerably. As a result, much of the earlier literature on the issue reflects a great deal of uncertainty.[42] A recent large study *suggests* again a positive association between the consumption of chlorinated drinking water and cancers of the bladder and rectum.[43]

There is a general perspective that the occurrence of the halomethanes in drinking water must be blamed on the spills and effluents from the chemical industry. While that is true in a few isolated cases, the widespread occurrence of chloroform in municipal water supplies is mostly due to the chlorination process. As a result of the uncertainty of the risk estimates, alternative methods have been sought.

A frequently mentioned replacement for chlorine in water treatment is ozone. Ozone is toxic, both to micro-organisms and to humans, in relatively small amounts. It is a gas that is formed by exposing the oxygen in the air, or pure oxygen, to electric discharges. That is why ozone is readily formed during lightning storms and in photocopiers. It is not very stable. It falls apart very easily to give back the oxygen from which it was formed. It cannot be made in one city and transported to another in gas cylinders, like other gases. It must be made where it is used. This increases the amount of handling by operating personnel, which, in view of the toxicity of ozone, is a disadvantage of this method. In terms of air pollution the amount of ozone released by a water purification plant is small compared to the amounts that result from other pollutants such as automobile exhaust. Nevertheless, for very large plants in areas with heavy air pollution it would make sense to do the arithmetic on the atmospheric impact of an ozone-based water purification plant.

The use of ozone suffers from several shortcomings. It is a very strong oxidizing agent that does not only kill micro-organisms, but also reacts chemically with many soluble organic chemicals in the water. The short lifetime of ozone leaves no residual protection in the distribution system.[44] This may explain the unfortunate experience with ozone in the city of Tucson, Arizona. In 1990 this city planned to build a new ozone-based water treatment plant.[45] It would have a capacity of 150 million gallons a day and would require about 6000 pounds of ozone per day. The plant was built, and used to treat the CAP (Central Arizona Project) water that came by canal from the Colorado River, and supplied it to part of the population of Tucson. The resulting poor taste, unpleasant smell, and unattractive colour of the treated water caused such an uproar that Tucson City Council decided to close the plant temporarily. Tucson City Council must now try and decide whether to spend or save the one million dollars it would cost to "mothball" the plant.[46]

Ozone treatment does have advantages too. For example, chlorine kills most micro-organisms in water, but not all. The *Cryptosporidium* parasite is not easily destroyed by chlorine and when water supplies get infected with this bug there is trouble. The city of Milwaukee, Wisconsin experienced an outbreak of *cryptosporidiosis* recently. A large number of people got sick and there were a number of fatalities. The tiny parasite is common in livestock and apparently it can easily make the trip from the cattle ranch to a city's water supply. It does not survive ozone treatment of the water.

Another alternative for the purification of drinking water is irradiation with ultraviolet light. Exposure to intense ultraviolet radiation is deadly for micro-organisms, for plants and for animals, including humans. There seem to be no direct environmental or human health disadvantages, except that this method, like the others, consumes electrical energy. Recent improvements in the reliability of UV radiation equipment has made this option more attractive and UV-based water treatment plants are gaining popularity in Europe.[47] In terms of health effects, the emphasis, as was the case with ozone, is on the safety of the operating personnel.

In summary, two factors are of importance in providing alternatives to chlorine treatment of drinking water. Both ozone and ultraviolet have potential effects on operating personnel. This could mean that an additional occupational health factor is added to the water purification process. The other factor is the considerable cost of retrofitting existing water treatment facilities to utilize the alternative methods. A better delineation of the risks and benefits with respect to public health and environmental health must become available before individual communities can decide on the implementation of one of the alternative technologies.

Fit for drinking

Notwithstanding controlling legislation, industrialized areas and agricultural communities alike often have a water supply that, although not legally toxic, has a poor taste. The qualification "legally non-toxic" means that legal limits for listed contaminants are not exceeded. There is good reason to believe that such limits are reasonable and do protect public health. They do not necessarily protect the water's taste. A few alternatives are available for those who don't like the taste.

A limited number of people chose to move out of the cities and live in a rural area. Rural living often means relying on ground water in areas where agricultural chemicals are in use. Fertilizer contains salts like nitrates, which are very soluble and readily dissolve in ground water. A well, drilled or dug where fertilizer is frequently used, can easily produce potable water of good quality, but with a high nitrate content. In most legislations the permissible level of nitrate is set at 10 ppm. Higher levels do not readily affect adults, but can cause a peculiar problem to young infants. Their intestinal microbiology and less acidic digestive liquids causes the nitrate to be reduced, *in vivo,* to nitrite. Nitrite can cause a condition known as *methemoglobenia,* in which the oxygen-carrying capacity of the blood is reduced. The affected baby will get a bluish skin colour and start to suffer from slow asphyxiation. The problem is limited to the first ten weeks of life and occurs only with bottle-fed babies.

Another alternative for people who are dissatisfied with their water supply is to buy bottled water. Water from many a "natural source" or "spring water" is available in bottles all over North America and Europe. Depending on the source and the bottler, it may or may not be carbonated. The growth of the bottled water market suggests that many people have become less confident about their drinking water supply. Concerns about public water supplies include perceptions concerning chlorine and chlorine compounds, objections to fluoride addition, increasing lead concentrations, contamination by micro-organisms, and chemical contamination of ground water. For many it is simply a matter of the taste of tap water.

The safety of bottled water is regulated in most jurisdictions. However, the consumer has little or no information on analysis or water quality, the aquifer that supplied it, and hygienic conditions in the bottling plant. Unless such information is available, there is no reason to assume that water from a "natural source" is somehow better or healthier than tap water. Whether the water comes out of the kitchen faucet or from a bottle, the only criterion is trust in the supplier.

Yet another popular alternative for dealing with an unsatisfactory water supply is filtration. Many devices on the consumer market are aimed at purifying the water by filtration as it comes out of the faucet. Most of these filters are based on activated charcoal. The device can be connected in-line, attached to a faucet, or it can be part of a specially constructed water pitcher. It is intended to remove insoluble particulates as well as dissolved impurities. The filtration of particles depends on their particle size and the pore size of the carbon packing. Coarse, loosely packed charcoal is not effective. Removal of dissolved chemicals depends on the nature of the chemical to be adsorbed, and the capacity of the charcoal bed. In the adsorption process chemicals *compete* with each other. Some, like the chlorinated organics, are preferentially and strongly adsorbed, others, like water itself, are not adsorbed very well. Once the limited adsorption capacity of the filter cartridge has been reached, no further adsorption takes place, the filter has become useless until the charcoal is replaced. As with bottled water, the amount of peace of mind purchased depends very much on the trust placed in the manufacturer of the charcoal cartridge.

The case for frugality

In addition to the conditions and factors that affect water supplies at the local level, the adequate supply of potable water also has a global dimension. For people in regions blessed with abundant water resources the idea of scarcity is difficult to comprehend. The maps of Europe and North America show the hundreds of thousands of lakes and rivers available to a relatively small part of the world population. When *renewable water resources* are measured in terms of total precipitation, in cubic meters per capita per year, the numbers show significant scarcity in Northern Africa, the Middle East, and the Arabian sub-continent.[48] In those regions the amount of renewable water is around 1000 cubic meters per year, an amount taken to mean scarcity. Much of mainland China and Western parts of South America have only borderline levels of renewable water resources of between 1000 and 4000 cubic meters. Using renewable water resources emphasizes that ground water is a renewable resource only to the extent that it is replenished by adequate precipitation and that is protected from contamination. The "mining" of ground water without replenishment leads to depletion and promotes pollution of the underground supply.

The treatment of both drinking water and waste water is expensive. Although industrialized countries have paid little attention to water costs, modest moves toward conservation are becoming noticeable. The reasons are decreasing availability, energy costs, and demands for more stringent sewage treatment.

The purification of drinking water is greatly influenced by the quality of the source, the lake or river or aquifer from which the water was taken. Often that source is the same lake or river that receives a city's dirty water. For example, the vast drainage basin of the Rhine includes parts of Switzerland, Germany, France, and the Netherlands. In 1985 the Rhine transported pollutants from this heavily industrialized area to the North Sea that contained over a million tons of chloride, 3500 tons of phosphate, 450 tons of copper and about 10 tons of cadmium per year.[49] The quality of drinking water for a city on the banks of the Rhine depends on who and what is upstream. A similar situation exists for the Mississippi River in the USA, where water is taken, treated, and via a sewage treatment facility returned to the same river. A few miles downstream another town or city repeats the process.

Similar natural limitations apply to the use of ground water. An example is the extensive use of underground water resources in Mexico City. Sewage is disposed into a canal system that allows contamination of the underlying aquifer.[50] Overuse of the groundwater supply is affecting a resource that is not easily replenished but which is easily polluted. Even in an apparently water-rich country like the Netherlands, there are concerns about falling groundwater levels. In the southern part of that country enormous amounts of ground water are withdrawn for irrigation. Much of the irrigation water then evaporates or becomes run-off into the rivers and the groundwater resource shrinks.

When the dirty water, the city's sewage, is returned to the river it must be purified before it is released. This sewage treatment often leaves much to be desired. The first step is the so-called *primary treatment*. It involves little more than letting the sewage sit in a tank or reservoir for some time. Suspended solids fall to the bottom, grease and oil are skimmed off the top, and the water is discharged. Anything that remains in solution goes with it.

With additional expense for better equipment, the waste water can undergo a *secondary treatment*. Here bacteria are used to biochemically break down dissolved organics such as detergents, shampoos, and cleaning agents, and suspended materials that did not settle out in the first step, such as faecal matter and food residues. This is the same biodegradation process we discussed above. Because of the large quantities of biodegradable organics, treatment plants must add extra oxygen to the water for the bacteria to work. The liquid sewage is therefore aerated by blowing air through it. This consumes energy for pumping water and air.

The more biodegradable waste there is in the sewage, the more energy is used. Therefore, the typical North American gadget called a "garborator" is a burden on the treatment plant. The garborator, rarely seen in Europe, is an electric cutting device mounted under the kitchen sink. Vegetable waste and potato peels are washed down the sink, and with the water running, the machine is switched on. Rather than placing kitchen waste in a composter, many North Americans use water and electrical energy to add to the BOD burden of the sewage treatment plant.

Secondary treatment produces a cleaner waste stream for discharge. A semisolid bacterial mass, not unlike the yeast sediment from a beer or wine making operation, remains behind. It is called *sludge* or *activated sludge*, because the bacterial culture has to be actively growing for the process to work. The sludge is wholly organic and includes valuable nitrogen and phosphorus containing nutrients. It is often used as a soil conditioner or fertilizer. However, concerns have been raised in recent years about this practice. Depending on its origin, the sludge may contain unacceptably high concentrations of heavy metals such as cadmium, chromium, or mercury.[51]

Water resulting from secondary treatment is relatively clean, but it still contains residual micro-organisms. It is often disinfected with chlorine or ozone before it is discharged. This is another point in the process where organochlorine chemicals can be formed. At this point the waste stream still contains a variety of inorganics such as phosphorus and nitrogen compounds. These are not particularly toxic, but they are also good nutrients for plant growth. That is bad news for the receiving lake or river.

Biologists recognize two kinds of aquatic plants. Attached to the bottom with their roots are the *benthic* plants like water lilies and underwater grasses. Floating free in the water is the *phytoplankton*, mainly algae. Nitrogen and phosphorus compounds act as fertilizers for both kinds of plants; the more fertilizer, the more growth. Both kinds of plants also need sunlight. Here the phytoplankton has the advantage because it inhabits the upper layers of the water. Fertilized with the nitrogen and phosphorus from the secondary treatment sewage, the algae grow vigorously and an *algal bloom* occurs. The mass of algal material cuts off the light for the benthic plants below, which then start to die off. As the benthic plants die, their biomass decomposes and releases more fertilizing waste, which further promotes the algal growth. As algae keep growing part of them also die. Dead benthic plant and algal material further increase the biological oxygen demand (BOD). The water becomes turbid and the lower layers become deprived of oxygen, which in turn kills fish and other aquatic life. The process is called *eutrophication*. A eutrophic lake is characterized by profuse algal growth, reduced or absent benthic plant growth, increased plant debris on the bottom, and endangered fish stocks.

Eutrophication of lakes and ponds can take place naturally. If natural causes are temporary, the lake will eventually recover. If they are permanent, like a change in climate, the lake dies, dries up and becomes another feature in the landscape. Accelerated eutrophication caused by human waste can only be prevented at the source by removing the excessive nutrient load. This can be achieved by decreasing the amount of waste in the water at the disposal point, for example by removing phosphate salts from synthetic detergents and by cutting down on vegetable waste and shampoos. It can also be done by *tertiary treatment*. Tertiary sewage treatment consists of a number of actions aimed at the characteristic problems of a specific waste stream.

Phosphate can be removed from the sewage, after secondary treatment, by adding calcium carbonate (lime), which precipitates water-insoluble calcium phosphate. Other treatment methods remove excess nitrate and other nutrients. Sewage treatment, especially up to the tertiary level, has two principal consequences. First, the inevitable result of these treatments is the formation of additional sludges, such as calcium phosphate, which must be disposed of. These sludges are semisolid or solid materials and the disposal of solid waste is already a widespread problem in most industrial nations. Second, full sewage treatment is expensive. Considerable capital costs are involved for construction and operation. During recent discussions of clean-up plans for the Great Lakes, the costs for upgrading the sewage treatment plant of the city of Toronto were estimated at $1.7 billion (Can.) over about twenty years.[52]

These costs increase as the population increases. In this context we should note that the average North American uses more than 70 times as much water as the average person in Ghana. Even if water were abundantly available, the costs of a large waste water burden are a strong argument for practicing frugality with water.

Environmental arguments in industrialized countries tend to concentrate on issues of phosphate-in-detergent and the polluted rural well. The phosphate issue is a good example of how easy it is to introduce errors in the environmental balance sheet. When phosphate-free and phosphate containing detergents are compared in terms of energy consumption and resource extraction for making a given amount of each finished product, the difference in total environmental impact of the two is just about zero.[53] The phosphate-free detergent is easier on the lake, does not promote algal growth, but costs more energy and petroleum products to make.

The problems in poor countries are mostly seen in terms of disease, poverty and food shortage. Better water supplies would bring immediate improvement for all these problems by providing sanitation and increased agricultural productivity. Thus two main features of the global water problem are the over-use of water in the rich countries and poor sanitation conditions in poor countries.

Efforts to supply water to regions with scarcity problems include the diversion of water from its natural course. Dams and canals are common tools for letting the water flow where it is needed but where it did not flow naturally. Much of the diverted water is then used for irrigation and several problems usually follow. Normally, water from precipitation is partly taken up by living organisms and another part evaporates. However, ground water contains the mineral salts that were leached from the soil on the way down. When this mineral rich water is used for irrigation, evaporation and uptake by living organisms leave the minerals behind on the soil. About a million hectares of agricultural land per year are affected worldwide by this *salinization* process.[54]

A representative example of the problems of irrigation can be found in the state of California. An irrigation scheme called the California Central Valley Project was started in 1935. A system of dams, canals, two tunnels and several generating stations converted a semi arid area into very productive agricultural land. The irrigation water began to dissolve enough minerals from the formerly dry soils so that it became toxic to crops. An underground clay layer subsequently trapped the mineral rich irrigation water and the water table started to rise. Another canal, the San Luis Drain, was supposed to allow excess water to run North into the Sacramento river delta and San Francisco Bay. The canal was never completed and the water presently runs into a catchment area called the Kesterson Reservoir in the San Joachim Valley. Parts of this valley are particularly rich in selenium, a very toxic element. The water, travelling an unnatural course, leaches selenium out of the soil which concentrates in the Kesterson Reservoir. As a result, animals in a nearby wildlife refuge show signs of selenium poisoning. The San Joachim Valley problem is a clear illustration of the complex ecological problems that can result from water diversion.[55] It is a textbook example of Commoner's technological flaw.

Modern water management makes extensive use of large scale computer modeling for planning future water availability.[56] Such management planning must also take into account the political implications. On the North American continent there is the continuing question of water diversions from the North to the Western and central USA. California has its eyes on a supply of water from the North via an ambitious diversion plan. The issue will almost certainly remain on the agenda of Canada–US relations for the foreseeable future, and could become more urgent if climate change begins to dry out the central part of the continent.

Other contentious water issues are associated with big, international rivers such as the Nile in Egypt and the Ganges on the Indian subcontinent. The conflict between Jordan and Israel over the West Bank of the Jordan river is probably as much a conflict over water as it is about territory.[57] Sooner or later some unorthodox solutions to these political water disputes must be developed.

Of the many solutions proposed for solving the world's water problems, one that makes eminent sense is to concentrate on more efficient use of the available supply rather than to try and increase it. For example, on average less than 40 percent of irrigation water is actually taken up by crop plants. The rest runs off, evaporates or contributes to topsoil salinization. More efficient irrigation techniques could stretch the supply and prevent salinization.

In industrialized countries the irony of water supply problems is that only a tiny fraction of the water is actually used for drinking. By far the major part of all this pure water is used for flushing toilets, washing the family car, and watering the front lawn. For the industrialized, wealthy countries the solutions to water and sewage problems have little to do with science and technology; they are a matter of lifestyle.

CHAPTER 10

Food for eating

Together with air and water, the third vital ingredient for life is food. Originally, food was collected or hunted in the fields, or caught in a stream or lake. With agriculture came the cultivation of vegetables and animal husbandry. This level of control decreased food scarcity and populations began to increase. When the development of science, particularly medical science, allowed population numbers to rise again, increased food supplies were needed. Technological methods were introduced to protect crops against insects and micro-organisms. Transportation of food became critical for survival.

The technological approach to the food supply has allowed for further increases in population. So far this has not led to the crises predicted decades ago. While a few vegetable foods like wild berries, mushrooms or wild rice, are still collected as they were in prehistoric times, nearly the entire food supply is now dependent on technology. There is one exception. Fish is still mostly "harvested" with little or no regard for survival or extinction. Commercial fishing is still a textbook example of Hardin's Commons.[1] Since the fish belongs to nobody, it belongs to all. The "hunting" of fish with sophisticated technological equipment is bringing the depletion of major fish stocks in sight.

Food became an industry many years ago. It has become part of the environmental scene more recently. In this context the most important aspect of food is that it *consists* of chemicals. Benarde's book on the subject expresses that in a slightly provocative title: *The Chemicals We Eat.*[2] The point is further emphasized in another book on the subject: "Food is a gross mixture of carbohydrates, fats, proteins, oils, flavours and aromas, amino acids and minerals. Most are essential for life and some are innocuous bulk. *All are harmful* if taken in excess" (emphasis added).[3]

Our daily bread

Under the law of "eat or be eaten," plants and animals evolved by developing defenses against their enemies. These defense systems are generally of two kinds. They are mechanical, such as thorns, hard leaves, sharp claws or fast legs, or they are chemical. Skunks and stink bugs (*Scutelleridae*) treat their enemy to unpleasant odours, snakes and wasps inject their poison into their attacker, and many plants make chemicals that are toxic and taste bad. Even micro-organisms biosynthesize chemicals that are toxic for their enemies. Nature contains a rich variety of these so-called *allelochemicals*. Plant eaters beware.

The elders knew, sometimes

For prehistoric hunters, strategy and stamina determined the amount of available meat. Collecting edible plants needed other aptitudes. Taste and experience were the tools for learning which plants were edible and which were poisonous. *Ethnobotany* traces how different cultures have used plants and plant products. The interaction between people and plants determined which plants were used, and also the way the plants changed in the process. An up-to-date treatise on ethnobotany is Timothy Johns' book *With Bitter Herbs They Shall Eat It.*[4] Although the main theme of the book is the story of the common potato, it gives a good general picture of this fascinating area of biology.

The origin of the potato is Northern South America. The tubers are rich in carbohydrate, and they contain a significant amount of protein. The potato plant also has powerful defenses against predators. Several natural chemicals in the tuber are very toxic to insects and mammals. The bitter taste of these chemicals taught the predators to leave the plant alone. However, the human predators were persistent, particularly when they learned to use fire to cook their food. Occasionally they found potatoes that were not quite so bitter and these plants were preferentially cultivated. Over many years, potato varieties with less bitterness evolved until the current, sweeter tasting cultivars remained. In this evolution-with-human-assistance it was not the fittest that survived, but the sweetest tasting.

The potato story has an interesting technological twist. When it was recognized that plants make these toxic chemicals as a defense against insects and fungi, attempts were made to cultivate, by cross-breeding, "improved" potatoes that had better resistance to such spoilage. When these attempts did indeed produce a more resistant cultivar, it was too toxic for human consumption.[5]

Most animals, including *Homo sapiens*, select their food on the basis of taste, preferring sweetness and avoiding bitterness. Where animals simply avoid a bitter taste, humans have learned to associate the bitter taste of plants with other useful properties. Medicine men and shamans used bitter plants either to kill or cure. Some plants were used to repel insects or evil spirits, others were weapons in hunt or war, still others were used as aphrodisiacs. Not surprisingly, chemists, biologists, and medical scientists have studied the medicinal and toxic properties of plants almost since the beginning of modern science. Several books on the topic are available and some details are relevant to this chapter.[6,7]

While the potato was "made to evolve" into less toxic varieties, other foods could not be altered in this way. Then a variety of pre-treatments could still make the food edible. Soaking in water or treatment with clay or charcoal could partially remove toxic ingredients.[8] Control over fire greatly improved the human food supply since many toxic chemicals in plants are readily destroyed by cooking. In other cases attractive looking fruits with deadly toxic ingredients just could not be made edible. The urgent warning "Man of God, do not eat, death is in the pot" applied to the good looking, but deadly bitter cucumbers, the "wild gourds" of the scriptures.[9]

Toxicity is not restricted to vegetables. Animal meat and fish can also contain toxic ingredients. Quails, when their alternative food sources are scarce, can survive on hemlock seed. The toxic hemlock does not affect them, but their meat becomes toxic to humans. That was how the Jews, on their move through the desert "died with the flesh between their teeth."[10] In contrast, central American hunters could kill their prey with poison and eat the meat with impunity. Their curare tipped arrows killed the game, but this poison does not pass the intestinal barrier. The meat was safe to eat.

Plants that were unsuitable as food were often used in primitive forms of medicine as teas, poultices, and similar concoctions. The vast majority of these "natural medicines" have been found, after careful investigation, to be useless. Many so-called folk medicines are at best physiologically inactive; at worst they are toxic. Nevertheless, a significant number of plants have been found to contain chemicals that can affect human biochemistry in a positive sense. From India came reserpine, an important antihypertensive drug found in species of *Rauwolfia,* the wolfsbane. From South America came the curare alkaloids and quinine and from Europe came the heart drug digitalis. When humans learned to distinguish between bad food and good food, some of the bad food became medicine. Good tasting plants continued to be eaten without further questioning, because they were *generally regarded as safe.* Knowledge about vitamins, essential amino acids, and protein deficiency disease came many years later.

This system of selecting food worked well for hunting and gathering societies. Agriculture brought fruits and vegetables that were cross-bred to produce more abundantly or grow under less favourable conditions. Animal husbandry provided an abundant supply of protein and dairy products. Notwithstanding the alarms sounded by the Malthusians, and more recently by Paul Ehrlich and the Meadows group, the world's food supply kept up with the population increase. Several recent famine conditions in underdeveloped countries are of a political making or are the direct result of mismanagement of available resources. The African area known as the Sahel contains examples of both.[11] Nevertheless, concerns about the world food supply have resurfaced.[12]

In the global context it is important to note that many poorer countries rely heavily on protein from vegetable sources. Producing protein by means of livestock animals is a very inefficient process. It requires a much larger surface area for a given amount of protein. The environmental impact of clearing this land is a substantial problem in countries like Brazil and in central Africa. Since wealthier nations get most of their protein supply from animals, our meat eating habits have a negative effect on the environment in our own and in other countries.

Food for eating

Intentional food additives

For many big city dwellers the connection between food and farming has dropped to the subconscious, something out of sight. Food has become a commodity that has to do with production. It is the output of an industry. Processing and transport over large distances cause delays between production and consumption. The logistics of the food supply require an appreciable shelf life.

Eating cherries once meant finding a cherry tree and picking the cherries; now the industry brings the cherries to the consumer. When cherries are not available in winter and tropical fruit cannot be had in polar regions, the modern food industry provides these privileges; at a price. The food has to be manipulated, cleaned, stems and skins removed. Fish and meat have to be cleaned, packaged and stored cold.

Removing the stem of a fresh cherry leaves a "wound," a tiny patch of fruit unprotected by skin. This exposed spot, containing sugar, vitamins and other nutrients, is an ideal growing place for bacteria and fungi. They will rapidly colonize the fruit and make it unfit for human consumption. Meat and meat products are also susceptible to bacterial spoilage. Food has to be protected from spoilage. Just boiling is not always sufficient. *Clostridium botulinum* is an anaerobic bacterium that produces the much feared botulinus toxin. The dangerous toxin is readily destroyed by cooking but the spores of the bacterium are resistant to heat. To protect prepared food after canning needs more drastic methods. However, some older methods of food preservation are still used frequently.

One of these involves smoke. Ever since humans started using fire, smoke has been used both to enhance flavour and to preserve. Smoke contains a huge number of chemicals and many of these stay in the food. Therefore smoke is officially considered a food additive in most jurisdictions. Other older methods, still in use, consist of adding substantial amounts of sugar to fruits, or salt to meat, fish, or vegetables. High concentrations of water soluble chemicals like sugar and salt, literally pull the water out of bacterial cells and the micro-organisms die. Similar damage to the cell walls also softens the fruit and the meat.

At very low temperatures, growth of micro-organisms slows down almost to a standstill, and the frozen food section of modern supermarkets keeps growing. Preserving food this way relies on chlorofluorocarbons (CFCs) or effective replacements thereof. The grocery industry accounts, on average, for about 20 percent of CFC use in North America. Besides the pollution potential of CFCs, preserving food by freezing also consumes large amounts of electrical energy, both for storage and for refrigeration-equipped transport. Thawing frozen food also damages cell walls and softens fruit and vegetables.

Another method for preserving food is to add small amounts of chemicals that are toxic to micro-organisms, so that bacteria and fungi cannot grow. Since micro-organisms are in many respects different from mammals, it is often possible to find chemical preservatives that are not very toxic to humans. In some cases it is advantageous to add preserving chemicals not to the food itself, but to the packing materials used during transportation. The addition of chemicals to food for preservation is not free of problems. Sometimes the safety of the chemicals is questionable. Some hyper-sensitive individuals cannot tolerate specific chemicals. For example, the use of sodium sulphite or sulphur dioxide as a preservative in fresh salads is quite safe for most people. It can cause allergenic reactions in asthma patients.

When prepared food is packed in tin cans, or in glass bottles with metal lids, the acidity, particularly in fruit, can dissolve minute quantities of metals such as iron, lead, and zinc. The toxicity of iron and zinc are relatively insignificant; lead could have serious health effects. A major problem is the disastrous effect these metals have on colour and taste. Many a home-made wine has been spoiled by the use of the wrong metal utensils. Metals must be prevented from reacting, in a chemical sense, with the food. The answer to the problem is to use only stainless steel or to add *sequestering agents* to canned food to prevent spoilage by metals.

Notwithstanding the problems inherent in all methods, the case for preservation processing in the food industry remains strong. Many consumers live far away from where their food is grown, and protection during storage and transport is essential. The need for this protection is probably less obvious for people living in rural areas than for those in big cities. The importance of food protection against bacterial spoilage was described recently in considerable detail.[13] The statistics mentioned in this report are reason for concern. The number of food poisoning cases in England and Wales has increased almost fivefold in the past ten years, and between 200 and 500 Americans die each year as the result of their insufficiently cooked hamburgers. The author mentioned increasing meat consumption and a rising demand for fast food as two important causes among many.

Food additives are also used to enhance nutritive value. An old example is the lemon juice that was supplied to the British navy around the late eighteenth century. The vitamin C in the fruit was an effective weapon against scurvy. Some natural components of food, particularly some vitamins, can be damaged or lost during food processing and are added afterward from other sources. Typical examples of nutrient additives are vitamin D in milk, iodine in salt, and niacin, one of the B vitamins, in flour and breakfast cereals.

The case weakens significantly for another class of food additives. We will stay with the cherries for a while. Preserved in a tin can or in glass, the addition of a sugar syrup changes the appearance of the cherries. The chemical in the cherry skin that is responsible for the bright red colour dissolves in the syrup, which becomes pink. The cherry is left with a pale, fleshy, unattractive colour. The consumers' preference for good looking, bright red cherries taught the producer to restore the colour of the final product by adding a dye. A *cosmetic food additive* was invented.

Colour, texture, consistency and flavour are so important in prepared food that an industry has developed for addressing these issues, mostly in terms of additives that are not naturally occurring in the plants or animals from which the food product was made. Yet many of these chemicals are derived from other plants or animals. The well-known red food dye cochineal is made from dried insects and the thickening agent carrageenan is made from a seaweed. Other additives are wholly synthetic. Opponents of the practice argue that the addition of cosmetic additives is consumer deception and an unnecessary use of chemicals.[14] The food industry, no doubt armed with consumer surveys, will counter with the argument that they are merely responding to consumer demand, and that the additives are an additional expense they would prefer to avoid.

A special place in this picture is for the so-called convenience food, the ready-made meal that moves from the freezer to the dinner table in minutes, via the microwave oven. Food that is cooked is more readily attacked by micro-organisms than raw food. Texture and colour are particularly important for these products. They are displayed prominently in large frozen food storage freezers, and the carefully crafted frozen pie should not collapse as soon as it is thawed out. Therefore the labels on the packaging of convenience food products often read like the inventory of a laboratory store room.

This intensive manipulating of prepared foods needs a large number of chemicals that are deliberately added to food, a long list of *intentional food additives*. Chemicals are added to this list or removed from it on a regular basis. New items that are added must undergo extensive toxicity testing before they are approved. Occasionally, when new evidence shows that an additive may adversely affect public health, it is removed from the list. Amaranth, also known as Red #2, was removed from the list of food additives in the USA by the Food and Drug Directorate in that country. Red #2 was the subject of many controversial studies but eventually it was shown that this dye was indeed a carcinogen in laboratory animals.[15] One of its replacements, Red #40, is a close chemical relative.

Sometimes toxicological testing leads to a caution alert instead of a removal from the list. For example, saccharin and sodium cyclamate are synthetic sweeteners that are both "rodent carcinogens." That does not mean that there is any evidence that these synthetic sweeteners cause cancer in humans, but it is a good reason to *control* their use as food additives. Conversely, for some people the consumption of table sugar has serious health effects. Those persons must weigh the *calculated* risk of consuming the synthetic chemicals against the *real* risk of eating sugar. There are many such risk–benefit choices in the food we eat.

Descriptions of individual food additives are outside the scope of our discussion, but such information is readily available from the literature.[16] The list is very long indeed. Opponents of the use of food additives frequently point out that the vast majority of the chemicals on the food additives list have not been adequately tested in terms of toxicity, especially the kind of toxicity that can lead to cancer. Therefore, they say, a comprehensive testing program is mandatory to protect the public from the consumption of toxic chemicals in food.

The problem with comprehensive toxicity testing is with the long list of additives. Comprehensive testing of all these chemicals is not a realistic proposition. It is the kind of trans-science problem based on impossibly large numbers. That is why this list is known as the GRAS list, a list of food additives that are *Generally Regarded As Safe*. The chemicals on this list have been used so long in human nutrition, without observable adverse effects, that there are no cases of adverse health effects on our records. It is exactly the same mechanism that all of humanity used for selecting acceptable food in the hunting and gathering days. They had no methods for testing their food other than taste and experience. In the days of food additives, taste and experience are no longer the only tools, but they are still important for measuring acceptability.

There is one special issue in the considerations of intentional food additives, that is the Delaney amendment of the US Food, Drug, and Cosmetics Act. The amendment states that substances that have been found to have carcinogenic potential in laboratory animals, may not be used as food additives. It only concerns intentional food additives and does not address unintentional residues. Neither does the amendment have anything to do with naturally occurring carcinogens. These distinctions are not always clear. In one essay we find:

> for some substances, for example carcinogens, it is widely accepted that there is no safe threshold level of exposure. Reflecting this, the Delaney clause of the US Food, Drug, and Cosmetics Act stipulates that *no substances found to be carcinogenic will be allowed in food,* drugs or cosmetics in the United States (emphasis added).

The author does not acknowledge that there are many areas where the no-safe-level hypothesis is not "widely accepted." Moreover, the Delaney clause does not stipulate as quoted by this writer. Americans are allowed, fortunately, to eat all the naturally occurring carcinogens in their black pepper and peanut butter and broccoli.[17]

Today's preoccupation with food additives also has to deal with a terminology problem in the food industry. We already looked at the candy bar that needed to have a vanilla flavour. The unfortunate mix-up of words led to the introduction of the idea of "artificial chemicals," where a simple descriptive terminology like "synthetic flavour" would have solved the issue. The otherwise informative review mentioned above even goes so far as to state that "vanillin is the imitation of vanilla."[18] We can put terms like imitation or artificial in another perspective by measuring the acute toxicity of vanillin, which is low but not "zero." Tested on rats, the LD_{50} of vanillin is 1580 mg/kg, only slightly less toxic than the pyrethrum pesticides which have an LD_{50} of 1200 mg/kg. The rats do not distinguish between natural and synthetic vanillin. In both cases, once the LD_{50} test is over, half of them are just as dead.

Unintentional food additives

Apart from the chemicals that are intentionally added to food, many chemicals end up in food by accident. Plants are continuously exposed to whatever falls from the atmosphere, precipitation if it is wet and fallout if it is dry. Naturally occurring acid rain would have little impact on the product because of the small quantities of acid involved. Acidity in rain can affect the health of the crops and the appearance of the fruit. Incomplete combustion, from industry or from volcanos, can deposit particles of soot, fly ash, heavy metals, chlorinated dioxins, and PAHs on the surfaces of produce. Accidental fires involving industrial chemicals and their combustion products could add to the particulate fallout.

In the field, plants are desirable food for other creatures: birds, rodents, insects, and micro-organisms. Before agriculture, these other organisms hardly competed with humans. Growing the same crop on very large areas, so-called monoculture, made life easier for the insects. They could multiply at a tremendous rate, became competition for humans and were called pests. When agricultural practice allowed insects to proliferate, it also introduced chemicals to fight them. A huge number of synthetic chemicals were developed to kill insects before they could destroy the human food supply. Pesticides allow for increased crop yields, protect the nutritional value of crops, and make prolonged storage and year round availability possible. There are many estimates of the amounts of food that would be lost without the use of pesticides. The economic and nutritional benefits of pesticide use are enormous. Pesticides are also enormously controversial.

Opponents of the use of pesticides tend to emphasize the negative impact of pesticide use in terms of human health and environmental damage. Qualitatively benefits and risks have been well documented. The benefits of larger, healthier crops and year round availability must be compared to insect resistance, damage to non-target organisms, applicator accidents and consumer exposure either directly or through the food chain. Unfortunately, in a risk-benefit analysis neither the benefits nor the risks are easy to express in numbers and the debate is not making much progress.

At least one important point must be part of such an analysis. After the initial alarm was sounded by Rachel Carson, several events took place. Many so-called persistent pesticides have been phased out, and present practices are controlled by a vast amount of legislation. Many studies have shown that virtually all pesticides are broken down by biodegradation and photodegradation, some faster and some slower. Several references to those studies were presented in the preceding chapter and many more are on the record. The danger of discussing today's problems in terms of the thirty-year-old scenario of *Silent Spring* must be avoided. That book was written when persistent pesticides like DDT and dieldrin were in extensive use. The leftovers from that time, the residues of these pesticides, are now common contaminants in many places, but the amounts are slowly decreasing. They are residues from a bygone era and should not be labelled as pesticides in the 1990s. Practices from thirty years ago are not good reasons to condemn current agricultural methods. A more recent perspective for understanding the current pesticide-in-food picture is provided by a series of essays published under the title *Silent Spring Revisited*.[19] The presentations in that book contain some important points we will summarize in the following few paragraphs.

One particularly relevant review alleviates some of the present day concerns. Davies and Doon cite convincing scientific studies that show that the usual health effects of pesticides found in *occupational* exposure and in animal studies do not apply to *public* exposure.[20] The reason they do not apply is entirely a matter of dose, that is the amount per kg bodyweight. Another up-to-date discussion is somewhat more technical, but it also includes a realistic discussion of cancer risks in relation to pesticide use.[21]

An alternative to synthetic pesticides is "natural pest control" or pest control with "natural chemicals." The bacterium *Bacillus thuringiensis*, which is specifically pathogenic toward the Spruce Budworm, a major forestry pest, is an example of these methods. Another often cited example is the anti-nematode action of several species of marigolds (*Tagetes* spp.). A review of the literature on so-called natural pest control includes many examples of the use of marigolds.[22]

The principal advantage of many natural methods is their specificity for one target insect. Such assurances must be treated with caution however. The claim that *B. thuringiensis* is specific for the Spruce Budworm was recently questioned by an Oregon entomologist who reported on the negative effect of this "non-chemical" pesticide on butterflies and other non-target insects.[23] *B. thuringiensis* is often referred to as a biological or natural pesticide or a "non-chemical" pesticide.[24] However, it is a chemical that does the job. The *toxin* secreted by this bacterium has been isolated in pure form and can be used by itself in the absence of the bacterium. This toxin, *thuringiensin*, is effective against several insect pests and its name has become a frequent key word in *Biological Abstracts*.

One problem that applies to both the conventional methods and the alternative natural methods is the issue of insect resistance. Among an insect population there are always a few individuals, with a slightly different genetic make-up, that survive the pesticide treatment. These survivors continue to breed and pass the "resistance genes" on to their offspring. There is no reason to believe that such an evolution into pesticide resistant strains will occur only with synthetic chemicals and not with the naturally occurring chemicals that come from a biological source.

On the other hand, these biological agents, although perhaps not as specific as was first believed with respect to other insects, have little effect on other wild life. They are relatively safe for humans although the emphasis has to be on *relative*. *B. thuringiensis* is a cultured bacterium and bacterial cultures of all kinds have been known to become contaminated with competing micro-organisms. There is good reason to treat the so-called biological insecticides with the same respect as is due to the synthetics.

An alternative to pest control that must be considered with great caution is no pest control at all. The removal of fungicides, for example, would drastically increase the potential for public exposure to aflatoxins. Any method of growing food will lead to a disturbance of the "natural" system and have undesirable side effects. It is only the nature and the severity of these side effects and the identity of the victims that will be different, regardless of whether the "assault" on the insects is chemical, biochemical, or genetic. Humans will remain in competition with other occupants of the planet, a competition that will only become more intense as our numbers continue to increase. There are many good reasons to continue research on the kinds of pest control that have a minimum environmental impact. The proper perspective for such research must be the awareness that zero impact does not exist.

An application closely related to pesticides is the use of chemicals to speed up or delay the ripening of fruit. The classical example is alar, which caused a major upheaval in North America in 1987 and which has almost disappeared from use. Alar has mistakenly been called a pesticide, although it is actually a growth regulator.[25] It was the Natural Resources Defense Council that initially spread the news that alar would cause thousands of cancer cases among apple juice drinking children. The public reaction caused a major media campaign led by film star Meryl Streep. Although the furore was condemned in the scientific literature, alar has disappeared from the orchards.[26] A few years later a retrospective view stated that the alar risk was not what it had been made out to be, and described the calmer European perspective on alar.[27] Both articles were followed by several letters to the editor, pro and con. The alar case is now history, but it teaches a useful lesson.

One letter writer pointed out that it does not matter how large or how small the actual risk was, the very fact that there was *talk* of a cancer risk was enough to transfer the problem from the scientific sphere to the emotional.[28] This author stated: "the fact that alar is present in apple products without their [the consumer's] consent or knowledge and that consumers can do nothing to remove it, makes this sort of risk *outrageous*, whether it is a tiny risk or not."

The perspective for the "outrageous risk" came from Bruce Ames, also in a letter to the editor. The formal chemical name of alar will help to appreciate Ames' comments.

That name reads: *succinic acid 2.2-dimethylhydrazide.* This chemical is readily changed in the body into *unsymmetrical dimethyl hydrazine* or UDMH. This UDMH is also a common trace impurity in the manufacture of alar. And every chemist and toxicologist knows that hydrazine and many of its derivatives are carcinogens. However, ...?

As Ames pointed out, mushrooms, including edible mushrooms, contain a variety of such hydrazine derivatives.[29] On a comparative index scale the cancer risk from eating one mushroom a day for life is about a thousand times higher than the risk from one daily glass of apple juice that contains traces of alar. The carcinogenic hydrazine derivatives are in the mushrooms without the consumers' consent or knowledge and they can do nothing to remove it. Nature does present us with some outrageous risks indeed. We will return to the issue of moral outrage later in our discussion.

Although pesticides have a prominent place in the ranks of unintentional food additives, there is more to food contamination than pesticides. We considered the fly ash from coal-fired power plants in an earlier part of our discussion. That fly ash contains many potentially toxic elements such as mercury, cadmium, arsenic, and lead. Contamination of the external plant structure is possible. Plants that are not normally subject to much washing during processing or preparation can easily be a dietary source of these contaminants. Tobacco plants, which are legally not food plants, have large, fuzzy leaves that are perfect traps for airborne dust, and they are not washed before use. Some researchers believe that the natural radioactive fallout on tobacco leaves, particularly polonium-210, is a more potent carcinogen than the chemicals that result from the combustion of the tobacco. Others have indicated that cadmium, a highly toxic metal, is a common contaminant on tobacco leaves.[30]

Yet another class of unintentional contaminants in food comes from animal health products: pharmaceuticals, biologicals, and feed additives. Many such products are used for maintaining the supply of milk and meat. Some are essential, others are merely cosmetic. Feed additives that affect the animal's growth and meat production are a natural result of the large demand for cheap meat. Particularly controversial is the use of estrogenic hormones to promote meat production. Seen in perspective, the amounts of estrogenic chemicals in meat are a tiny fraction of those naturally made in the human body or voluntarily consumed in the form of birth control pills. Whether or not the removal of such chemicals would be a public health benefit is a trans-science question, impossible to answer by scientific investigation. Such a removal would also increase the cost of the meat.

On the other hand, the prevention and treatment of mastitis, a common mammary gland infection in cattle, with antibiotics is essential to safeguard the milk supply. In Britain, the Swann Committee led to a legislated decrease of antibiotics in livestock in the early 1970s, but levels have now increased again until almost the pre-1971 situation.[31] A comprehensive discussion of animal health products in food is beyond the scope of our present discussion. Those readers interested in the details will find a recent summary in the literature.[32]

In an earlier chapter we considered the contamination of drinking water and the possible health effects on humans. Pollutants in drinking water can also affect farm animals. Those farm animals eventually become human food, and that food is often transported over long distances. This is how water pollution in one part of a continent can contribute to the contamination of food in another.

Industrial contaminants can also enter the human body, indirectly, via contamination of livestock. This group of contaminants includes pesticides, used around the animals, cleansing agents, solvents, packaging materials, machine lubricants, and natural and anthropogenic radioactive fallout.[33] In contrast to fish and shellfish, farm animals do not readily accumulate industrial contaminants from water or feed. Bioaccumulation is not a problem in the meat supply. Further sources of food contamination are the packaging materials used for prepared foods. Particularly plastics can contribute various potentially toxic contaminants to food.[34]

Food irradiation

Our discussion of food protection would not be complete without mention of food irradiation. Treatment of food with ionizing radiation is a controversial and emotionally charged subject. The controversy will likely last, but a few common misconceptions can readily be removed. The process has been in use since the early 1950s and is relatively simple.[35] Food is exposed to high energy radiation, usually from a radioactive source such as cobalt-60. This source is enclosed and is never in contact with the food. The radiation passes through the food in the same way diagnostic X-rays pass through a patient's body. No "residual radioactivity" is left behind in the food, for the same reason that no radioactivity remains after a medical X-ray. The "contamination of the food with radioactivity" because of the treatment does not take place. Any radioactivity contamination that was present before the treatment will, of course, remain.

Closely related to the irradiation of food is the use of radiation for the sterilization of medical equipment and supplies. Bandages, plasters, syringes, and a host of other medical consumables can be conveniently wrapped in paper or plastic wrappers without too much attention to sterile conditions. Once the packages are sealed, a dose of radiation kills whatever micro-organisms may be inside, and the product remains sterile until it is opened by the user. These sterilization methods for medical supplies have been in practice for many years. There is no evidence of the formation of toxic chemicals inside such wrappings.

The objective of food irradiation is the same. The high energy gamma rays are fatal to microbes and the process eliminates, up to a point, the need for refrigeration or chemical food preservatives. The result is a much longer shelf life, particularly for fresh fruit and packaged raw meat. Food irradiation is very important for protecting public health in countries where refrigeration is not readily available in either supermarkets or private homes. As with virtually all technological practices, there is always a price tag.

The high energy of the radiation can break chemical bonds in some components in the food. This leads to the formation of smaller molecules that are not normally present in the food. These chemicals are formed by the lesion, or lysis, of food molecules and are therefore called *radiolytic products*. A persistent concern with food irradiation is that some of these radiolytic chemicals could be toxic.

Many jurisdictions require toxicity testing on irradiated foods. This is difficult to do with conventional toxicity assays. Since the exposure time to the radiation is usually very short, radiolytic products are only formed in extremely small amounts. These small amounts make it almost impossible to use conventional animal tests. For the same reason, toxic effects on human consumers, if any, would also be extremely small. Because of today's preoccupation with absolute safety and the technical difficulties of testing very low exposure levels, the question of the toxicity of radiolytic products has not been settled after some thirty years of research. However, there is another approach to this question.

Radiolytic products, which may or may not be toxic, represent a very small fraction of the total of many toxic substances formed by other processing methods, or occurring naturally in food. More specifically, many of the changes that can be caused by irradiation are of *the same kind* as the changes caused by other processing methods. Changes in food and food products caused by processing techniques are well documented. Heat treatment causes changes in vegetable oils, which tend to form peroxides and polymeric compounds that can have adverse health effects. Solvent extraction, used to remove unwanted components such as caffeine, can leave traces of solvent behind. Various foodstuffs undergo treatment with alkali, and fumigation is a common treatment for stored grains. Many cooking oils are obtained from their vegetable sources by extraction with solvents such as hexane. Finding traces of hexane in the cooking oil is simply a matter of instrumentation. Heating of many foods also causes the so-called non-enzymatic browning. Several of the breakdown products formed during this browning are toxic. Browning of protein containing food can lead to formation of carcinogenic nitrosamines. Much of the literature on this topic was well established in the 1960s and 1970s.[36] Ames, in a recent review, elaborated further on the carcinogenic potential of these browning products.[37] He pointed to the gram amounts of this "brown matter" we consume daily in the crusts of bread, toast, fried meat, and numerous other browned food. In many instances food is treated with chlorine or chlorine dioxide and organochlorine chemicals formed in these processes have been demonstrated to remain in the product as impurities.[38]

All these processing methods leave contaminants behind in the final product. It is not reasonable to accept that all of these contaminants would be found harmless in toxicity testing. Nevertheless, they are GRAS, generally regarded as safe, with a few exceptions. Methionine, an amino acid that is part of the protein in every living organism, can easily be converted into a strongly mutagenic organochlorine compound.[39] The commercial processing of a Japanese fish with nitrite and common salt readily forms this chemical, which may be an explanation for the high incidence of stomach cancer in areas of Northern Japan. Living organisms contain many essential chemicals like methionine, and their interaction with common preservatives is only to be expected. There must be many more of these unsafe, "non-GRAS" processing products waiting to be discovered.

Both theory and practice in the organic chemistry laboratory show a close relationship between the chemical events that take place under influence of heat and those that are promoted by radiation. This supports the observations that changes caused in food by radiolytic processes are very *similar* to those caused by food processing that involves heat treatment. Of course, radiation treatment of food has not been unequivocally established as being "completely safe." It never will be, because that is outside the capabilities of science. The present state of knowledge does not support the notion that food irradiation is a potential threat to human health. Conversely, the alternative to processing, completely untreated food, is most definitely a threat to human health.[40]

Those who remain unconvinced may find solace in another aspect of food irradiation. The biochemical events associated with the ripening of fruit are quite complicated and not well understood in many cases. Exposure to high energy radiation can interfere with this biochemistry in an unpredictable way, slowing down the ripening process in one product and accelerating it in others. The softening of the food is bad for texture; consumers prefer firm, crisp fruit. In some cases softening also robs the produce of its natural defense against spoilage. In potatoes, high doses of radiation lead to increased spoilage, the opposite of the desired objective. Other problems are the logistics of many high level irradiation sources and potential hazards to operating personnel. It seems unlikely that food irradiation will *replace* chemical additives. However, it is a useful tool to reduce the use of chemical additives. In North America consumer resistance is a bigger obstacle than in Europe, where food irradiation is a part of the mix of methods used by the food industry.

Of vegetables and medicines

The strong emotional resistance to food processing and intentional and unintentional food additives has promoted two developments that we must consider in some detail. The trend to do away with pesticides and preservatives altogether, led to the birth of the health food store. One step further is where the belief in the "good of nature" makes food into medicine.

Nature knows best

One of Barry Commoner's statements that assumed the status of dogma was that "Nature knows Best." Commoner himself called it a "law of ecology." The Earthmother, or Nature, has only the very best in mind for all her creatures. The corollary is that when things go wrong, it cannot be her fault. When cancer and heart disease strike in every family and industrial chemicals are everywhere, the conclusion must be obvious. Cancer, in Ralph Nader's words, is an industrial disease.[41]

Our discussions on cancer led us to explore one part of the scientific literature that demonstrates the flaws in this vision of the law of ecology. There are so many natural phenomena, from radiation to fallout to carcinogens in almost every plant, that have the potential to cause cancer, that the wishful thinking that "nature wouldn't do it to us" clearly does not hold. Now that we specifically look at food, what more does nature have in store for us?

The emphasis on pesticides that started with *Silent Spring* and subsequent literature on the topic had a profound effect on the way people think about food. A very large body of knowledge about food and food chemicals has almost been forgotten or is ignored.[42,43,44] The monograph by Liener contains some three thousand references to the literature on thousands of chemicals isolated from common vegetables, that have been shown to have the potential to cause negative effects on human and animal health.

In addition to the examples of naturally occurring carcinogens, many of these naturally occurring chemicals can cause liver disease, destroy red blood cells, damage kidneys, or cause a host of other health problems. Naturally, all these effects are dependent on the dose.

Bruce Ames called these toxic allelochemicals the "natural insecticides" made by plants.[45] Allelochemicals are used by living organisms to communicate, to attract or repel other organisms. Insect sex attractants are an example of the first, self-defense with the help of toxic chemicals is of the second kind. In his review Ames listed mostly examples specifically related to cancer. The other sources cited tell us about many other kinds of toxic action.[46] A few of these are suitable as examples.

Probably the best known are the solanum alkaloids in potatoes and tomatoes, and the bitter principles of some wild cucumbers; we mentioned both earlier. Cyanogenic glycosides are another group of toxins that are very widespread in the plant world. These chemicals consists of combinations of carbohydrates with small molecules called cyanohydrins. Enzymes, both in the plant and in the plant eater, can break these chemicals down into smaller fragments, and one of the fragments is hydrogen cyanide. Hydrogen cyanide blocks one of the enzymes vital to cellular respiration and small amounts are fatal to humans. It acts very fast and that is the reason it has been used for executions in "the gas chamber." Cyanogenic glycosides are present in some 2000 plant species belonging to over a hundred different plant families, including bitter almonds, wild cherries (leaves), apricots and peaches (seeds), sorghum, bitter cassava (tubers), and many kinds of Lima beans.

The pyrrolizidine alkaloids, also known as the *Senecio* alkaloids, are a large family of naturally occurring chemicals with about 250 members. They occur in as many as 6000 plant species, or about 3 per cent of all flowering plants, many in the *Compositae* and the *Leguminosea* (legumes) families. Poisoning by these plant toxins in rodents, domestic animals, and humans primarily affects the liver and continued exposure can lead to chronic liver failure. Species of *Heliotropium* and *Crotalaria* containing these toxins are weeds that readily contaminate grain crops. Pyrrolizidine alkaloid poisoning is common in poor countries where primitive harvesting methods lead to human consumption of these toxic weeds. There are case histories implicating the use of alkaloid containing plants in herb teas in cases of liver disease. Some common examples of natural plant toxins are listed in Table 10.1.

Rachel Carson referred to synthetic pesticides like the chlorinated hydrocarbons, and chemicals like parathion and malathion as "the synthetic creations of man's inventive mind... having no counterparts in nature."[47] However, the "counterparts in nature" are abundant and well known in the chemical literature. For example, the red seaweed genus *Plocamium* produces several chlorinated and brominated hydrocarbons with strong insecticidal action.[48] Chemicals like parathion inhibit an enzyme that is vital to the transmission of nerve impulses; their toxic action is the same as that of nerve gases like sarin, developed for chemical warfare. They too have their counterparts in nature. Solanine, present in potatoes, is a common example of a natural *cholinesterase inhibitor.* Other poisons of this kind have been identified in broccoli, sugar beet, asparagus, turnips, radish, celery, carrots, oranges, apples, and raspberries.

The quantities of these chemicals in a balanced diet are small and there is little evidence of a human health threat by ordinary food plants. Human metabolism can

Table 10.1 Some examples of natural toxins in food plants

Plant	*Chemical*	*Health effects*[a]
Alfalfa sprouts	L-canavanine	systemic *lupus erythematosus*
Black pepper	safrole[b]	carcinogen
Edible mushrooms	various hydrazines[c]	carcinogens
Spinach, rhubarb	oxalates	corrosive gastroenteritis
Nutmeg	myristicin	hallucinogen
Broad beans	vicine, convicine[d]	hemolytic anemia or favism
Potatoes	solanine	cholinesterase inhibitor
Celery, parsley	furanocoumarins	carcinogens
Lima beans	cyanohydrins	cyanide poisoning

Notes: [a]Only in toxic doses
[b]Also in nutmeg
[c]Close relatives of Alar
[d]Probable causative agents
Source: B. Ames, "Dietary carcinogens and anticarcinogens," *Science*, 1983, vol. 221, September 23, pp. 1256–64 and I. E. Liener (ed.), *Toxic Constituents of Plant Foodstuffs*, New York, Academic Press, 1980

detoxify many of them; others are sensitive to heat and are destroyed by cooking. In fact, the human species is testimony that not only survival, but proliferation is eminently possible in an environment saturated with naturally occurring vegetable poisons. Our very evolution on this planet is a strong empirical argument for chemical hormesis. However, there are two realistic scenarios that can present a significant risk from the poisons in our food. Both have to do with the well-known danger of an overdose.

An overdose can easily occur because of the abundance of one food source and the scarcity of another. The *Cycadicae* or cycads are a plant family containing over a hundred species. Particularly the tubers are an important food source in many countries. In addition to many carcinogens, these plants also contain a non-protein amino acid called beta-N-methylamino-L-alanine or BMAA. Monkeys given repeated doses of this amino acid developed symptoms similar to those of Parkinson's disease. This study of the cycads was prompted by a serious health problem on the island of Guam (Micronesia), where the consumption of these plants is common.[49]

There are many other cases of mass poisonings by plant chemicals. The literature reviews listed in this chapter contain abundant examples. Some authors agree that these naturally occurring toxic chemicals are indeed present in our food, but they suggest that they are "safe" by the reversal of a common argument. For example, Culliney *et al.* stated that some of these substances have proven to be carcinogenic in animal studies, but "the carcinogenicity of these compounds to humans has not been established."[50] These reviewers did not mention that this also holds for almost all other carcinogens, including pesticides and growth regulators like alar. It is the very argument against the "no-safe-dose" hypothesis. The law of ecology that says "nature knows best" is more about the plants protecting themselves from plant eaters, than about nature protecting the plant eaters. Or, in another language, the idea that a benevolent nature "would not do it to us" is a purely anthropocentric view of ecology.

Vitamins, cult or cure?

Another food issue deals with the recent surge of interest in vitamins and mineral supplements. Evidence is slowly accumulating to support the notion that modest nutritional supplements can have beneficial effects. However, the medical community remains sceptical on this issue.

Intensive reading of the literature on plant toxins and naturally occurring carcinogens can quickly spoil the appetite. If this is real, do we stop eating altogether? Of two possible answers to this question, the first is surprisingly simple. We have already looked at Bruce Ames' summary of "anticarcinogens" in food. We have also reviewed several aspects of the defense mechanisms of the human body, not the least of which is the chemical "obstacle course" for a chemical on its way to a target cell. The second answer is more subtle.

Notwithstanding many questions about the use of vitamin and mineral food supplements, there is a definite trend developing on the borderline between food science and medical science. It started many years ago with an American chemist and a Scottish physician. The publications of Linus Pauling and Ewan Cameron on vitamin C have been around for decades, but they were largely ignored.[51] More recently, however, the possible link between diet and disease prevention has received considerable attention. A conference of the American Institute for Cancer Research was held on the issue and the proceedings of that meeting were published in book form.[52] About the same time the National Heart, Lung, and Blood Institute (US) held a workshop on antioxidants and heart disease.[53] In 1990 the National Cancer Institute held a conference on vitamin C. Noteworthy about that conference was the almost complete absence of representatives of the medical press, with the journal of the American Medical Association as the only exception. The American Society for Clinical Nutrition also devoted its 1990 conference to antioxidant vitamins in disease prevention.[54] So far, the medical establishment is clearly not impressed by the interest in vitamins.

These meetings covered an enormously wide area, and a detailed discussion would be beyond our present scope, so we will restrict our discussion to a few key aspects. The general tone of vitamin research so far is supporting the notion that diet and health are related in such a way that further important research must be done. The associations between antioxidant vitamins and cancer are being pursued along two different lines, prevention and cure.

A detailed review lists at least thirty studies involving human subjects which show that vitamin C supplements did offer a statistically significant protection against cancer.[55] Here also, a consensus seems to be building that antioxidant vitamins can play a role in prevention. Not surprisingly, the majority of the medical community strongly objects to suggestions that vitamins, particularly vitamin C, could actively cause the remission of a cancer. Another study, in which the researchers used cells of hamster embryos, demonstrated that several of the presumed inhibitors seem to work in concert. The protective effect against the results of radiation of a combination of vitamins C and

E was considerably larger than the sum of each vitamin separately. The authors were very careful not to extrapolate their results from the cells in a petri dish to humans.

Tannenbaum *et al.* studied the chemistry of vitamin C with respect to one specific interaction.[56] Nitrite, which is a component of so-called butcher's salt, is added to virtually all prepared meat products as a preservative and to improve the colour of red meat. Nitrite is also formed in the human digestive system by the reduction of nitrate, which occurs naturally in almost all green vegetables. The interaction of nitrite with many amines, which are natural components of food and of human tissue, produces the strongly carcinogenic nitrosamines as well as the above mentioned mutagen in some fish processing. The chemical interaction is called nitrosation, and it is inhibited by vitamin C. In this specific instance vitamin C is clearly playing the role of an anticarcinogen.

Another study, this time not related to cancer, also links the actions of vitamins C and E.[57] People who took larger daily supplements of these vitamins experienced a significant reduction in cataracts. These results, which the authors suggested be followed up with further study, agree with Ames' thesis that several of the agents that play a role in cancer are also important in other age-related diseases. The common thread that runs through all these studies on vitamins is that although preliminary results look promising, much further research is needed. As yet, vitamins are no wonder drugs.

The foolish immortals

Mass production, monoculture, and the logistics of distance will remain an integral part of the world's food supply, making the complete elimination of synthetic chemicals from human food improbable. However, the information age has drastically increased public awareness of the flaws in the system. The apparent increase of dread diseases afflicting aging populations has turned awareness into fear, and fear turned into defensive action. As a result, there has been a growing demand for food that is not contaminated with "all these chemicals." Inevitably, someone coined the term *health food.* Specialty shops that fill this need became known as health food stores.

Health food is subject to regulation in many jurisdictions. It cannot contain any residues of the synthetic chemicals used in modern agriculture. Only what is natural has a place in health food and the unfortunate term "organic gardening" entered the green language. Unfortunate, because organic gardening helps to increase the confusion. Gardening is inherently organic, inorganic gardening is a contradiction in terms. Terms like "natural gardening" or "gardening without synthetics" would have served better to promote an understanding of the practice. In view of the fact that *food is chemicals,* the perception of food "free of chemicals" is nonsensical, expensive and it can be dangerous. We will again use our favourite fruit, the "organically grown" apple as our example.

The attractively coloured fruit, preferably without blemish and with a soft shine, is the symbol of good nutrition and health. It appears to be fashionable to sit waiting for a bus or a train while scanning the headlines and munching away on a beautiful, crisp apple. With the red skin on it, of course. Let's have a close look at this apple. Misunderstanding about pesticides, ripening agents, and additional treatment particularly applies to apples. In some of the popular literature, the writers claim that pesticides like captan can be removed from fruit by washing.[58] Since captan is virtually insoluble in water but quite soluble in wax, it will end up embedded in the natural wax layer on the skin of the apple. It could only be washed off with the help of a strong detergent, hot water, and much elbow grease. Not so for the "organically grown" apple, or so the proponents say.

Organically grown or not, any apple grows with natural wax to help its skin keep the juice in and the rain out. That wax traps water insoluble contaminants in the air, such as PAHs from passing trucks, water insoluble chemicals sprayed on the apple, and dust, including natural radioactive fallout. Of course, an apple grown without the use of synthetic chemicals is less likely to contain pesticide residues. Nevertheless, those concerned with "chemicals" should consider peeling off the apple's skin. Enough is now known about the environmental history of this apple to discard the old fashioned notion that the skin is "the best part." The motto: "what you can wash off won't harm you, and what can harm you cannot be washed off" applies particularly to apples, whichever way they are grown. On the other hand, the evidence for chemical and radiation hormesis suggests that it does not make any difference which way we eat the apple: "you gotta eat a speck of dirt."

Health food is not limited to vegetables. Many prepared foods on the market are claimed to be free of additives. Protection for the consumer of this category is still in its infancy. Several states in the USA issue a state administered certification for "organically grown" produce, but this kind of protection is not as readily available for packaged and prepared foods. In any case, processing inherently introduces unintentional contaminants and some foods hardly exist in an uncontaminated version. The natural, made-on-the-spot peanut butter may be made from organically grown peanuts, but it is almost impossible to obtain peanuts that are completely free of aflatoxins. Incidentally, this is another example where zero is never zero. All that is needed to find aflatoxin in "clean" peanuts, is next year's more expensive, more sensitive instrument. The occurrence of the *Aspergillus* fungus on human foods is almost universal.[59]

There are two problems with fresh produce grown without synthetic chemicals: supply and long distance transport. The supply problem is obvious from comparing the relative sizes of regular supermarkets and health food stores. At present, the latter can supply only a tiny fraction of the population. Should all of society insist on the availability of naturally grown food, an instantaneous supply problem would be inevitable. Naturally grown food is a specialty product. Feeding mass populations requires mass production. The long distance mass transport requires a technological approach to prevent mass spoilage.

This still leaves the question whether health food is actually better than conventional food or vice versa. Formal studies on this question are not readily available, and attempts to answer the question by epidemiology would be riddled with confounding factors. For example, a recent article addressed the relationship between poverty, nutrition and disease, and concluded that in many places the war on cancer is really a war on poverty.[60] Since the consumers of health food must be sufficiently well-off to pay the higher prices, it would be difficult to measure if wealthy people who eat health food are healthier than wealthy people who eat a balanced diet from the supermarket. So far the health food store is mostly for the economically privileged, and only two benefits have been definitely established: peace of mind for the health food consumers and the bottom line of the storekeepers' balance sheet.

For the ordinary occasional visitor there is little danger in the health food store. It can be different for those who seek "unconventional" therapies for their ailments, and who believe that health food and herbs can accomplish what conventional medicine cannot. These health food consumers are not aware of the extensive literature on toxic chemicals in plants. Their interpretation of the "nature knows best" law is stuck in the idea that nature knows the difference between good and bad. Good and bad for humans, that is. This outdated view could lead to dangerous overdose situations.

The Russian Comfrey, *Symphytum officinale*, contains several of the pyrrolizidine alkaloids mentioned earlier.[61] Yet, comfrey appears in several cook books as a culinary herb, and it is a common ingredient in herb mixtures and teas. A dictionary of herbs sings the praise of this valuable medicinal plant in over three pages without so much as mentioning its toxic properties.[62] Chronic overexposure to herbs like comfrey could have very unpleasant results.

Similarly, wholesome alfalfa sprouts contain a curious, highly toxic amino acid, called L-canavanine. This amino acid does not occur in the normal proteins found in our food. It is a relative of arginine, another amino acid that does belong in protein. Canavanine can "fool" the cell into making a wrong, malfunctioning protein. A condition similar to human systemic *lupus erythematosus* (SLE) can be induced in Macaque monkeys by feeding them alfalfa sprouts.[63] While the many valuable nutrients in alfalfa sprouts are important ingredients in a healthy diet, overdosing on alfalfa sprouts, or extracts thereof, is not to be recommended.

Broccoli is mentioned in so many health-related magazines and books that it has almost become a cure-all. This vegetable is rich in several common nutrients. It has been reported that broccoli contains a chemical called sulforaphen, which can activate certain enzymes that are thought to play a role in detoxifying carcinogenic chemicals.[64] Sulforaphen also occurs in radishes.[65] Broccoli is also rich in potassium, an essential mineral that has also been investigated with respect to cancer prevention.[66] Consider now the following scenario.

The innocent believer in the nature-knows-best ecology has found some good-looking capsules of concentrated broccoli extract, and is ready to add a few to the usual breakfast-time collection of pills. One of the other pills, a prescription drug, is an enzyme inhibitor that lowers the blood pressure. Its side effects can be increased blood levels of potassium. Upsetting the potassium sodium balance in the blood can be very dangerous. How much potassium is there in the extract of the potassium-rich broccoli? It does not say on the label. Moreover, with the potassium from the broccoli comes, unavoidably, the naturally occurring, radioactive isotope potassium-40.

Paul Gallico, a well-known-fiction writer of the 1950s, told his own vision of the "Elixir of Life" in a novel *The Foolish Immortals*. It was about a rather odd little group of Americans travelling to Palestine to find the secret of why Methuselah lived so long. They too believed that they could live longer if only they found the right elixir. Unfortunately, the laws of ecology forbid us to believe in the Good of Nature without accepting also the Bad of Nature. Wherever "The Elixir" is to be found, it is not in our food.

CHAPTER 11

Waste not, want not

The Oxford Dictionary describes waste paper as "paper regarded as spoiled or valueless." Something that has become valueless has to be thrown away. Already the environmental literature of the early 1960s made it clear that we could run out of space for throwing things away. Waste was one of the first issues that entered public awareness with Vance Packard's book *The Waste Makers*.[1] His emphasis was quite different from today's waste problems, but he added "waste" to the environmental language.

Waste management

We have in the meantime learned that waste cannot just be thrown away, it needs to be managed. Sometimes it must even be treated as a resource. Waste management is an industry, and the language contains words like hazardous waste, industrial waste, household (domestic) waste, solid waste, waste stream, and waste exchange. These terms do not apply to a natural, balanced ecosystem. In such a system waste is always a resource. In a balanced ecosystem there is no waste, because everything is *recycled*. Without recycling nature would come to a standstill.

Because it means following nature's example, recycling has become an important buzzword. The automatic assumption is that it is the right thing to do. Our brief examination will show that there other laws of nature that must be obeyed as well.

Waste and the laws of nature

All living organisms consume matter. They ingest mineral and biological material from which they extract nutrients. Then they get rid of the parts they cannot use. The material they cannot use has often been modified so that now other organisms can use it. The waste of one species has become food for another. When the organism dies, its entire body becomes biomass for consumption by others. On the molecular level, the bits and pieces of one organism are put together, in a different arrangement, by another. *Recycling* makes the "waste" of one available to another. Humans fitted into this scheme for most of their existence on earth. We ate plants and animals; human waste was part of the biomass used by other organisms. Under the laws of nature, waste management is perfect, provided that certain rules are obeyed. One of the rules is about the limits imposed by the *carrying capacity* of the ecosystem.

When humans invented industry, they started manufacturing things. Initially, manufactured things were made of stone, or biological materials such as wood and leather. When they wore out and were no longer useful, they were thrown out. The wood and the leather rotted away, were *biodegraded*, recycled into bacterial cells, insect bodies, and plants. Then manufacturing began to include metals, other minerals, then coal, and still later petroleum. When these manufactured things "died" their mass could not be eaten by anybody. The natural recycling of mineral materials occurs on a different time-scale. When manufactured things became "spoiled and valueless" they became waste. The manufacturing process itself also produced waste. This kind of waste is unique to industrialization and manufacturing.

Human waste in surface waters is a problem in many countries, but there is, at least theoretically, a simple natural solution. Proper treatment of sewage or a drastic reduction in the number of people producing the waste, or both, would make the problem disappear. Not so with manufacturing wastes.

Initially, industrial waste was treated the same as ordinary sewage. Rivers, lakes, and oceans were seen as the unlimited sink where waste just disappeared. Human life expectancy was short enough so that nobody paid much attention to potential long-term health effects of improperly disposed waste. There was not much concern for wildlife or nature. Then populations grew, industrial output increased, and so did the waste output. Awareness of human and environmental health increased and waste became a problem.

In the mining industry they say: "If you can't grow it, you have to mine it." Materials for manufacturing are mined, they are not grown. This includes building materials that contain wood components. Harvesting wood requires manufactured heavy equipment and vast quantities of petroleum products. Composite wood products contain glues made from petroleum. The use of cotton and wool similarly involve the use of industrial equipment, petroleum-based dyes and lots of energy. Most of industry needs the energy from mined fuel.

Mining means extracting substances from the crust of the earth and subjecting them to change. Iron ore has to be treated with coal in order to make iron. The wastes are slag, ashes, and smoke. When iron is shaped and machined, turnings and filings become part of the waste. Petroleum refining processes leave a host of unwanted materials behind before the fuel or the plastics can be used. When wood is made into paper, only part of the wood, the cellulose, is used. The other part, the lignin, becomes waste.

Other materials are temporarily needed in many manufacturing processes. Salt, a natural product from the oceans, is used to make chlorine and hydrochloric acid and caustic soda. Sulphur, a mining product, is "burned" to make sulphuric acid. Sulphuric acid and caustic soda are used for petroleum refining, but they do not become *part* of the products. The acid and caustic used in the process become waste. Chlorine is used for making paper, but it does not become *part* of the paper, it becomes part of the waste. During these processes acid and caustic must be washed out of vessels and containers and pipes with water. The resulting dirty water is now also part of the waste. All these wastes are *processing wastes*. They are called *liquid waste*, and, since most of these liquids are toxic, they are often also labelled as *toxic waste*.

Eventually, manufactured products themselves also become waste. Petroleum-derived fuel is burned in a car engine, and waste comes out of the exhaust. After ten years the whole automobile becomes waste. When the plastic toy breaks, it becomes "valueless." Yesterday's news is no longer news and the newspaper becomes waste.

Goods and commodities are transported by truck or by rail. Eventually the batteries, tires, and the steering wheel become waste. Cardboard boxes, glass bottles, plastic packing materials, become waste. And yes, even the most sophisticated stereo system or computer or refrigerator eventually all become part of a waste stream that is uniquely human. Unique because most of this waste does not fit into the ecological biomass cycle any more. It has to be dealt with in a different way.

When Vance Packard raised the issue in 1960, he was mainly concerned with the economic aspects of wastefulness. His main interest was in human behaviour and economics. He coined terms like "planned obsolescence" and attacked the economics of waste. His concern was with the cost that packaging added to the product, and the cost of replacing items for the sake of fashion. Packard had little to say about the disposal of waste. Landfill capacity, the space to put the garbage, had not yet become a problem. Incineration was small scale and was not seen to hurt the environment.

In the thirty years since Packard's book, little changed in the way waste was generated, but the emphasis changed. Populations increased and individual wealth and consumption expanded. Landfill sites started to fill up, space available for new disposal facilities became scarce as environmental constraints began to take hold. Environmental effects and human health became linked to waste disposal. As a result, the disposal of all kinds of waste is now a major predicament. The two parts, the disposal of domestic waste and the management of industrial wastes are two separate issues.

Domestic waste

Most domestic waste consists of packaging material, paper, glass, wood, plastics, discarded consumer products, and some vegetable material from food waste. A significant contribution, although strictly speaking not "domestic," is made by construction activities. The waste management industry uses the term *solid waste* for domestic garbage or, in other jurisdictions, rubbish. Most industrial waste contains water, oils, and organic solvents. Waste managers call it *liquid waste*. Solid waste can most easily be stored in a large hole in the ground, a landfill site, or more formally, a *sanitary landfill*. In some rural areas control over who dumps what and how much may be fairly relaxed. Near cities and towns the landfill site is usually chosen with some care, and certain conditions must be met. As environmental legislation proliferates, these conditions become stricter.

Rain water can dissolve some components in the waste and contaminated water can percolate through the waste and surrounding soil and enter the ground water in the area. Therefore the geology, and particularly the hydrology of the site are important considerations. Most landfill sites are surrounded by monitoring wells, so that the condition of the ground water can be watched. In some facilities, contaminated *leachate* can be pumped out if something should go wrong.

Another option for domestic waste is *incineration*. Most of the waste will burn in such a process. Unless removed beforehand, glass and metals remain behind as a slag. Incineration is not just burning waste; it occurs in a closed vessel, at very high temperatures, and with a carefully controlled supply of oxygen or air. The gaseous combustion products go through an elaborate system of scrubbers and ash precipitators. The scrubbers remove acidic gases and vapours and the precipitators retain solid ash particles. The properly operated incinerator for solid waste emits almost nothing but carbon dioxide and water vapour. Solid ash and waste from the scrubbers must still be disposed of, but this volume is much smaller than the original feed of the incinerator. Eduljee *et al.* studied an incinerator in Scotland in detail for the emission of chlorinated dioxins and furans.[2] They looked in the soil, in the stack gases and particulates, and the ash. They even examined the tissues, particularly the livers, of a few cows in the area. They did not find any.

Of course, that was in 1986, and what was zero then is no longer zero today. As was the case with water pollution, today's more sensitive instrumentation probably would find offending emissions from that Scottish incinerator. If we were to spend additional effort to reduce these emissions to today's notion of zero, we would find our efforts wasted when in a few years time a better instrument again finds the offending chemicals in still smaller amounts.

A third option, composting, is only available to the vegetable part of the waste. Vegetable (and garden) waste, separated from other solid waste, contains many naturally occurring bacteria and fungi. Under suitable conditions of temperature and moisture these organisms, with help from added vitamins, *partially* biodegrade the biomass. The resulting *compost*, a fibrous, partly decomposed vegetable material, is a good soil conditioner and fertilizer. None of the three methods of disposal are free of problems.

Sanitary landfills are unsightly, they cause odour problems and can attract scavenger animals. These problems are partly dealt with by covering each day's load with a thin layer of soil. The serious problem is that landfill sites tend to fill up. That means another place has to be found, and that may be difficult.

Even if they are built with proper monitoring facilities, landfills are not very good for containing materials that are soluble in water and which can leach out of the area. There are many such materials. Hearing aid batteries contain silver or mercury. They are small, but there are many of them. The landfill for a medium-sized city annually receives many pounds of mercury this way. Portable radios and children's toys often use nickel-cadmium batteries. Although these are often rechargeable, they eventually find their way to the landfill. In spring, when the lawns and gardens are prepared for summer, grass and plant clippings and left-over yard chemicals from last year also make their one-way trip. There is a long list of such water soluble, more or less toxic chemicals. Sanitary landfills are not the perfect solution to domestic waste.

Waste incineration also has its problems. Many incinerators are not really incinerators, they are just facilities to burn trash. Bad odours and smoke can be problems. When a real incinerator is operating properly, very little offense comes out of the stack. Very little, but not zero. The incinerator stack problem is directly proportional to the number of incinerators and the amount of waste they treat. Ultimately, the stack problem depends on the number of people. Much controversy about waste incineration is based on a failure to distinguish between burning trash in a little building with a smokestack and a true incinerator.

The composting option is a more recent addition to waste management. Compost is a useful material as long as it is free of offending chemicals. Garden and yard waste can contain a variety of agricultural chemicals that are freely for sale at the retail level. This can cause quality control problems in the compost trade, particularly since it was found that composting is one of the many processes that can lead to dioxin formation.[3] Also, large scale composting needs temperature control. If it is too cold, the micro-organisms go on strike. In very cold climates in the northern regions of Canada and Europe composting may require an input of energy that must be entered on the debit side of the environmental ledger.

Diverting the compostable materials from the landfill is not good for the landfill's containment characteristics. As with surface waters, the leaching of soluble chemicals is limited by their adsorption on soil particles, particularly those of vegetable origin. This adsorption therefore depends on the amounts of soil used for daily cover and the vegetable waste present in each daily load. When that vegetable material is diverted to a composting facility, the probability for the chemicals leaching from the landfill increases.

Although the risk of contaminated ground water of most sanitary landfills is small, it has become such a prominent issue in waste disposal that the site selection for new landfill facilities finds a major obstacle in the NIMBY (Not In My Back Yard) syndrome. Finding new landfill capacity is running headlong into the curious phenomenon of people collectively producing waste and at the same time using their collective political clout to block the disposal of that waste.

Industrial wastes

Some wastes from industry are solid, cuttings and filings, others are semisolid sludges, or liquid. The liquid waste may be part of an *effluent stream,* or it can be the kind of waxy, oily, greasy material that is kept in drums and, at least initially, stored. Disposal options for industrial waste are in a sense similar to and yet quite different from those of domestic waste. While vast quantities of industrial wastes are solids with little toxic potential, the emphasis is often on the sludges and liquids. The reason is that these often contain potentially toxic materials and are easily dispersed. As a result, industrial liquid waste is often called toxic waste.

The similarities are in the use of landfill facilities and incineration, as with domestic waste. The differences are in the way the processes are controlled. This control includes careful screening and monitoring of the nature of the waste, particularly with respect to human and environmental toxicity. A landfill for toxic industrial waste must be a so-called *secure landfill*, a facility that is carefully constructed to contain the waste indefinitely.

A secure landfill is constructed with liners of impermeable clay, plastic sheeting, and in a carefully chosen geological formation. It is divided into many cells or compartments to keep incompatible chemicals separated from each other. A detailed system for monitoring and, if necessary removing, ground water is used to protect the surrounding area. In many jurisdictions all these precautions also apply to sanitary landfills. Such facilities are expensive to build and maintain. Their operation now also requires large insurance coverage. Building new ones is next to impossible, not for technical reasons, but because of fear, perception, and regulations.

It is not long ago that industrial wastes were disposed of in the equivalent of an ordinary sanitary landfill. When some of these proved inadequate because of leaching and leaking, major problems surfaced. The classical example is the area near Niagara Falls (New York) known as Love Canal. That 1976 drama has been described in so many books and from so many angles that just a few brief comments will be sufficient here. After the initial complaints from residents about "chemical odours," a number of investigations were carried out. A study reporting chromosome damage among residents triggered fear and panic in the area. That study was subsequently found, "in the opinion of many experts to have close to zero scientific significance."[4] A later report in *Science* concluded that "Data from the New York cancer registry show no evidence for higher cancer rates associated with residence near the Love Canal toxic waste burial site in comparison with the entire state outside New York."[5] Details on the Love Canal affair can readily be found in the news sections of *Science, Chemical and Engineering News*, and similar publications. A detailed description is also presented in Whelan's book.[6] The overall picture that emerges from these various reports on the Love Canal tragedy is that legitimate complaints about the leakage of chemical waste did not get answered with practical remedial action. The fear and panic that were produced instead, led to serious psychological trauma for the residents of that community.[7] Another product of Love Canal is the question that remains, in the rest of the USA and elsewhere, about other improper storage sites with waste chemicals.

Incineration can also be applied to industrial wastes. Incineration demands very tight control on operating conditions such as temperature and oxygen supply. Stack emissions are so rigidly monitored that they have provided the figures for numerous articles, studies, and reports. The reason that such studies are needed is because waste incineration is one of the more controversial issues in waste management.

The lack of recognition of the difference between domestic waste and hazardous industrial waste is one of the major problems in the waste disposal business. Incineration is a sophisticated operation and a huge volume of technical literature exists on the design and workings of modern incinerators. That literature shows quite conclusively that the amounts of hazardous materials that escape the stack of an incinerator are very small. They are not believed to present a public health hazard, but of course, they are not zero. Many other methods and devices for the destruction of hazardous waste exist. None of these processes are one hundred percent efficient and ever-improving capabilities in analytical chemistry can always detect the offending chemical somewhere. This has left the unfortunate perception that most approaches to hazardous waste management are flawed. However, the only alternative to that problem is no hazardous waste at all, and that utopian view does not fit the industrialized society we live in.

In North America the use of incinerators has been under attack for several years.[8] Surveys of the technical literature on the topic have been presented as a rebuttal of this criticism.[9] The problem has become a trans-science issue that has more to do with politics than with technology. The political aspect is clear from the fact that critics on either side offer no realistic alternative. Incineration is common practice in the more densely populated countries in Western Europe, where landfill capacity is scarce because of a lack of available land. In North America waste disposal is more complex. Political activism in terms of the NIMBY syndrome acts against new landfill sites and incinerators.

An important feature of industrial waste is transportation. Facilities for hazardous waste disposal, be they landfill or incineration, are not as abundant as for the management of domestic waste. Therefore hazardous wastes must frequently be transported over long distances. As "sins against the environment" are precipitating larger and larger fines, even imprisonment in many jurisdictions, the transportation of hazardous materials, including waste, is becoming a risky trade. It is heavily regulated by governments and fortunes are paid for insurance.

It follows, then, that disposing of waste has become a serious problem for the industrialized world. It is significant that the promoters of the resourceful earth ideology consistently ignore this side of the environmental problem.[10] Their reassurances concentrate on the plentiful resources of the planet and the irresponsibility of the doomsday prophets. The waste problem, if at all mentioned, is dismissed as a simple problem to be solved by technology, but the details of these solutions are usually missing.

Rs and myths

The approaches to the waste problems of the western world are deceptively simple. The amounts of waste can be reduced drastically by modifying manufacturing processes, using less packaging, re-using packaging, and recovering materials. If the packaging boxes cannot be re-used, maybe the cellulose fibre is still usable. The *material* of the old box, the cellulose, can be *recycled* and used to make new cardboard. The concepts of reducing waste, re-using materials, and recycling ingredients quickly became a slogan. Reduce, Re-use, Recycle. All levels of government, desperately squeezed between the lack of disposal space and political pressure against other disposal methods, vigorously promote the three Rs.

The three Rs

Reducing one's individual waste output is a tricky problem. It is easy to reduce waste with a re-usable grocery bag. Europeans learned this years ago, North Americans still find it difficult. Plastic and paper bags for the weekly haul of groceries are just too easy, and the disposable bags are recyclable anyway. In any case, disposable grocery bags make up only a tiny portion of the load that goes to the landfill, or so the argument goes.

Paper, on the other hand, makes up a large part of municipal waste. Reducing one's output of paper waste would be more effective. But retailers have demonstrated convincingly that advertizing does bring customers. We do buy from the flyers and catalogues and the newspaper ads. Reducing our consumption of paper would require a major change in lifestyle. It would mean drastically reducing the size of Sunday newspapers in New York and Toronto and London, for example.

In the area of "big" waste, like cars, washing machines, and television sets, not much has changed since Vance Packard. Built-in obsolescence in consumer appliances is still with us and reducing waste of this kind is very difficult. There is no economic or other incentive for the maker of new refrigerators to accept the return of the old ones. However, even in this area there is a glimmer of hope.

Another fertile area for argument is packaging material. Most of that problem is in the hands of wholesalers and retailers. Although some attempts have been made to reduce the amount of plastic bottles, that impact on the disposal facility is fairly small. Plastics make up only about 7 percent of municipal waste, so no great savings are to be realized there.[11] There is also the trend of buying food in bulk. It applies the "economies of scale" to the product and it saves on packaging material. The system works well as long as we don't pay attention to our fellow shopper coughing and sneezing all over the bin with bulk breakfast cereal.

And so, is the ultimate solution to recycle? No more three Rs? Will one very big R do? The notion that recycling will save the earth is embraced by all colours of the political environmental spectrum. National, provincial, and state governments are showing their voters how much waste can be "diverted" from the landfill or incinerator.[12] Environmental activists urge us to recycle to save the trees.[13] Even the bountiful earth promoters claim that resource scarcity is a doomsday scenario that will be proven wrong because "ultimately everything can be recycled."[14] Recycling seems an important weapon in the war on environmental degradation. It is worth a closer look.

Recycling is an industry. An industry aiming to recycle newsprint must "harvest" its raw material, old newspapers, just as trees must be harvested for virgin newsprint. To recycle aluminum cans we must "mine" the aluminum from the domestic waste, just like bauxite must be mined for making virgin aluminum. Recycling requires collection, transportation and, at the recycling plant, industrial scale processing. Transportation and processing mean energy consumption as well as processing wastes. The recycling industry is an industry and it inevitably produces *its own industrial waste.*

This means that arithmetic must be done to determine the debits and credits on the environmental ledger. Unfortunately, this kind of environmental bookkeeping is not so rigorously applied to the recycling industry. If it was, the idea that recycling is universally friendly to the environment would need some readjustment.

Some of these aspects of recycling and details on the "energy and pollution costs" of the recycling industry were discussed in a recent review.[15] First there is the issue of the collection of recyclable materials. For the consumer it is little effort to take empty containers and waste newspapers to a centrally located collection facility. It just takes public education to convince people of the need for waste management. Therefore most jurisdictions have chosen this central collection depot style. The big steel containers, clearly marked for paper, colourless and coloured glass are found all over Europe and parts of the North American continent, with one curious exception.

In some parts of Canada and the USA local governments have chosen to collect recyclable waste at the curbside. The householder separates the materials in blue boxes. These plastic boxes are made from petroleum, a non-renewable resource. Although they are reasonably durable, some amount of turnover is inevitable. This turnover is on the debit side. The content of the blue box is collected by additional vehicles. Their fuel consumption and carbon dioxide production are also on the debit side. Sometimes special, compartment type trucks are built for the purpose. The entire system is expensive in energy, and operating costs are so high that virtually all blue box operations lose money on a big scale.[16] The blue box collection system is clearly an operation aimed at saving the aging landfill with the absolute minimum effort on the part of the public. The impact on the environment is clearly on the debit side.

Another question deals with the effects of recycling on resources and energy. In the early 1970s there was no waste crisis. There was an energy crisis, or so we thought. Would recycling cost less energy than making new? If so, could we stretch the energy budget? Several studies provided results that are useful for a rough bookkeeping exercise on recycling. The energy crisis is gone, but we can still use the numbers.[17]

Figure 11.1 Depot collection of recyclables – avoiding the environmental insult of curbside collection

It makes a great deal of sense to recycle materials that already "contain" a lot of energy. For example, substantial energy savings can be made by re-using waste iron. The old trade in scrap iron came about not because of a scarcity of iron ore, but because it realized substantial savings in energy. Particularly the size of old automobiles and refrigerators makes "mining" of scrap iron reasonably efficient. Making aluminum also needs lots of energy, specifically electrical energy. Remelting used aluminum costs only about 5 percent of the energy needed to make aluminum from the natural resource bauxite. Although bauxite is plentiful in the earth's crust, the energy bookkeeping makes recycling aluminum almost mandatory. Recovering this kind of material clearly makes environmental as well as economic sense.

When distances between collection and remanufacturing facilities are short, many other materials can also be recycled with a net environmental gain. In Western Europe distances between cities are relatively short and population density is high. Transport of recyclable materials is not a problem. However, geography creates exceptions. There are many areas in North America, Asia, and Africa where distances can be very large, and some of these areas have low population densities. There transportation energy can make recycling an outright insult to the environment. Glass and paper are typical examples.

The ingredients for making glass are mainly sand and potash, raw materials that are abundant enough to be considered inexhaustible. With certain assumptions, we can calculate that making glass bottles from old bottles saves about 4 percent of the energy that is needed for making glass bottles from the raw materials.[18] One of the assumptions is that the raw materials for the glass are available within a few hundred kilometres of the glass foundry. Another assumption is that the waste glass to be recycled is also "mined" close to the foundry. The energy costs of the actual collection system are not included in the calculation, since they are considered equivalent to the alternative transport of the waste glass to a local landfill.

The conclusion is straightforward. If the distances are short, there is a small energy saving of about 4 percent in exchange for the personnel costs of collecting, sorting, and transport. The impact on the environment is near zero and the justification for recycling glass is to save landfill space. When the distance from waste glass collection to glass foundry is more than about 500 km, environmental cost of the transport begins to overtake energy savings. With increasing distance *the net impact on the environment becomes increasingly negative.* Instead, waste glass could be crushed and used locally for road fill or aggregate in asphalt. But what is the point, economically speaking, if gravel is freely available? Much better environmental and economic savings could be obtained by the use of *locally* refillable bottles.

For the recycling of newsprint, the collecting and sorting stage is followed by the removal of ink and repulping the fibre. This means paper mills have to invest in a de-inking plant. If they have to buy the de-inked, repulped fibre from another supplier, there may be additional costs for transportation energy. Two factors affect the de-inking and repulping.

Removal of ink *inevitably* causes a loss of fibre of between 10 and 20 percent. This means the process produces a sludge of ink and fibre which is waste, and which has to be disposed of. The sludge can be burned to recover some energy. That process produces carbon dioxide and also leaves ash. The ash, or the sludge itself, can be used as soil conditioners, but their usefulness depends on the amount of soil that needs to be conditioned. For paper recycling, the de-inking operation is the first obstacle to "zero waste."

De-inking and repulping include a mechanical treatment. This damages the cellulose fibres. They break and the average fibre length decreases. Fibre length has a significant effect on the quality of the paper, therefore the fibre cannot be re-used indefinitely. There does not seem to be a consensus on the number of times a fibre can go around, but eventually all cellulose fibres become waste. Much less waste than the whole newspaper. Nevertheless, this is a second obstacle to zero waste.

A large part of domestic waste is old newspapers, so recycling of newsprint can be a major step toward solving disposal problems. Therefore many jurisdictions, particularly in the USA, require by law that newspapers be printed on stock that contains X percent recycled fibre, where X is determined by the appropriate state or provincial government. Since a substantial part of the US newsprint supply comes from Canada, Canadian paper mills need a supply of used newspapers in order to meet the US legislation. Because of differences in population density, this supply cannot be met locally. It must be imported, mostly from the eastern states, a thousand miles away at a substantial cost of diesel fuel, waste tires, and batteries. This is a third obstacle to zero waste. *Waste reduction* makes good sense, zero waste violates the laws of nature.

North American paper is exported to other countries that have few trees and overloaded disposal facilities. There it makes good sense to recycle newsprint. In much of North America itself, however, newsprint recycling is promoted, particularly among schoolchildren, with the emotional appeal for "saving trees." In reality, the insistence on newsprint recycling is a desperate attempt to save the overflowing landfills of New York and Chicago and Toronto.

There is also a conflict between political pressure and economics. The landfill problem forces, by political means, a recycling process that is also dependent on market conditions. Lave *et al.* demonstrated the complexity of this problem.[19] These authors used 1988 figures to argue that the price of a mere twenty dollars per ton of used newsprint did not adequately cover the collection cost. In the few years since, the price of newspaper waste has skyrocketed to well over a hundred dollars per ton, causing mills in central Canada to cut back on recycling because of the costs.

Why do trees need to be saved? Modern forestry has it that trees can be cultivated and harvested, just like vegetables. When the arithmetic is done the list of recycling expenses is quite impressive. After the diesel fuel is burned, the carbon dioxide and nitric oxides produced, we have waste tires, waste batteries, and waste trucks at the end of the line.The question that needs to be asked is, what is more important, to save trees that can be replaced, or save the diesel fuel, which cannot be replaced? The answer is subject to a few further questions. Are we really "farming" with trees? If not, why not? And how essential is a three pound newspaper?

In the end, there is little difference between the recycling of glass and the recycling of newsprint. Both save on disposal capacity, but neither is particularly "environmentally friendly." Under favourable conditions they can be environmentally neutral, under less favourable conditions of distance and collection they can be environmentally hostile. Scarcity of resources is not the issue for either glass or paper. The only real issue is disposal capacity. However, back to the trees.

Trees as a renewable resource are not quite as straightforward as it would appear. The world's total consumption of wood is enormous. At a total of about 3.5 billion m^3 annually, we consume about as much wood as we eat food. Only, wood grows much more slowly than food, and supply and demand are badly out of balance. As demand keeps increasing, particularly in the form of fuel in less industrialized countries, global wood resources will not be able to supply that demand without compromising plant biodiversity.[20] The time lag between investment in plantations and the harvesting of the wood makes for uncertain ventures in investment. The supply and demand situation for wood, on a global scale, is almost certain to require environmental concessions. As long as we have not solved the economic part of "tree farming," newsprint recycling does make sense. Transporting waste across continents does not.

A difficult category for the recycling option is plastic. Notwithstanding the obvious disadvantage of consuming petroleum resources for making it, plastic has entered our lives in many ways. The reasons are its light weight and resistance to breakage. Plastics offer no great promise for lightening the burden on disposal facilities. On the other hand, some of the pigments used for colouring plastic contain cadmium, a very toxic metal. It makes sense to keep that out of the landfill and the incinerator. A difficulty with plastics recycling is the great chemical variety of these materials. The easiest to deal with are the relatively rigid containers used for food, household cleaners, and beverages. A coding system to help recyclers separate the different types is now in general use by the industry. Even so, collection of used plastics is a highly inefficient "mining" of resources.

Plastic objects can often be made into other objects. Ingenious technologies for impregnating wood with molten plastic yield plasticized wood that is virtually weather proof and insect proof. Used plastic jars can be made into garden furniture. Both are examples of re-use, not really recycling. Eventually the fence post and the garden chair will end up in the landfill or incinerator. Their birth from plastic waste was a once-only make-over.

On the other hand, plastic shopping bags can indeed be recycled into other plastic shopping bags, theoretically forever. Usually the second time around the shopping bags don't look so great; controlling colour is a common problem in plastics recycling. For shopping bags that is not a real problem because re-usable shopping bags will win the durability contest anyway. Plastic soft drink bottles must be kept carefully separated by colour for direct recycling into new plastic bottles. Polyurethane plastic, the kind that is found in furniture upholstery, can be reprocessed in several different ways. In general, as was the case with other materials, the arithmetic on the environmental impact of plastic recycling must include such variables as transport energy, collection and sorting energy, and the subsequent transport of the recycled products. It is very difficult to find the ledgers of this environmental bookkeeping in the literature.

The myth of biodegradation

In the meantime somebody remembered the classics. That is, the classic writings of the early warriors, Commoner, Carson, Packard and others. Commoner often wrote about how detergents did not break down in the environment and caused water pollution.[21] Much of Carson's message dealt with the persistent pesticides that would stay around forever.[22] In the past thirty years, we have made good progress with these two problems. Synthetic detergents have become almost fully biodegradable, and so are most of the modern pesticides in use today.

Since nature does such a fine recycling job, why not follow this good example? Natural biodegradation could get rid of much of our waste. Shampoo, soap, and detergents readily biodegrade, so why not paper? Paper is a vegetable material and is fully biodegradable. Perhaps we could make biodegradable plastics? Unfortunately, in this case physics and biochemistry offer no help. In fact, they stand in the way.

When water-borne organic chemicals such as detergents and vegetable waste biodegrade, the water contains the needed oxygen in solution. On land, biodegradation by oxygen-using bacteria, aerobic biodegradation, is slower. Biodegradation on land occurs mainly in the surface layers of the soil. In the compacted layers of a sanitary landfill, oxygen is almost absent and biodegradation is accomplished by *anaerobic* bacteria. Anaerobic biodegradation is much slower, and organic materials in a landfill do not break down very quickly.

The ultimate end product of anaerobic biodegradation is methane. Some bacteria can use the oxygen from nitrates and in doing so they convert nitrate salts into nitrogen gas, ammonia, and nitrous oxide. These gases rise through the soil and the waste to the surface of the landfill and become part of the lower atmosphere. The nitrogen is a normal part of the atmosphere and it does no harm. The physical behaviour of methane makes it a very efficient greenhouse gas, and the ammonia and nitrous oxide are common air pollutants. In short, anaerobic biodegradation in a sanitary landfill produces air pollution and greenhouse gas.

This is textbook theory, and it is readily confirmed by experiments. Archeology is an activity that conjures up images of dusty people digging in dusty sand for the leftovers of long gone civilizations. However, Rathje and Murphy practiced some unconventional archeology to uncover interesting information about our own, current civilization.[23] Their archeological searches through sanitary landfills showed clearly that newspapers survive, quite legibly, for decades without biodegrading. The condition of the old newspapers served them as a dating tool to estimate the age of the layers in the landfill and date the other treasures they found. This is how they found that bread also survives almost unchanged for many years.

The measuring and weighing and counting by these archeologists revealed more interesting details. One year's subscription to the *New York Times* weighed as much as eighteen thousand aluminum cans, and the city of Tucson, Arizona, threw out 350,000 bottles of nail polish every year. If the chemicals needed for making this nail polish were to be packed in fifty gallon drums, they would have to be classified as hazardous waste. Many people believe that disposable baby diapers are a big problem. In reality the disposable diapers used by the victims of incontinence, particularly in hospitals, are by far the biggest contributors to the diaper "problem." Taken all together, disposable diapers are a minor part of the landfill problem. When the waste is classified in terms of social status of a neighbourhood, the most frequent throw-aways in low income families are car care products. Higher income neighbourhoods throw out the most lawn and garden products. So far for modern archeology.

Although anaerobic biodegradation is very slow, the generation of methane in a large and deep landfill can be significant. Thus, there are a few facilities of this kind where drilled wells allow the recovery of enough methane, for fuel, to make the effort economically viable. However, the Rathje and Murphy study is an experimental confirmation of biochemical theory and shows that biodegradation does not play a significant role in the disposal of solid waste. More specifically, in a landfill not equipped for the recovery of methane and other gases, biodegradation causes air pollution, is a distinct nuisance, and one of the least wanted conditions.

This is often not understood. In the USA some local governments have gone so far as to legislate a requirement for biodegradability of certain materials, particularly packaging materials. They are thereby worsening the conditions in landfill facilities in terms of localized air pollution, which in turn strengthens the public's opposition against the siting of new facilities.

Although biodegradation in a sanitary landfill is both slow and undesirable, the biodegradable plastic garbage bag attracted much attention. Many attempts have been made to make biodegradable plastics. An initial approach was to incorporate small particles of starch into the plastic. Micro-organisms would then consume the starch and the plastic garbage bag would disintegrate. However, the plastic itself remained non-degradable and the process merely changes a plastic garbage bag into microscopically small pieces of plastic, that would readily blow away in the wind, get into water supplies and cause health hazards to both humans and wild life.[24,25] For similar reasons an early attempt to make a *photodegradable* plastic bag had to be abandoned, because of the potential hazards to the environment.[26] However, truly biodegradable plastics are on the way. A Dutch research institute is developing biodegradable plastics based on wheat starch and on fermented rape seed oil. The potential of such experimental products to compete with petroleum-based plastics is still an economic unknown. In the landfill the biodegradable plastic would still be subject to the same limitations as other biodegradable waste.

Vegetable kitchen and garden waste form another aspect of biodegradation. There are good reasons for treating this type of waste separately. Since it contains a fair amount of water it would be a burden on the energy needs of an incinerator. A simple solution is to collect this waste separately and by a segregated biodegradation process, convert it to compost. Keeping it out of the landfill would help to slow undesirable biodegradation. It would also inhibit the landfills' capacity to retain leachable chemicals by adsorption.

In the end, the economics and quality control of the resulting compost and the energy input must be part of the calculations. It makes sense to keep compostables out of an incinerator. Whether or not they should be diverted from a landfill is questionable.

Vanishing resources

An earlier term for recycling was *resource recovery*.[27] Although the most compelling reason for recycling and composting is the pressure on disposal facilities, this is not usually given as the reason. The idea that recycling helps to conserve scarce and valuable resources is a stronger emotional argument that helps to promote the practice.

In the debate about resources the "catastrophists" argue that the resources of the planet will soon be exhausted by a human population with a growth rate that is out of control. The opposite view of the "cornucopians" holds that for all practical purposes, non-renewable resources are available in unlimited amounts.[28]

Malthus' early views were revived by Packard and Ehrlich thirty-five years ago. Packard's worry about North America running out of fossil fuel energy was based on his conviction that nuclear energy is just not economically viable. His prediction that "Americans will be trying to mine old forgotten garbage dumps for their rusted tin cans" suggested that we were about to run out of iron. Considering that 5 percent of the earth's crust is iron, that seems rather unlikely. Packard's dire predictions have not come to pass except one. The supply of fresh water has indeed become a global problem.[29]

The modern version of the catastrophist view is best represented in the well-known *Limits to Growth*.[30] The Meadows group argued that if present rates of population growth and resource consumption continue humanity will cause a catastrophe of such a proportion that our species may not survive. The methods used in the original Meadows study have been severely criticized. The sponsor of the study, the Club of Rome, reversed its position and in 1976 actually went on record *promoting* growth.[31] Aurelio Peccei, the originator of the Club, was reported to have "explained" the exaggerated views in the book as part of a deliberate strategy to "jolt people from the comfortable idea that present growth trends could continue indefinitely."[32]

Philip Abelson, editor of *Science*, criticized the methods used in *Limits* because of the errors in averaging conditions and variables.[33] In attempting to discuss a truly gigantic problem, the Meadows group viewed the earth as one single system. Population growth, for example, is treated as a single variable although actual growth rates in one region may be as much as six times larger than in other regions. Then Abelson cautiously joins the other side of the argument by suggesting that the supply of resources will be improved by new technology. More recently part of the Meadows group repeated the warnings in an updated version of *Limits*.[34]

The other side is represented by those that call themselves the Cornucopians. Their view that the planet's resources are inexhaustible, with very few exceptions, has been presented by authors of several disciplines: scientists, philosophers, and economists. In 1976 *Science* devoted an entire issue to the question of resources. Most of the articles in this issue were cautiously optimistic. One speculative discussion suggested that the supply of nearly all non-renewable resources is so vast that it would be impossible to exhaust the supply.[35] Moreover, the authors point out, many materials now in use can be replaced with others, provided that energy is cheap and readily available. The picture is very attractive and it is not surprising that the paper by Goeller and Weinberg is frequently quoted in the plentiful-earth type of literature.

In 1981, Julian Simon published *The Ultimate Resource*, also reassuring us about the availability of renewable and non-renewable resources.[36] His optimism about cheap energy is likewise based on technologies that have made little progress so far. Simon, an economist, confidently states that pollution is not all that bad either: "life expectancy, which is the best overall index of the pollution level, has improved markedly as the world's population has grown." It is strange that the health care costs of that improved life expectancy are not a part of this economist's argument. Many industrialized countries are now discovering that their huge national debts are closely tied to overburdened health care systems.

Another highly optimistic view of the technological society was written more recently.[37] This book, *Trashing the Planet*, is sufficiently recent to address global problems like acid rain and ozone depletion. In keeping with the cornucopian viewpoint, the authors present evidence from reputable scientific literature to suggest that, yes, indeed, science and technology do have a positive contribution to make. One example of such contributions is the earlier quoted report on the "benign design" approach that is becoming part of synthetic chemistry.[38] Another example of a technology-based approach is the "lease rather than sell" trend we see in Western Europe, where manufacturers are taking back not only packing material, but also the actual product when it has reached the end of its useful life.[39] That this is no panacea is clearly demonstrated by the economic disasters that dominate much of the European recycling industry. These reviewers also mentioned, incidentally, that in the USA, municipal solid waste increased by almost 60 percent between 1960 and 1990.

The cornucopian view does acknowledge acid precipitation and ozone depletion as problems, but assumes that these can be readily solved with cheap and plentiful nuclear energy. As recently as 1992 this debate was still in full swing. Leaders of the opposing sides debated the issue of scarcity versus abundance and published the discussion in a book reviewed by Stuart Pimm.[40] The views of Julian Simon, the economist, and Norman Myers, an Oxford biologist and environmentalist, are as far apart as ever. There are a few flaws in the cornucopian view that we must address.

The requirement for plentiful, cheap energy is one of these. Nuclear energy is indeed plentiful, particularly if we include nuclear breeder reactors and nuclear fusion in the mix. However, fusion is still far in the future. The trans-science problem of nuclear waste has not been solved satisfactorily, and that makes nuclear energy expensive. The technologies of solar and wind energy have made significant progress in recent years, but the actual application of these alternatives is still in its infancy.

Also, these authors usually do not address the basic physics of energy consumption in the sense that waste is an intrinsic part of energy conversion. Generating energy produces wastes like carbon dioxide or nuclear fuel bundles or worn-out windmills. Consuming the generated energy creates industrial processing wastes and waste consumer products. The cornucopia of cheap and abundant energy cannot exist if there is no cheap and easy way to manage the associated waste. The cornucopians have not offered solutions for these problems.

These snapshots of recycling and resource availability show some of the complexity of the waste problem. There are many difficulties with both recycling and disposing of waste. We will leave the domestic waste problem with a simple common-sense suggestion. On a per capita and per year basis, Europeans generate only half as much garbage as do US residents.[41] There is no reason to believe that Canadians behave any better in this respect than their southern neighbours. Clearly, in North America more attention should be given to the first of the three Rs.

Special embellishments

There are a few special situations that make life a little more challenging for the waste manager. Some of these deal with special kinds of waste, others deal with special challenges in terms of disposal.

One of the more complicated mixtures of materials is the automobile. The steel carcasses have been recycled by the steel industry for decades. A more recent problem is the mixture of weight saving plastics in modern vehicles, introduced as a result of legislation that demands better fuel consumption. Little progress has been made with the re-use of these materials, although several jurisdictions have instituted recycling taxes or levies on parts like tires. The Dutch have introduced a "recycling levy" on the sale of new cars. Supposedly these special levies will be used to dispose of the eventual waste in an environmentally appropriate way. Common political cynicism leads us to believe that these taxes are more likely to disappear into a big pot called general revenue.

A waste stream not unlike that derived from the automobile comes from the leisure time industry. Psychologists and other social scientists have written volumes on the "problems" of leisure time. In our context the problems of leisure time have to do with the technical consequences. For example, taking family pictures led to thousands of small operations called the "One-Hour Processing Lab." Without special precautions, these small photo processors can collectively contribute significantly to the toxic load for a city's sewage treatment plant.

The availability of travel time resulted in many small "houses on wheels." They used to be called trailers or caravans or motor homes, now they are recreational vehicles or RVs. Together with speed boats and snowmobiles, the RVs help us to burn millions of gallons of non-renewable, non-replaceable petroleum *just for fun.* There is also a large waste stream both from the manufacturing of these leisure time toys, and from their eventual disposal. This waste stream includes many potentially toxic materials such as coloured plastics, paints and varnishes and cadmium-plated metal parts. On the positive side, the bulky, high volume plastic waste from discarded automobiles and the RV industry are easier to handle for the recyclers.[42] This fast growing industry does cast a special light on the notion of "sustainable development."

Another special problem is formed by used tires. They are a heavy load on landfill facilities, and they present a serious fire hazard when insecurely stored. Notwithstanding a few disastrous fires in used-tire storage facilities, Canadian progress with tire disposal has been almost non-existent. Caught between political pressure "to do something" and the NIMBY syndrome, governments are paralyzed into procrastination. A few small facilities convert old tires into floor mats or pelletize the rubber for use in asphalt roadbeds. That raises questions about the re-usability of the asphalt, which is normally torn up from old roadbeds, re-pulverized and used again. However, of the twenty-four million tires discarded annually in Canada, less than 10 percent is recycled, about 6 percent is burned for energy recovery, and about 16 percent is exported to the USA.[43]

A promising approach to the management of industrial waste is the multiple waste exchange or the waste network. The concept is based on the fact that what is waste for one company can be an ingredient to another. This approach not only deals with environmentally responsible waste management, it also has significant economic advantages. A metal plating operation usually has a waste stream containing mineral acids used for "pickling" the objects to be plated. If this industry can make contact with another operation, for example leather tanning, that produces a strongly caustic effluent, the two wastes may be combined, and the resulting, neutralized mixture could be disposed of much cheaper, because it is no longer an environmental hazard. There are many other examples. The waste heat from electrical generating stations can be used for heating public buildings and private dwellings. Thousands of tons of fly ash from the combustion of coal are used in road construction and in cement manufacturing.

The key to these trades in waste and combinations of waste streams is a supporting information network. Many such networks are in operation in Europe. The Ontario Waste Exchange, operating under the auspices of ORTECH International, is a North American example. The primary driving force for this form of waste management is environmentally induced economics. When legislation makes it increasingly expensive to dispose of industrial waste, innovative waste management has a direct effect on profitability.

No waste management story would be complete without the saga of the coffee cups. Since North Americans are major consumers of coffee, the addictive "drug" for which rain forests are sacrificed, the coffee cup is an important tool on this continent. Much of that coffee is slurped "on the run," on the way to work, or from work, or in the car. More often than not the coffee cup is made to be thrown away. Other products of the fast food industry operate on similar principles so the coffee cup scenario also applies to the packaging of other fast food.

Disposable coffee cups are usually made of paper or of foamed polystyrene. The latter has the advantage of good insulating properties to keep warm food warm and cold food cold. The common reasoning goes like this: for foaming the plastic we need CFCs, they damage the earth's ozone layer, and the plastic cups are not biodegradable so they "stay around forever." The paper coffee cups are biodegradable and therefore they are intrinsically good. Sometime in 1990 somebody questioned this reasoning.[44] A year later somebody else did a more detailed analysis and came up with some interesting numbers. The author did not offer solutions, but the data were nevertheless surprising.[45]

The comparison is based on two observations. One is that few manufacturers ever used CFCs for making food containers, and those who did phased them out soon after the Montreal Protocol of 1988. Several other blowing agents are presently in use. Pentane, a flammable hydrocarbon, is often used because it can easily be recovered and re-used repeatedly. A second point is that neither kind of coffee cup is *biodegradable*. That is an advantage for the landfill, it is a disadvantage when the coffee cup gets thrown at the side of the road. For the incinerator it makes no difference. Then follow the numbers.

A paper cup, impregnated with paraffin wax, weighs about 10 grams, a styrofoam cup weighs 1.5 grams, therefore disposing of paper coffee cups adds six times more weight to the landfill or incinerator than the plastic cup. Making paper coffee cups consumes about 25 percent *more* petroleum products, such as diesel fuel, than the styrofoam cup. The paper cup needs about forty times more energy and costs twelve times more steam. Making a ton of paper produces between 50 and 190 m^3 of waste water; a ton of polystyrene, which yields six times as many cups, gives only a few cubic meters of effluent. Clearly, the choice between paper and styrofoam is not nearly as straightforward as it appears. Thus we have to wonder about one recycling effort, which involves washing the cups and pelletizing and melting the polystyrene "for future use."[46] With this approach, once we account for transport, water, effluent, and energy, we could well end up on the negative side of the environmental ledger.

It is somewhat unsatisfactory to end this chapter without pointing to answers and solutions. Waste management is probably one of the most challenging problems of our industrialized society. On an environmental list of priorities it could share a place with the world's water supply problems. On the positive side, while waste can never be eliminated, our brief exploration of the problem showed that there are many opportunities for making the problem, literally, much smaller.

CHAPTER 12

Organochlorines – killer chemicals?

> There is something nonbiological about halogenated organics (excluding iodinated compounds).... Chemicals that do not occur naturally ... are often persistent, since there are often no natural biological processes to metabolize or deactivate them.

This was written by the Science Advisory Board of the International Joint Commission on the Great Lakes in their report in 1989.[1] This joint American–Canadian organization was formed to manage the waters in the Great Lakes Basin. Presumably on the basis of this advice the Commission recommended the complete phase-out of chlorine and chlorine containing chemicals as industrial feedstock.[2]

The element chlorine is involved in about two-thirds of all chemistry practiced in laboratories and chemical plants. Directly or indirectly, chlorine plays a role in the manufacture of some fifteen thousand chemicals. Chlorine chemistry represents an enormous part of the economies of industrialized countries. The complete phase-out of chlorine would create huge disruptions of these economies. Apart from the need to find substitutes for tens of thousands of chemicals, there is another side to the chlorine problem: the manufacture of sodium hydroxide, also known as caustic soda. Sodium hydroxide is, literally, the other half of the same process that produces chlorine from salt. To stop manufacturing chlorine means to stop the supply of sodium hydroxide. That will paralyze another major part of the industrialized economies. Oil refineries, the food industry, and many other industries depend on the availability of sodium hydroxide.

Chlorine and chlorine containing chemicals, including PCBs and dioxins and furans, are also very much an integral part of nature. There are well-defined processes for metabolizing them or deactivating them, and the biochemical processes for their formation have been studied and are well understood. These controversial chemicals deserve a special place in our discussions. We will find that the Science Advisory Board to the International Joint Commission was badly in error.

Chlorine

Chlorine is one of the fundamental substances we called elements. With fluorine, bromine, and iodine, it belongs to a small sub-family of elements, the halogens. This name comes from the Greek and means salt formers. The halogens themselves, as free elements, are very different from each other. Fluorine and chlorine are gases, bromine is a liquid and iodine is a crystalline solid. Yet their chemical behaviour is very similar. These four elements, particularly chlorine, are part of the molecular make-up of many thousands of chemicals, organic as well as inorganic, naturally occurring as well as synthetically made.

Some of these halogen containing chemicals are vitally important to our physical well-being, and indeed the physical well-being of plants and animals. Thyroxine, a hormone made in the thyroid gland, is an organo-iodine. Valium, the well-known tranquilizer, is an organochlorine. None the less, in the glossary in an ecology textbook we read:

> Chlorinated hydrocarbons: Synthetic organic poisons containing chlorine that are often used as pesticides. They tend to be persistent and to move around in the environment; they often accumulate in fatty tissues of organisms that feed high on food chains. DDT is the best known example; PCBs are also chlorinated hydrocarbons.[3]

These descriptions have attracted much attention during the past decade or so. At the same time, as this example shows, beneficial effects and the natural occurrence of these chemicals tends to be ignored. Organochlorines have become a subject of major controversy and the beneficial properties of many organohalogen compounds have become overshadowed by the hazardous properties of a few. Debates on chlorine have taken on the appearance of a war. The chlorine industry, the associated thousands of jobs, and the economies of cities and countries are on one side. The environmental concerns and personal fears of thousands of people are on the other.

This debate is hindered by the size of the territory it must cover. There are so many chemicals and processes involved that it is easy to lose track. The problem is further aggravated by the extraordinary publicity that a small group of these organochlorines has attracted. Positive news about halogen chemicals has had a very small share of the media's attention and the occurrence of organochlorines in nature is one of the better kept secrets. A bird's eye view of the territory is therefore our starting point.

The halogens

Halogens occur in nature nearly always in combination with other elements, that is, as molecules. The most abundant are the combinations of the halogens with different metals in the form of inorganic salts. Of these salts the most abundant and best known is sodium chloride, table salt or rock salt. Halide salts occur abundantly in the crust of the planet and in the oceans.

Solutions of these salts in water conduct an electrical current and therefore they are also called electrolytes. Salts are important in electrical systems like batteries, electrical cells, and other electrical conductors. They also conduct electrical signals in a living organism. The concentration of electrolytes in body fluids is literally of vital importance. Chlorine in its mineral or inorganic forms, such as sodium chloride or potassium chloride, is generally not very toxic to most living organisms, provided the amounts remain within certain limits. In small quantities, mineral salts are essential nutrients for virtually every form of life.

When a salt is melted in the absence of water, the hot, molten mass also conducts electricity and the electrical current causes the salt to change into the elements that formed it originally. The process is called *electrolysis*, a lysis or lesion or breaking of a molecule by electrical energy. Electrolysis of sodium chloride gives chlorine gas and sodium metal. Sodium is an alkaline metal. It interacts violently with water to form strongly alkaline solutions of sodium hydroxide. Factories for producing chlorine and sodium by electrolysis are therefore called chloralkali plants. Often the electrodes for the process contained mercury. Sludges from the process, improperly disposed of and containing appreciable amounts of mercury, have caused major pollution problems in many areas.

The electrolysis of iodine and bromine containing salts gives us access to iodine and bromine. Sodium hydroxide and the free halogens are very reactive. They readily combine with other chemicals. Occasionally, like in volcanic explosions, free chlorine gas can form without the help of electricity because of extreme conditions of temperature and pressure. In the laboratory or in industry these elements are occasionally used in their free (elemental) form. In this form the halogens are very toxic and are corrosive to animal tissue. They deserve considerable respect. For example, fluorine can etch glass. Chlorine can bleach various materials such as paper, flour, and laundry. Chlorine gas is also used for destroying micro-organisms in swimming pools and in drinking water. Particularly the bleaching of wood pulp for the manufacture of paper is a major source of controversy.

Chlorine does not only the bleach the paper and make it bright white, it also helps to break down the wood structure. It helps to remove the lignin, a component of the wood that holds the cellulose fibres together. In the process, many components of the wood interact with the chlorine to form a variety of other organic chemicals that are then called organochlorines. We will look at these chemicals in more detail below.

The inorganic chemistry of the halogens is quite extensive, but we do not need it for our further discussions. The *organic* chemistry of the halogens is vital to an industrialized society. It is also one of the biggest controversies in the environmental debate. The free halogens are frequently used for combining them with organic chemicals. As before, combining has a chemical meaning, not just a mixing of chemicals, and organic is used as distinct from inorganic. These chemical interactions produce different molecules in which one or more chlorine atoms are attached to carbon atoms. These reaction products are collectively known under several similar sounding names: organochlorines or organochlorides or chlorinated organics. If the other halogens are involved, we also use the terms organohalides or organohalogens.

Organochlorines – killer chemicals?

Organochlorines

A molecule that contains only carbon and hydrogen is a hydrocarbon. If one or more hydrogen atoms are replaced by other atoms the molecule is no longer a hydrocarbon. A molecule containing chlorine is not a hydrocarbon, and *chlorinated hydrocarbon* is not a correct chemical term although it has found general acceptance in the environmental language. Organochlorines, organochlorides or, more generally, organohalides, make for better language.

Since the organochlorine family contains so many chemicals, we need to bring some order in the topic. Many different classifications are of course possible. Since we will concentrate on environmental issues we will choose the criteria accordingly and recognize three broad groups:

1. Industrially *made* organochlorines (synthesized).
2. Accidentally *formed* organochlorines (byproducts of processes).
3. Naturally *occurring* organochlorines (biosynthesis and other processes).

This classification will become easier to handle as we continue and add a few subclassifications and some familiar names. It also applies to organobromines, iodines and fluorines. The number of known organohalides is very large.

These chemicals have many different applications. We find organochlorines in the pharmaceutical and cosmetic industries, agricultural industry, plastics and surface coatings to name but a few. To get some sense of numbers we can look at a few standard reference books, found in chemical laboratories and in libraries.

The catalogue of a leading chemical supplier, Aldrich Chemicals of Milwaukee, USA, lists about 17,000 chemicals, both organic and inorganic. About fifteen hundred of these, about 9 percent, are organochlorines. Adding also the bromine, iodine, and fluorine compounds brings the total for organohalides to over 10 percent of the entire catalogue. These chemicals are for sale, by the pound or per gram, for use in chemical plants and laboratories. A chemical supplier's catalogue is not the most exciting reading but it does give an idea of the numbers.

More interesting is *The Merck Index*, called by its publisher "an encyclopedia of chemicals, drugs and biologicals."[4] It has been published since 1889 by Merck & Co., another leading chemical supply company. The *Index* places more emphasis on pharmaceuticals and gives details on the uses of the listed chemicals and their source or origin. For each entry it refers to the literature reports of who synthesized the chemical for the first time, or who identified it in a plant or animal.

The *Index* contains 10,000 entries. Most are organic and about fifteen hundred contain fluorine, chlorine, bromine, or iodine. Many organohalides listed came from the laboratory or the chemical industry. Many others were found in plants or in micro-organisms. More than 70 percent of all the organohalogens listed are pharmaceuticals. These include anabolic steroids, anticonvulsants, antidepressants, antihistamines, antihypertensives, beta-blockers, diuretics, tranquilizers, vasodilators, and many more. Both librium and valium are organochlorines. Most of the fifteen hundred are organochlorines, but we also find organo-iodines, used as X-ray contrast media, and organobromines and fluorines. We also find, of course, well-known names like dieldrin, 2.4-D, and DDT.

Since these chemicals were *designed to interfere* with the chemistry of living cells, virtually all of them can be toxic, depending on the dose, of course. A brief look back to our discussion on the relationship between dose and effect, acute and chronic toxicity, detoxification, and no-effect levels may be useful at this point. *The Merck Index* gives limited information on acute toxicity (LD_{50}). Reference books more specifically aimed at human health give more detail on the toxicity of organohalides. A very comprehensive reference book is Irvin Sax's *Dangerous Properties of Industrial Materials.*[5] Hodgson's *Dictionary of Toxicology* contains more detail on the physiological effects, but is not as comprehensive.[6] The perspective in Sax is on occupational health, Hodgson contains listings of both industrial and medicinal chemicals. These reference books are indispensable to specialists, for our present discussion they can be useful to answer an occasional question.

Some organochlorines are found in agriculture mostly for use as pesticides. A pesticide is an agent used to eradicate pests. Organisms, insects, fungi, or plants that compete with humans, mostly with respect to food, have come to be known as pests. Some pests eat our food, others eat us, or part of us, like mosquitos and biting flies. This terminology could be fertile ground for ethical discussions, at this point we will just use the commonly accepted definitions. Thus, a chemical used to control insects is an insecticide, for a fungus we use a fungicide, and for weeds we use a herbicide. A difficult problem with pesticides is the persistence of several factoids about them.

For example, the *Basic Guide to Pesticides* contains some classifications that confuse rather than explain.[7] This book labels the insecticide Aramite as "organic," the herbicide Benefin as "a dinitroaniline", and the common antiseptic camphor is classified as "biological." In reality, these three pesticides are all organic chemicals. Although camphor was originally found as a plant product, over 75 percent of camphor now sold is synthetic. The classifications in this book are a little like classifying a Golden Delicious as "an apple" and a Red Delicious as "a fruit."

Another factoid has it that most pesticides are organochlorines. However, in North America and most of Europe the majority of chlorinated pesticides have been phased out. Very few, such as the insecticides methoxychlor and dicofol and phenoxy herbicides like 2.4-D are still in use.[8] Almost all pesticides today are either organophosphates or carbamates. These are chemicals with a much higher acute toxicity, but which break down into much less toxic degradation products very quickly.

Another factoid about pesticides relates to the question of persistence. Pesticides are generally perceived to stay around in the environment almost forever. However, studies showing the ready biodegradation of many organochlorines had started well before the writing of *Silent Spring.*[9,10] Most organohalides *are not persistent* and are readily broken down, either by photodegradation or biodegradation.[11,12] The persistence problem applies to a small number of organochlorine chemicals, the most persistent ones are no longer used, and eventually even they will also degrade.[13] Nevertheless, we do have a legacy of these earlier years, and some persistent organochlorines are an inevitable part of that legacy. Many documents still refer to these residues because environmental monitoring must still continue.

In addition to the chemicals used to control insects and unwanted vegetation, many organochlorine pharmaceuticals are intended for veterinary use. When normal agricultural practices are followed, feedstock animals metabolize these chemicals just like humans metabolize their librium. Residues in the meat of healthy animals do not come from bioaccumulation, but they can result from poor agricultural practices.[14]

A third important group of synthetic organohalides is found in the plastics industry. This group does not contain a large number of different chemicals, but the quantities involved are enormous. Obvious examples in this category are teflon (a fluorine containing polymer), saran (cling wrap) and polyvinyl chloride, vinyl for short. Vinyl consuming industries make upholstery, building materials, automobiles and many consumer products. In Canada the amount of polyvinyl chloride used in 1993 was 271,000 tons, up from 250,000 tons in 1988 and expected to rise to 300,000 tons by the year 2000.[15] A proportional estimate for the much larger markets in the USA and Europe would yield the staggering numbers that would apply to the task of finding chlorine-free alternatives.

There are several lesser known but very important halogen containing polymers that find application in many consumer products. Neoprene, a polymer of chloroprene, is a synthetic "rubber" that is resistant to gasoline and many petroleum-based solvents. Neoprene tubing is essential for the automotive industry. Grease and oil resistant soles of safety boots often consist of neoprene or one of its close relatives. Fluorothene is a fluorine containing polymer used in wire and cable insulation and in surface coatings and films. Although the variety of chemicals is not as large in halogen containing plastics as it is in pharmaceuticals, the volume of these industries is very large.

A small but important group of industrial chemicals that does not readily fit in a general classification are the polychlorinated biphenyls or PCBs. These chemicals form a sub-family of closely related compounds. The name means that a molecule called biphenyl may be altered to contain many (poly) chlorine atoms. It is possible to synthesize a biphenyl that contains only one chlorine atom, or two or three, up to a limit of ten. When there are two chlorine atoms, they can be in several different places with respect to each other and *within the same molecule*. This means that there can be several different *dichloro*biphenyls and quite a few *trichloro*biphenyls. Chemists call these close relatives *congeners*. In the PCB family there are a little over two hundred of them. Many of these have been synthesized in pure form in research laboratories to study their chemical behavior and toxicology.

PCBs were made on a large scale because they have properties that were considered highly desirable. They are very good electrical insulators, and they are not readily broken down by chemical or biological processes. That makes them ideal as cooling fluids in large electrical transformers. PCBs were also used in the ballasts of fluorescent light fixtures and in insulating material for electrical cables. Other industrial applications included hydraulic fluids, fire retardants, and, mixed with plastics, in sealants, surface coatings, printing inks, and carbonless copy paper.

PCBs are not single, pure chemicals. The industrial process results in mixtures of several dozen congeners, although a few usually predominate in a given mixture. Mixtures of differing composition have different properties such as viscosity. They vary from sticky and syrupy to thin and free-flowing. Transformer oils containing PCBs had to be sufficiently mobile so they could move through a system of tubes. Sticky, tar-like oils in capacitors or fluorescent light ballasts were formulated so they would not leak from the equipment. PCBs could be mixed with other chemicals to produce different formulations for different purposes.

In North America the leading producer of PCBs was the Monsanto Chemical Company which used the trade name Aroclor. A four-digit number after the name indicates the approximate composition, and thus the average chlorine content, of the mixture. Aroclor 1221 contains about 21 percent chlorine and Aroclor 1260 about 60 percent. This in turn would tell the user about the product's electrical behaviour or its viscosity. Bayer in Germany used the trade name Clophen and Kanegafuchi in Japan called its product Kanechlor.

The very properties that made PCBs suitable for these applications, their resistance to degradation and their electrical properties, were the cause of trouble. PCBs are mildly toxic to humans, and their presence in the environment can cause problems. Manufacture and use of transformers, disposal of old equipment, improper handling of equipment and of PCB containing fluids, all led to the dispersion of these chemicals into the environment. The contamination of soil, micro-organisms, and animals with PCBs has led to a variety of concerns.

The manufacture of PCBs is now banned in most countries, but these materials remain in use in existing, operational electrical equipment. There are also large quantities in storage from electrical equipment that has been taken out of service. Both PCB materials in storage and in equipment that is still operating are controlled by a tight web of legislation in most jurisdictions. Chlorine-free alternatives have been developed and PCBs are now gradually being replaced.

It would be interesting to speculate how industrialized countries would have developed without the highly efficient distribution system for electrical energy that we have today. This infrastructure, which depended on the use of PCBs for many years, is only one part of the heritage. The other part is the presence of these chemicals in the environment. The problems that result, although sometimes serious, are not quite as catastrophic as they are often made out to be.

It has been suggested that chlorine as a stock chemical should be banned completely or, as second best, the industrial use of this element should be drastically curtailed. Although fluorine, bromine, and iodine are not normally mentioned in such suggestions, toxicological information suggests that organic chemicals containing these other halogens are also toxic. As expected, almost all of these chemicals can cause cancer in laboratory animals.[16] Presumably this would mean that a ban on chlorine would have to be extended to the other halogens as well. This important and logical extension is usually omitted from discussions on the organochlorine ban.

Nature's own

There is a general perception that organohalogens are strictly anthropogenic, that they are exclusively made, or caused to be formed, by humans. This may have its origin in Rachel Carson's reference to "the synthetic creations of man's inventive mind... having no counterpart in nature," which we encountered earlier.[17] We can only speculate why

Carson, a trained biologist, ignored the existing literature on a special group of organic chemicals containing halogen atoms.

These are the organohalides that occur in nature. They were not spilled or accidentally formed by industry, but they were *biosynthesized*, by plants and micro-organisms. Truly natural, organically grown, organochlorines and bromines and iodines, and even fluorines are common in nature. Particularly among fungi and lichens and marine organisms there are many species that can readily incorporate chlorine or other halogens into complex organic molecules. These organisms obtain their chlorine supply from chloride salts, the chlorine containing minerals, in the soil and in the oceans. Examples are readily found in the chemical literature. Many of them were well known before 1962.

The family of the *Aspergillus* molds is probably best known for one member, *Aspergillus flavus*, which produces the dreaded aflatoxins. The aflatoxins are not organochlorines, but other *Aspergillus* species biosynthesize several different organochlorines. As early as 1936 two British chemists found that *Aspergillus terreus* produces two chemicals, called geodin and erdin, which are organochlorines.[18] The fungus *Drosophila subatrata* makes a phenol with four chlorine atoms in its molecule.[19] This naturally occurring mold product is such a close relative of the infamous synthetic pentachlorophenol that the two could almost be twins. A literature survey of mold metabolites by Turner, now twenty years old, contains information on some six hundred mold products known at the time.[20] Many of these, such as the tetracyclines, are important antibiotics. Several dozen organochlorines are included in this survey and many of these were well known before 1962. Fungi are not the only makers of organohalides. In the 1940s sheep farmers in the Karoo desert, in the northern part of South Africa's Cape Province, had a poisoning problem among their sheep. After grazing in areas where one particular plant was abundant, the animals would get ill and, literally, drop dead. The plant was appropriately named "gifblaar" which, translated from Afrikaans, means "poison leaf." Botanists call it *Dichapetalum cymosum*. The leaves of *D. cymosum* were found to contain a relative of acetic acid that contains fluorine.[21] Mono-fluoroacetic acid is truly a deadly toxic chemical.

In a more recent literature study by K. Engvilt we find many higher plants that also biosynthesize organochlorines.[22] Some of these are common vegetables or weeds. Plants like safflower, *Carthamus tinctorius*, the common bracken fern, *Pteridium aquillinum*, and artichoke, *Cynara scolymus*, biosynthesize a variety of organochlorines. A perspective on the numbers of natural organochlorines follows from a recent review of the literature.[23] This collection of recent discoveries particularly underlines the abundance of these chemicals in marine organisms. The estimated number of naturally occurring chemicals that contain chlorine, iodine, bromine, or fluorine in their molecular make-up is now close to two thousand. Since a relatively small proportion of organisms has been investigated to date, many more of these chemicals are waiting to be discovered.

Always depending on the dose, some halogen containing natural products have medicinal properties, others are toxic to both humans and animals. When cattle graze in fields that are heavily infested with the fungus *Sporidesmium bakeri*, they develop a facial eczema. The problem is caused by several chlorine containing fungus products called the sporidesmins.[24] In addition to the effects of monofluoroacetic acid and the sporidesmins the toxicity of the common Bracken fern is well known.

The detailed searches for naturally occurring organohalogens sometimes yield unexpected results. The vast quantities of simple halogenated hydrocarbons produced by marine algae, and the chlorofluorocarbons formed in volcanos, must be included in further forecasting of stratospheric ozone depletion. The PCBs, specifically synthesized for industrial use, were thought to qualify eminently for the epithet "no counterparts in nature." It is an illustration of the universality of chemical interactions to find that several pentachlorobiphenyls have been found in the ash from the 1980 eruption of Mt. St. Helens in Washington State.[25] Careful analytical chemistry showed that these PCBs did not come from environmental pollution. A probable explanation for their formation was the incomplete combustion of plant material in the presence of the chlorine gas emitted by the volcano.

Organohalides in higher animals are not very common, although an obvious example is the biosynthesis of thyroxine, the iodine containing thyroid hormone. In summary, the variety of organochlorines made by marine organisms, micro-organisms, plants, animals and in natural processes like volcanic eruptions, adequately dispels the notion that there is "something nonbiological" about them.

The unwanted

So far our classification has dealt with several groups of organohalides that are manufactured and another group that occurs in nature. There may be differences of opinion on whether all the manufactured chemicals are indeed useful, but that debate is beyond the scope of our discussions. Even without a consensus on usefulness, the fact remains that these chemicals were made for practical reasons. They were wanted, they were intentionally synthesized, and they were used.

The chlorinated organic pesticides were made *for the purpose* of spraying them on the land and even on people. PCBs were also intentionally made, but not for release into the environment. PCBs were extremely useful chemicals that were not easy to replace. There have been transformer fires because of the higher flammability of some PCB replacements.[26] PCBs became an environmental problem because of handling procedures that led to *unintentional release* into the environment such as spills and improper disposal. Incidentally, waste transformer oils and hydraulic fluids were used as a dust suppressant spray on many unpaved secondary roads in Canada and the USA. The transport of soil particles by spring run-off is one of the mechanisms for PCB contaminated material to get into lakes and rivers.

Intentional introduction of organochlorines into the environment and accidental spills are processes that can be prevented, at least in theory. This is not as easy as it sounds, but at least DDT and PCBs are on their way out. When we continue our classification scheme, we find some situations where the solution is not so straightforward.

The "chlorine free" that isn't

There is one group in our classification that is usually ignored, both in the environmental literature and in the chlorine debate, because it is invisible. This group is of special importance just *because* it is so often ignored. It is the chlorine containing intermediates and precursors from which other, non-chlorine containing products are made. Because chlorine is so reactive it is often used in chemical processes in many different ways *without becoming part of the final product.* A typical example is ABS plastic, the black plastic pipe and fittings used by plumbers. The acronym ABS means *acrylonitrile butadiene styrene* resin. Its manufacture starts with a stock chemical called ethylene chlorohydrin. This, in turn, results from the action of chlorine and water on ethylene. So, although ABS plastic does not contain chlorine within its own molecular make-up, and is not an organochlorine, chlorine gas and an organochlorine are essential in its manufacture. The manufacture of ABS plastic inevitably produces a waste stream that contains organochlorines.

There are countless other examples where chlorine or bromine is needed in a process for a consumer product, a building material, or a drug. It is highly probable that substitutions will be found in *some* of these processes. A recent report summarized the considerable progress in this area.[27] This report does not only address the question of the use of organohalides, but deals in general with the use of more environmentally friendly methods in chemical processes. Alternatives for heavy metal catalysts are other examples of this approach. However, these developments move slowly, and a simplistic assumption that chlorine can be completely banned from industry "because there are alternatives" is premature.

We looked earlier at the issue of chemical waste or, as it is often called, hazardous waste. After the synthesis of a chemical is complete, the next step is a *purification* process to obtain a homogeneous and chemically pure product. The removed impurities become part of the waste stream. Waste from the chlorine using chemical industry unavoidably contains organochlorines. Sometimes these can be re-used in the process. Improving re-use makes sense environmentally and economically.

Industrial waste that contains organochlorines is not any different from other types of waste. There are waste streams that do not contain chlorine compounds that are just as hazardous, for example wastes containing heavy metals like chromium or lead. Legislation and good corporate citizenship usually ensure the proper management of these wastes to prevent environmental problems. Usually, but not always. However, it is not always improper management that causes problems.

One group of these manufacturing impurities, specifically related to the organochlorines, has a particularly unpleasant reputation. This is the family of the chlorinated dibenzodioxins and the closely related chlorinated dibenzofurans. The names are often shortened to dioxins and furans. The very names strike fear in the heart of plant managers and environmentalists alike.

Chlorinated dioxins and furans

Few other groups of chemicals have been as intensely studied as the dioxin and furan families. Any on-line database will return many entries on the key word "dioxin." The names dioxin and furan are misleading simplifications which refer to two closely related *groups* of chemicals, the polychlorinated dibenzodioxins and the polychlorinated dibenzofurans. An important characteristic of these chemicals is that they are not manufactured for any specific purpose. They are coincidentally formed and become merely undesired impurities.

Many detailed reviews are available for readers who desire more information. A book, written for the "educated lay person" specifically reflects the perspective of the American experience in Vietnam, where dioxins were involved as impurities in the defoliants used for military purposes.[28] Other technical details of the dioxins and furans are in reports prepared for the US Environmental Protection Agency and the Ontario Ministry of the Environment.[29,30] These reviews contain in-depth discussions of the origins and sources, toxicities, and analytical problems of these chemicals. These details are useful for those with the relevant background, they are not essential for our discussion.

As is the case with the PCBs, two groups of *polychlorinated* chemicals are involved. This time the maximum possible number of chlorine atoms in both groups is eight. If we have between one and eight chlorine atoms per molecule, there are seventy-five possible congeners for the polychlorinated dibenzodioxins (PCDDs) and 135 combinations are possible for the polychlorinated dibenzofurans (PCDFs). Thus the simple designation "dioxins and furans" refers to two closely related groups containing 210 *different* chemicals. The individual chemicals are subtly different from each other, and these differences can easily be proven in a chemical sense. For our discussion the pronounced differences in toxicity are more important.

PCDDs and PCDFs are undesired byproducts in many manufacturing processes. They can be formed in many different ways. Chlorinated dioxins can be formed from the pentachlorophenol in treated wood under the influence of sunlight.[31] They frequently occur as the result of the combustion of many organic chemicals, including wood. By the nature of their formation, the dioxins and furans occur in such small amounts that it is very difficult, frequently impossible, to remove them by common purification methods. It is very difficult, but possible, to remove trace amounts of these impurities from a chemical such as trichlorophenol, in a laboratory. It is impossible to remove PCDDs and PCDFs from the smoke of a combustion process.

The combustion example is important because it is now generally believed that combustion is a principal source of chlorinated dioxins and furans. Inorganic chloride salts, which are ubiquitous on the entire planet, are a sufficient source of chlorine for the formation of the PCDDs and PCDFs. Therefore virtually every combustion process is a potential source of these chemicals. They have been found in cigarette smoke and car exhaust, and many studies have confirmed that PCDDs and PCDFs are routinely found in the fly ash of municipal incinerators and wood fires. A few examples of these studies will serve to set the scene for the dioxin problem.

Czuczwa and his co-workers at Indiana University reported in *Science* on their analysis of the bottom sediment of Siskiwit Lake, on Isle Royale in Lake Superior.[32] The island is part of the state of Minnesota and Siskiwit Lake is some 17 metres *above* Lake

Superior. Only atmospheric input explains the presence of PCDDs and PCDFs in amounts similar to those found in the air of several big cities.

Siskiwit Lake is far away from industrial combustion sources. The nearest city, Thunder Bay in N.W. Ontario, does not operate a municipal solid waste incinerator, but prevailing West winds do bring the smoke of the big forest fires of Northern Minnesota and North-Western Ontario directly over Siskiwit Lake. Nevertheless, the authors of this study categorically reject the idea of forest fires contributing to atmospheric fallout of PCDDs and PCDFs because of the so-called 1940s horizon. Earlier records showed that PCDDs and PCDFs were virtually undetectable at sediment depths that correspond approximately to the year 1940. Some researchers took this as evidence that these environmental pollutants are strictly of anthropogenic origin.

More recent reports have removed this 1940 horizon. A sealed glass ampoule, containing a fifty-year-old sample of sewage sludge-based fertilizer was opened and the contents analyzed.[33] The fertilizer was found to contain PCDDs and PCDFs in amounts and proportions very similar to much more recent material. The sample had been on display at the Philadelphia World Fair in 1933. Chlorophenol-based pesticides only appeared by 1944. The authors suggested that naturally occurring phenols, chlorinated during drinking water purification, may have been the source of the contamination in this old sample. We now also know that chlorinated dioxins were around as early as 1860 in lake sediments in New York State.[34] Going further back in history, chlorinated dioxins have been found in eight-thousand-year-old marine sediments from Japanese coastal waters.[35]

This removal of the 1940s horizon reopens the question of forest fires as a source of PCDDs and PCDFs. The chemical architecture of some naturally occurring organochlorines, drosophilin A for example, perfectly fits the conditions considered as favourable for the formation of PCDDs and PCDFs.[36,37] Another report has also suggested that forest fires do indeed contribute to the environmental dioxin load.[38] Basic chemistry, the ubiquitous presence of natural organochlorines, and the physical conditions of a fire predict such a contribution. There is considerable circumstantial evidence for such a hypothesis.

Although analyzing the smoke of forest fires is obviously not an attractive activity, it has been shown that uncontaminated wood, burned in domestic fireplaces, does lead to the formation of chlorinated dioxins.[39,40] Further reports on the formation of dioxins and furans in the combustion of virgin wood were cited in recent reviews.[41,42] The quantities of materials involved in the smoke of forest fires is truly enormous. Relevant statistics indicate that in Canada alone the average annual area burned is about two million hectare.[43] Detailed analysis of this natural form of air pollution is long overdue.

The formation of dioxins and furans in many combustion processes, including forest fires, is reflected in another recent observation. These chemicals have been found to be so common in the environment that they can be found in many everyday materials like vacuum cleaner dust and clothes dryer lint.[44] The amounts of dioxins found in these materials are often ten or a hundred times higher than those found in paper products. The molecular make-up of these dioxin/furan mixtures suggests that they come from a common source. Their almost universal presence in fine, airborne dust suggests that smoke from combustion processes could well be that common source.

We have already looked at the natural occurrence of organochlorine and organo-

Figure 12.1 Where there is fire, there is smoke – but what's in the smoke?

bromine chemicals other than dioxins and furans. The abundance of these chemicals is important for two reasons. First, they affect the measurement, the analysis, of organochlorines in natural systems. We will discuss these analysis methods briefly below. Second, among these chemicals there are many that have a molecular architecture that makes them suitable as precursors for the formation of dioxins and furans. In other words, on the basis of the theories of chemical structure and of biosynthesis it would be reasonable to *predict* that dioxins and furans could be formed by natural processes other than fires or sunlight. Such a prediction means that dioxins and furans are not nearly as "alien to nature" as we have been led to believe.

Chlorinated dioxins and furans are generally considered to be truly and exclusively the result of human industrial activity. The suggestion that these chemicals can be formed by biological processes is rejected by most ecologists. We must therefore examine some recent scientific literature on this point. A key observation was the conversion of trichlorophenol into several different members of the dioxin and furan family with a *biological catalyst*, a peroxidase enzyme obtained from horseradish.[45] This is a very important discovery for three reasons. It represents a truly *biological process* for the formation of "the most toxic chemical ever made." The peroxidase enzymes are part of the biochemistry of many plants and animals, and naturally occurring chlorophenols are abundant in plants and marine organisms. For example, 2.4-dichlorophenol, the starting material for the manufacture of the herbicide 2.4-D, is also biosynthesized by a *Penicillium* species.[46]

Another significant observation resulted from a study of the composting of garden waste and the formation of sewage sludge. When otherwise unpolluted garden waste, containing traces of chlorinated dioxins, is composted, new dioxins with a higher chlorine content are formed.[47] This means that the proportions in a given mixture of congeners can change under the influence of biological, enzyme-catalyzed processes. This study also showed that the bacterial action in compost can convert naturally occurring and synthetic chlorophenols into chlorinated dioxins. The dioxins in the 1933 Philadelphia fertilizer sample could well have been formed by natural bacterial action.

Probably the most surprising recent discovery has been the natural formation of several brominated dioxins in the marine sponge *Tedania ignis*.[48] It is clearly only a matter of time until also the chlorinated and iodinated analogues of these chemicals are found in natural sources. In most of these recent literature reports on naturally occurring organochlorines we find that the researchers took extraordinary caution to ensure that they were not measuring artifacts caused by modern pollution.

A recent study, sponsored by the US Environmental Protection Agency, included detailed modeling calculations on the total environmental input (in the USA) of dioxins and furans. That input was estimated at about 80 kg/yr. A somewhat surprising result was the finding that only about 11 percent of that input could be accounted for by five major sources. Motor vehicles contribute approximately 5 percent, municipal solid waste incinerators about 4 percent, hospital incinerators less than 1 percent, residential wood burning about 1 percent, and paper mills about 1 percent.[49] The authors suggested that there is a large, unknown source of these contaminants, and once more forest fires were among the possibilities mentioned. In a recent US EPA report that stated that most dioxins come from incineration, it was also acknowledged that "there may also be unknown sources."[50]

Considering the size of the chlorine using industry, there is little doubt that many specific PCDD and PCDF problems are due to human activities. Nevertheless, some generalizations are clearly not justified. For example, the pesticide guide mentioned earlier states that dioxins and furans "occur in pesticides as contaminants," creating the impression that this *always* applies to *all pesticides*, while many pesticides have no association with chlorine at all.[51] With so many possible pathways for the formation of the dioxins and furans we must come to terms with the fact that these chemicals are a part of the environment that is unavoidable. A look at their behavior and their possible health effects is clearly our next step.

Organochlorine behaviour

We have seen that the behaviour of a chemical in the environment as well as its effects on a living organism depend directly on the physical characteristics of the chemical, such as its solubility in water or in oil. With respect to these characteristics we can make several useful observations among the organochlorines in general.

In water and soil and air

In terms of their chemistry and physical behaviour, the organochlorines are not much different from other chemicals. Some are very volatile, others are not. Some are very soluble in water, others dissolve much more readily in oil. Some are very toxic, others much less so. Therefore it is wrong to generalize. It is essential to make clear distinctions between the different groups of organochlorines.

Engine oil on the driveway or parking lot does not wash away when it rains. The oil is not only insoluble in water, it is also strongly attracted to the asphalt, to which it is chemically related. Water solubility and solubility in oil have an inverse relationship. A chemical that is very insoluble in water is usually very soluble in oils or fats and vice versa. Also, chemicals with a low water solubility are easily adsorbed on the surface of small particles of organic material. Such particles we find suspended in natural waters. They also make up the sediments on the bottoms of lakes and rivers. Water solubility, fat solubility, and adsorption on solid particles can be readily explained in terms of molecular architecture. This explanation is in any organic chemistry textbook. The details are not important for our discussion, as long as we are aware that many organochlorines behave this way.

We can use a simple experiment to shed some light on the environmental meaning of water solubility numbers as we find them, for example, in the *Merck Index*.[52] When we shake equal amounts of water and chloroform in a test tube, the two layers readily separate again and the first impression is that chloroform is "insoluble" in water. Yet we find that one half mL of chloroform will dissolve in one hundred mL of water, giving a concentration of one half of 1 percent. In environmental language we usually deal with much smaller concentrations and use parts per million (ppm) or parts per billion (ppb). Since "percents" are "parts per hundred," the solubility of chloroform in water is 5000 ppm or 5,000,000 ppb. This means that drinking water could contain enough chloroform to present a serious health hazard. Similar observations hold for many other chlorinated chemicals.

PCBs, the dioxins, and some chlorinated pesticides present a different picture. The water solubility of the insecticide chlordane is reported as six micrograms per litre (6 μg/L), which is 0.006 ppm or 6 ppb.[53] Thus chloroform, with its "low" water solubility, is roughly a million times more water soluble than chlordane. Trying to imagine what these numbers mean is more complex than our daily experience allows. It is also very difficult to measure such extremely small amounts.[54] This is why these numbers often need interpretation. Interpretation then opens the door to argument.

It follows that in terms of water quality, many organochlorines are not *dissolved*, but are associated with suspended solids in the water and with the sediments in lakes and rivers. This means that a simple filter, made with activated charcoal for example, which removes suspended matter, will also effectively remove PCBs, dioxins, and many other organochlorines from drinking water. The strong adsorption on soil particles combined with extremely low water solubility also means that these chemicals are not readily transported through the soil by the movement of ground water. They are not easily leached into underlying ground water or nearby surface waters when they are spilled on land. Such spills can threaten nearby surface waters when the contaminated soil itself is washed into lakes and rivers.

The extremely low solubility of some of these chemicals in water can also explain an aspect of pollution of the food chain. When fish absorb these chemicals, they mainly acquire that part of the chemical that is truly dissolved in water. Since fish pass their entire blood volume through the gills in a very short time, their uptake of these chemicals is very efficient. This explains the very large *concentration factors* between organisms like fish and the water itself.

This very low solubility in water does not apply to all organochlorines. Most of those made by the pharmaceutical industry behave quite differently. A 1962 patent by Ciba Geigy describes an antihypertensive drug called chlorisondamine chloride (trade name Ecolid), which contains five chlorine atoms in its molecular make-up and is fully water soluble. A drug like Ecolid would not normally be involved in a large scale spill, but a chemical with such a high solubility would readily percolate through soil into ground water or wash into surface water. If it did, it could pose a health hazard. In surface water it would rapidly biodegrade into less toxic degradation products, in ground water it could present problems.

Between these two extremes we find hundreds of organochlorines that have intermediate water solubility, where "intermediate" spans quite a wide range. These chemicals with intermediate water solubilities are also "more or less strongly" adsorbed on particulate matter.

A final point related to solubility deals with the matter of environmental persistence. Recalling our earlier discussions on metabolism and biodegradability, we note again that the chemistry of living organisms is a chemistry that uses water as the preferred solvent. It follows that chemicals with a very low solubility in water are not *readily* biodegraded in the environment. They generally stay around long enough to create problems, but most of them will eventually break down. PCBs *are* biodegraded in sewage waste water, pentachlorophenol *is* biodegraded by bacteria, and dioxins *are* biodegraded by fungi.[55,56,57]

Adsorbable organohalides

The presence of organochlorines can be significant for environmental health and human health, so it is important that their presence can be readily detected and measured. One common analysis method is based on the behavior of these chemicals in water.

Organochlorines can enter the natural environment from many sources. The plastics industries, paper mills, water chlorination, waste incineration, and the naturally occurring organochlorines all contribute to the mix. This is unusual in terms of analysis, where we normally seek to identify one chemical with clearly described properties. Analyzing mixtures is complicated. For the organochlorines we can use the fact that many are strongly adsorbed on solid particles of organic origin. This allows for the measurement of the total amount of all these chemicals in water without the laborious task of identifying each one separately.

To do the analysis, a water sample is treated with activated charcoal to adsorb the chlorine compounds. The filtered-off charcoal is burned in the presence of silver. This immobilizes the chlorine in its mineral, inorganic form, which can be easily measured. This amount of chlorine is *an indication* of the total of all the various organochlorines *that were present in the water,* and became adsorbed on the charcoal. Hence the name adsorbable organochlorines. The method applies also to organobromines and organoiodines.

If there is no need to indicate one specific halogen, chemists often use X to indicate any one of the four. The term adsorbable organochlorines is also used for adsorbable organobromines or iodines. The acronym AOX relates to all adsorbable organohalides and it also represents the analytical protocol used for the measurement.[58] Thus, AOX is the total amount of organochlorines and bromines and iodines that have high enough water solubility to be leached out of the soil or from consumer products, *and* a low enough water solubility so that *most* of the chemical is adsorbed on the charcoal. In environmental water samples, the AOX measurement would include the organochlorines that were *dissolved* in the water, but also those that were already adsorbed on particulate matter present in the water sample, unless the water sample was carefully filtered, *through a non-adsorbing filter,* before the charcoal treatment.

Some organohalogens that are very water soluble may be only partially adsorbed on the charcoal. This means that they may not be included or may be incompletely included in the AOX assay. We must emphasize that AOX is not an absolute measurement of total halogen compounds or total organochlorines. From the viewpoint of the analytical chemist, the AOX protocol is a major simplification of an otherwise very complex analysis. Keeping the limitations of the method in mind, the AOX analysis is an accepted, standardized method for quickly obtaining general information on water quality.

The development of the AOX analysis protocol has allowed for studies with interesting results. A Swedish group has shown, in several reports, that naturally occurring organochlorines are indeed very common in the environment.[59,60] They found these chemicals in lakes and peat bogs in many areas of the world.[61] These researchers also reported recently on a study of the molecular building blocks of the many chemicals that make up the AOX mixtures.[62] They found that the AOX in natural, unpolluted waters and those in paper mill effluents have the same molecular architecture.

Health effects

Toxicity does not require that a chemical contains chlorine in its molecular make-up. Many very toxic organic chemicals do not contain halogen atoms. On the other hand, the presence of chlorine, or another halogen, in a molecule nearly always increases toxicity. A chemical bond between carbon and chlorine can be broken by the energy in sunlight and by many micro-organisms. Mammals have more difficulty breaking down carbon-halide molecules. The toxic effects of the organohalides show much variation and are influenced by the same properties that affect environmental behavior, that is, solubility in water or in fats. Chemicals that are not very soluble in water, such as fats or oils, can be transported within a living organism, but that needs some chemical trickery. Some are emulsified with emulsifiers like the bile acids, which are made in the liver. Others may become associated with hydrophillic, water-loving proteins.

General toxicity

Some pesticides, such as the dieldrin–aldrin–endrin group, have a high enough *acute toxicity* to be of concern to both humans and wildlife. Most of the chlorinated phenols, used as wood preservatives or as precursors for herbicide manufacture, have a high acute toxicity. *Occupational* accidents with some of these chemicals have been reported in the literature. Rachel Carson's examples of the toxicity of these pesticides were mostly occupation-related. Some of the chlorinated dioxins have extremely high acute toxicity. In most of these examples, acute toxicity was measured as acute lethality by means of LD_{50} measurements with laboratory animals.

For many other organochlorines the acute toxicity is quite low. It was the lack of acute toxicity in DDT that was the very reason that pesticide was so successful. Even when acute toxicity is high, there are few situations where this leads to difficulties. Accidental spills, which may threaten both human health and environmental health, and industrial accidents are typical situations where acute toxicity is of concern. Most environmental concerns are related to chronic toxicity.

The detoxification of organochlorines involves chemical changes that remove chlorine atoms from the molecule in question. The altered molecule is often also more water soluble so that it is easily excreted. In mammals this detoxification takes place in the liver, often with some difficulty. When the liver becomes overburdened with this task the organochlorine acts as a liver poison. The damage to the liver can vary from an impairment of function to cancer or necrosis. Detoxification of organochlorines is often much easier in micro-organisms. The precise biochemical details of this microbial breakdown process have been unraveled and are quite well understood.[63] While the number of organochlorines discovered in nature grows, it is not surprising that more and more organisms are found that efficiently biodegrade these chemicals.

A high solubility of many organochlorines in fat is the reason that these chemicals are readily stored in the body fat or adipose tissue of many animals. PCBs, DDT, and dioxins can be stored in body fat for long periods of time. The relationship between fat solubility and bioaccumulation means that fatty fish from inland lakes, such as lake trout, can carry significant body burdens of organochlorines in subcutaneous blubber.

The list of potential effects of organochlorines on human and animal health is long and varied. We will examine a few aspects and begin with a general observation. With very few exceptions our knowledge of these potential and adverse health effects is derived from animal experimentation and industrial incidents. For example, several chemicals in this class can affect the immune system or growth and development of the test animals. In humans chronic exposure can cause a general feeling of tiredness, muscle pains and neurological problems. In severe cases a "wasting syndrome," resulting in weight loss, occurs. One of the fearsome stories about the organochlorines deals with chloracne, the formation of skin lesions that are often painful, disfiguring and last a long time. Relatively few chemicals have been found to cause chloracne in humans: PCBs, PBBs, *some* dioxins and furans, tetrachloronaphthalenes, and tetra-chloroazoxybenzene.[64] Chloracne is the result of acute, high level exposure sometimes found in industrial accidents. It has been observed in both humans and animals and it is virtually always temporary. Most organochlorines are potential carcinogens or teratogens. Not many of them are actually causing cancer. Much of the complex chemistry in a living cell can be affected by organochlorines. Only if the dose is high enough, of course.

Of particular interest is the issue of organochlorine toxicity in relation to the paper industry. Organochlorines in paper mill effluents have long been blamed for toxic effects on fish, and the controversy has been the main driving force for the trend toward unbleached paper and alternate methods of bleaching. After a long political battle the Canadian Environmental Protection Act now includes regulations that prohibit the discharge of dioxins and furans "at measurable levels."[65] In other words, the allowable limits have little to do with risk or toxicity. They depend on what is "measurable," that is they depend on instrumentation. The irony is that it has now been established that the toxicity of pulp mill effluent has little or nothing to do with the organochlorines. The effluent of chlorine free mills was found to have exactly the same adverse health effects on fish.[66]

All these health effects can be readily observed in laboratory animals. The interpretation of these animal tests and attempts to extrapolate them to humans is not straightforward. This is true for all toxicity testing and the organochlorines are no exception to this rule. In any case, the abundant presence of organochlorines in nature shows that the linear dose-response curve and the "no-safe-dose" hypothesis cannot be true. For chronic exposures the only true tests for humans follow from epidemiology. Specifically in the area of organochlorines many such studies have been done. For evaluating such epidemiological studies it is important to distinguish between health effects on the general population and those that result from *occupational* exposures.

We have already looked at the Finnish study that attempted to link organochlorines with breast cancer without considering dietary fat intake.[67] Several other studies showed results that are equally uncertain and do not indicate a *causal* connection. For example, Falck *et al.* found that levels of PCBs and a metabolite of DDT were higher in breast cancer patients than in the controls, but levels for several other organochlorines *were the same in controls and patients*.[68] These authors also had no information on the dietary habits of their subjects and stated that: "However, because other important risk factors for breast cancer were not studied, *there are uncertainties* in the interpretation of these findings" (emphasis added). Although some of these researchers mentioned the pseudo-estrogenic activity of many organochlorines, they did not acknowledge the *anti-estrogenic* and *anti-tumour* activity of some dioxins and furans.[69] Also missing from these breast cancer studies was the observation that lactoperoxidase, an enzyme in mammalian milk, can readily catalyze the formation of organochlorines.[70]

A detailed technical review of the toxicity of organochlorines emphasizes the danger of generalizing.[71] Although most organochlorines are toxic, there are significant differences. Some sub-families are potential carcinogens and others are not. In the benzene series for example, toxicity increases with increasing chlorine substitution, but the ability of benzene to cause leukaemia decreases with increasing chlorine content. None of the chlorinated benzenes are carcinogenic, but chlorinated phenols are. Both chlorinated benzenes and chlorinated phenols are found in nature.

The toxicity of synthetic organic chemicals containing chlorine or other halogens can be summarized in terms of what we have encountered before. They show the whole range of toxic effects that we have seen among thousands of other chemicals. Whether they are found in nature or made in a laboratory makes no difference.

The toxicity of PCBs

The difference between the PCBs and the chlorinated dioxins and furans is that the PCBs were deliberately manufactured, the dioxins and furans were unwanted impurities. As impurities, the dioxins and furans were most frequently found in the polychlorinated phenols and in the PCBs themselves. This places a question mark behind many studies on the toxicity of PCBs, because it is often not known if an effect seen in a PCB study was really caused by the PCBs or by the impurities. That is not of great comfort to the animal that died or the human who got sick. It does make a large difference for making regulations.

The acute toxicity of the PCBs is very low. Acute lethality in terms of the LD_{50} values for PCBs is similar to those of aspirin and valium and antifreeze. Human fatalities due to acute PCB poisoning are virtually unknown and are irrelevant to environmental discussions. On the chronic effects of the PCBs and the dioxins a vast body of literature is available. Many studies on laboratory animals have indicated the *potential* for health effects primarily in terms of reproductive effects and cancer. Epidemiological studies in this area are generally inconclusive.

Long-term feeding to laboratory animals can cause sub-lethal effects such as enlarged livers, changes in fat metabolism, changes in steroid hormone balance, and reproductive interference. Details of this kind of study are available in a slightly dated source, but are not essential for our discussion.[72] It is important to note that many of these adverse effects are not permanent. A sub-lethal effects study found that Rhesus monkeys that were fed PCBs in their diet developed several health problems, including reproductive failure.[73] When the monkeys were placed on a PCB-free diet for a year, there was "a decided improvement in their general health and reproductive capabilities."[74] As can be expected, in laboratory rats and mice PCBs can cause cancers, mostly of the liver.[75] The human toxicity of PCBs is not as straightforward.

It may have been this work on Rhesus monkeys that subsequently stimulated a study of newborn children by Jacobson and co-workers.[76] They measured birth weight and head circumference of the babies of mothers who were regular consumers of Lake Michigan fish. They concluded that PCB exposure of the mothers led to underweight babies with below average head circumference. Although the differences with non-exposed mothers were significant according to proper statistical methodology, they were extremely small. Gestational age was measured according to two criteria and one of these actually indicated a *higher* (i.e., better) number for the babies with the *higher* PCB levels in the umbilical cords. In a follow-up study the authors tested the cognitive functioning of the same children at age 4. They found that the *in utero* exposed children performed poorer on those tests than their non-exposed counterparts.[77]

These studies did address the problem of "confounding variables," that is other potential or contributing causes. The first report contains no information about which of these were actually measured, the second lists under "toxic exposure" only lead, DDT, polybrominated biphenyls, maternal smoking, and alcohol consumption. In a heavily contaminated lake like Lake Michigan one would expect an abundance of trace chemicals, including other heavy metals like cadmium and chromium, which could affect the health of newborns. Many reviews and reports mention this effect of PCBs on newborn babies and we easily get the impression that this is a well-established fact. A workshop on the general question of human toxicity of PCBs was held at Michigan State University a few years ago. The report on that workshop was summarized as stating that: "It was concluded that available evidence did not demonstrate serious adverse effects such as cancer, in exposed cohorts but did provide indications of possible neurobehavioural effects in children exposed *in utero*."[78] It is important to recognize that this repeated reference to *in utero* exposed children does not represent additional evidence. It is the same study that is cited by many other writers.[79] The work of the Jacobson group is obviously important and confirmation by further independent research is urgently needed.

In 1968 the so-called Yusho incident in Japan involved substantial PCB contamination of cooking oil. PCB consumption by many people was the result. A similar incident took place in 1978 in Yu-Cheng in Taiwan. The resulting chloracne, abnormal skin pigmentation, fatigue, and headaches have mostly disappeared in both cases. Long-term effects such as above-average cancer rates have so far not been seen. The effects on babies born to exposed mothers in Yusho and Yu-Cheng were similar to those seen by Barsotti in monkeys and by Jacobson in human babies. In both the above accidents as well as in the studies by Barsotti and Jacobson it is not clear whether the effects were due to the PCBs themselves or to the presence of the associated chlorinated dioxins and furans as contaminants. None of these studies addresses that question. Dioxins and furans are common impurities in PCB containing oils.

The carcinogenic potential of PCBs shown in animal studies naturally raises the question of human cancer. A report prepared for the Ontario Workers' Compensation Board involved the study of over 300 literature reports.[80] The authors found that "there is a *probable* connection between cancers of the liver, biliary tract and gall bladder and *occupational* exposure to PCBs." They also specifically mentioned that they did *not* find any probable connection between occupational PCB exposure and any other site specific cancers.

PCBs can have adverse effects on the health of *sufficiently exposed* humans and other primate species. Whether these health effects are due to the PCBs themselves or to dioxin impurities is often not clear, but the answer has little meaning to affected individuals. Although these chemicals have been phased out and do slowly biodegrade, they will be around for a long time to come. The quantities still in use should be treated with respect and be subject to careful management and continued monitoring.

Dioxin and furan toxicity

Health effects of the chlorinated dioxins and furans are similar to those of many other organochlorines with one notable exception. The acute toxicity of *some* of the chlorinated dioxins and furans is very high, about a million times higher than that of the PCBs. This extreme toxicity, measured as acute lethality, invariably stimulates human imagination. It is guaranteed to move a discussion on environmental matters from the rational to the emotional level.

In May 1987 a dioxin workshop was held in Quebec City for the Canadian Council of Resources and Environment Ministers. The participants were a mixture of experts and lay people. In an overview in the proceedings of that workshop we find: "the most common one [dioxin] is when you have chlorine at 2,3,7, and 8 positions, giving 2,3,7,8-tetrachlorodibenzo-para-dioxin or TCDD."[81] To most people "most common" would mean the one that is most commonly found, or is most frequently present. If that was the intent the statement was probably wrong. If the speaker meant to say that TCDD is the most *well-known* of the dioxins, he was probably right.

Chlorinated dioxins are *always* found as mixtures of many members of this family and the so-called symmetrical tetrachloro compound, also called 2,3,7,8-TCDD, or just TCDD, is usually *a minor component* of these mixtures. In the fly ash of municipal incinerators, TCDD is about 4 to 6 percent of the total dioxin content.[82] Why is this one compound in a family of 210 members so special? Of all these chemicals, toxicity measurements have been made on only a limited number. The results were enough to catapult "the symmetrical tetrachloro compound" right into the limelight. A comparison of the acute lethality (LD_{50}) of TCDD with some other chemicals, presented in Table 12.1, explains why.

When 2,3,7,8-TCDD is compared with several other "well-known poisons" it turns out to be about two thousand times more toxic than strychnine. Not only that, but its toxicity to guinea pigs was really phenomenal. The chemical was quickly dubbed "the most toxic chemical known" or "the most toxic chemical ever made." The first label was wrong, because botulinus toxin A is about a thousand times more toxic. The second label is half true because the dioxins are not really "made," they are coincidentally formed in both industrial processes and natural events. And all these numbers have little meaning with respect to the environmental impact of the dioxins and furans, for several reasons.

First there is the species difference that we see in the LD_{50} numbers in Table 12.2. TCDD is five thousand times less in acute toxicity to hamsters as it is to guinea pigs, and to dogs it is about three hundred times less toxic. Nobody knows what the *acute toxicity* to humans is. Also, these numbers are misleading because of the difference in toxicity between the members of the congener family. Another tetrachloro dioxin, the *unsymmetrical* 1,3,6,8-TCDD, is about *two million times less toxic* to mice than the symmetrical 2,3,7,8-TCDD (Table 12.1).

And the third reason for the limited use for these numbers? There is just no plausible way in which humans can ingest enough dioxin to cause *acute poisoning*. These chemicals are universally distributed in the environment, but they occur in such extremely small quantities, that they simply do not accumulate in one place in sufficient amounts to present an acute poisoning problem. There are no records of humans dying from acute dioxin poisoning. The issue of acute lethality of these chemicals is essentially irrelevant. It is, however, a popular way to steer environmental discussions from the rational to the emotional level.

Information on the long-term toxicity effects of these chemicals comes from two sources, animal tests and epidemiological studies on humans. The early observation of the extreme acute toxicity of TCDD toward guinea pigs was enough to start much research on animal toxicity, of both acute and chronic effects. Several industrial accidents provided useful information on the question of human exposure. Health effects on animals are naturally similar to those found for other organochlorines. They include cancer, teratogenic and reproductive problems, enzyme induction, problems of the lymph system, and pronounced weight loss due to the depletion of adipose tissue. The TCDDs and TCDFs also belong to the limited group of organochlorines that can

Table 12.1 Comparative toxicity of TCDD and some other well-known toxic agents (acute lethality for mice as LD_{50} in micrograms per kg bodyweight)

Tetanus Toxin[a]	1.5×10^{-10}
Botulinus Toxin (Botulin)[b]	2.0×10^{-3}
Symmetrical 2.3.7.8-TCDD[c]	1.4
Tetrodotoxin[c]	10.0
Curare (Tubocurarine)[b]	6.3×10^{2}
Strychnine[d]	5.0×10^{3}
2,3,7,8-Tetrachloro dibenzofuran[c]	$>6.0 \times 10^{3}$
1,3,6,8-TCDD[c]	3.0×10^{6}

Notes: [a]R. Rawis, op. cit.
[b]E. Hodgson *et al.*, op. cit.
[c]D. E. Laycock, op. cit.
[d]Merck Index, op. cit., in rats.
NB: the smaller the number, the higher the toxicity

Table 12.2 Relative toxicity of TCDD in different species (LD_{50} in micrograms per kg bodyweight)

Guinea pigs	1.3	Mice	200
Rats	33.0	Dogs	300
Rabbits	115.0	Hamsters	5000

cause chloracne.[83] A review of sub-lethal toxic effects has shown that many members of this chemical family have a fraction of the toxicity of TCDD, and most show no mutagenic effects in the Ames test. In terms of most toxic effects, the symmetrical TCDD is the exception in the family.[84]

Epidemiological information on human health effects has been obtained from the monitoring of *occupationally* exposed workers and observations on people who were accidentally exposed. Many such studies, some on quite large groups, have been published. The June 6, 1983 issue of *Chemical and Engineering News* was entirely devoted to the dioxin problem. A brief history of the dioxin problem contains a listing of significant industrial incidents, numbers on acute lethality, and comments from some leading researchers in the field.[85,86] Some of the accidental exposures, partially listed in Table 12.3, go back to the 1950s and 1960s, and at the time of that report (1983) no elevated cancer rates had been observed.

At that time, the most common health effects of workers exposed in industrial incidents were chloracne and liver and neurological damage. In the Nitro incident over a hundred cases of chloracne were seen, but the total number of deaths, thirty-four years later, was *lower* than expected and no excess cancer deaths were reported.[87] On the other hand, a more recent report indicated elevated cancer rates in the exposed population in Seveso.[88]

Table 12.3 Industrial accidents involving dioxins

Company	*Location*	*Year*	*No. of workers*
Monsanto	Nitro VA (USA)	1949	250
BASF	Ludwigshafen (Germany)	1953	75
Philips	Amsterdam (Netherlands)	1963	106
ICMESA	Seveso (Italy)	1976	156

Source: *Chemical and Engineering News*, June 6, 1983

One nagging question seems to have been lost in the literature on the Seveso incident. The actual product that plant was producing was trichlorophenol, the precursor for the herbicide 2,4,5-T. All literature on this event is about the several pounds of dioxin that were released in the explosion. However, the dioxins were an impurity in the product, trichlorophenol. None of the reports on Seveso, particularly those mentioning the damage to animals and plants, mention the hundreds of pounds of this highly toxic and corrosive chemical that must also have been part of that spill.

A major epidemiological study in 1991 included over five thousand workers in twelve US plants with potential exposure to TCDD.[89] The report reads: "the study did not confirm the high relative risks reported in previous studies." Although there was an excess, over the expected background, of various cancers, the numbers were small and exposure to other chemicals and tobacco smoke could not be excluded from the effects. Of course, this study was criticized by other researchers and defended by the original authors.[90,91] Notwithstanding the weak evidence from these epidemiological studies the US EPA has said recently that "dioxins probably cause cancer in humans." However, we must note that the EPA's statements are based on the hypothetical linear dose-response curve and the assumption that there is "no safe level" for dioxins, a view that is considered unrealistic in many European countries.[92]

Some *mammalian metabolites* of TCDD have been found to be substantially less toxic than the dioxin itself.[93] This indicates that even the "persistent" dioxins are biodegraded by mammals, not only by micro-organisms. This observation is interesting in connection with some related reports. The levels of PCBs in the milk of lactating Inuit women have been reported as "the highest ever reported."[94] A related study again commented on the high levels of PCBs and dioxins in the blood serum of Québec Inuit.[95] Organochlorines enter the human body almost entirely in the fatty parts of the diet, which in the case of the Inuit is seafood. Therefore the summary of this report is significant. It specifically states that "there was nothing [in the findings] to justify recommending limited consumption of marine staples."[96] Another report also mentioned high levels of organochlorines in the serum of human subjects. This report also contains no information on the presence or absence of any adverse health effects due to these extraordinary high body burdens of organochlorines.[97] This absence of adverse health effects in the presence of unusually high body burdens of these chemicals leads to a few important questions. What is the true nature of their toxicity for humans? What justifies the regulatory levels that have been set for "allowable" blood levels of these chemicals?

There is no question that many chlorinated organics, both of natural and industrial origin, can be toxic and can cause adverse health effects in humans and animals if the dose is high enough. A recent review clearly implicated several classes of chemicals, both chlorinated and *non-chlorinated*, as contributing to deformities in birds in the Great Lakes basin.[98] At the same time, finding many natural sources, biodegradability and sometimes high human body burdens raises questions about our collective handling of what has been called "the dioxin phenomenon."

Chlorinated dioxins and furans are not particularly "environmentally friendly." They are almost unique in that they are so readily formed in many situations. The picture of what these chemicals can do to animal health, in the amounts that are found in the environment, is far from clear. Other unanswered questions are about the quantities involved and their significance in terms of human health. These chemicals should be treated with due respect. Their abundant natural occurrence is no reason for us to add more. But there is no evidence or reason why we should pursue, at great expense, the fruitless goal of zero exposure. Nature has shown that to be an exercise in futility.

CHAPTER 13

Conclusion

Rather than a formal treatise of environmental science, ecology, and environmental ethics, we have now concluded a brief tour through several of the important areas of the environmental challenge. Although we were travelling light, in the sources that provided the directions for our trip there was a strong emphasis on the science literature. That emphasis was deliberate, because it is missing from much other literature dealing with the environment. We have not found miraculous solutions for the significant problems that we found along the way, but perhaps we now have a different perspective than that with which we started our tour.

The debates and arguments concerning the environment will still continue between the three main groups of participants, the catastrophists, the cornucopians and the pragmatists. However, perhaps our perspective of the debate will help us decide which group to join. When we summarize that perspective, we will stay focussed on the two major road signs that we followed throughout our discussions. One is to keep occupational health firmly separated from public and environmental health. The other was stated by H. F. Smyth, in his essay "Sufficient Challenge" in 1967:

> We should spend our time measuring something rather than exhausting ourselves in the necessarily impossible task of determining zero.[1]

Catastrophists, cornucopians, and pragmatists

The monkey on our back in the industrialized world is the combination of our lifestyle and its associated overconsumption, with the vast numbers of our species. The "coercive utopians" would have us go back to a simpler lifestyle.[2] But it is not possible to turn the clock back. The catastrophists believe that, since "nature knows best," an environmental calamity will turn the clock back for us. So far they have been wrong. And the cornucopians maintain that science and technology will solve all problems. So far they have not been right. What are pragmatists supposed to do?

A glance at the environmental literature shows that society has heard the wake-up calls from the early biologists-turned-writers. Perhaps it is not so obvious how much progress we have made since. The past thirty years have seen a revolution in the chemistry of pesticides and their use. The foaming rivers of earlier days are gone and biodegradable detergents and shampoos are commonplace. Infectious diseases can be readily controlled. Although heart disease and cancer are the two principal causes of death in many industrialized countries, progress in the fights against these diseases has been phenomenal. While the number of cancer cases is increasing as the number of older people increases, the *rates* for most cancers are close to being constant. In fact, so many people in the industrialized countries are getting older, that the World Bank has called it a crisis.[3] The industrial "fouling of our nest" has not caused a catastrophe in terms of human health and longevity. Yet, it is too early to celebrate.

While Ray and Guzzo described "how science can help us deal with acid rain, depletion of ozone, and nuclear waste," those three issues are still with us.[4] On population growth the argument between the catastrophist view of Ehrlich and the cornucopian picture of Simon is still continuing. The optimistic views of abundant and cheap energy have not materialized and progress with other sources of energy is painfully slow. While we continue our production of greenhouse gases there is little evidence of an increased acceptance of the nuclear option. The old promise of "affluence for ten billion" is all but forgotten.[5] The waste management problems of the rich countries have not been solved.

Although we had "the tragedy of the commons" described to us as early as 1968, we have seen several replays of that scenario.[6] After the catastrophe of the Sahel, recent examples are the international disputes about fish stocks. As Dales remarked, albeit in another context, "empirically it is clear that if an asset is depletable, it will be continuously depleted on the ground that everybody's property is nobody's property."[7] In a global sense, the atmosphere of the planet is also common property. However, as recently as April 1995 yet another world conference on climate change showed that the rich nations have little interest in cutting back on their comfortable lifestyle that is based on fossil fuel combustion. While many books have been written on how to "manage the planet," our management of the commons has not made much progress.

Our brief discussions have exposed a picture of what the words mean when we speak and write about ecology as a science, a philosophy, or a lifestyle. We have seen the thirty-year development of a hypothesis that allows us to view the entire biosphere as a single, self-regulating system. The Gaia hypothesis continues to inspire researchers that seek a better understanding of our planet and how we affect that complex system.[8] We had a detailed look at the important differences between the facts of science and the opinions of scientists. If science and technology are responsible for "the technological flaw," they are also an integral part of the solution towards fixing it.

We found that we can productively talk about chemicals and chemistry by merely gaining some familiarity with a few basic concepts and a handful of names. Our explorations of toxicology showed that effects of chemical and physical agents on our health and on the environment depend entirely on the amounts involved. Basic toxicology rendered the term "toxic chemical" meaningless unless we specify these amounts. Our manipulations of industrial chemicals inevitably include the potential hazards of occupational exposure and occupational disease. But to extrapolate the *occupational hazards* of radiation, asbestos, and carcinogenic chemicals to the entire population and make them into *public hazards* is wrong. It is wrong because most of these extrapolations are based on a hypothesis constructed from questionable assumptions.

At the same time, we developed a view of nature that is a little different from the popular "nature is benign" image. Our survey of naturally occurring, potentially toxic chemical and physical agents has shown us that our environment is not all that benign. This may provide some insight to the victims of cancer and Alzheimer's and Parkinson's disease. For many such patients the twin issues of guilt and blame often centre on the cause of the disease. There is the feeling that this could have been prevented "if I only had lived differently." The strong urge to blame somebody or something as the cause plays a key role in our personal and collective reaction to these diseases. Perhaps it helps to realize that by far the largest number of cancers are there because they are an inherent part of nature; potential carcinogens are an integral part of our planet. It may be easier to cope if we let go of the image of the inherent goodness of nature. Helen Hiscoe summarized the problem some time ago: "The underlying assumption of those who value things because they are natural, is their belief that nature is a benign force whereas human power tends to be evil. They equate naturalness with goodness."[9] Our brief surveys of naturally occurring toxic plants and natural radioactivity clearly show the fallacy of that belief.

Our exploration of the things-that-shouldn't-be-there in air and water and food showed that they are not that much different from many of the things that have a natural right "to be there." Biochemical processes do not readily distinguish between natural and industrial chemicals. The line between natural contamination and pollution is merely a matter of quantities. Our waste problems follow a similar trend. The quantities are directly related to population and consumption; none of these problems operate in isolation. To preserve clean air and water and healthy food is not a simple matter.

Thus we can build a pragmatic approach to environmental problems on our understanding of the words and the basic concepts of the environmental language. Many of these words and concepts follow from scientific investigations and their interpretations. However, we also encountered many examples of serious environmental questions that are outside the reach of science because they are questions in trans-science. It is important to recognize not only the capabilities of science, but also its limitations. That is where the environmental language must be expanded with the words of many other fields of endeavour. Those other words should not include the factoids of pseudo-science.

Factoids and pseudo-science

Although science and technology are as much part of the solution as they are of the problem, it would be foolish to leave all our problems in the hands of scientists and engineers. We have enumerated the many players in the environmental arena in some detail. We want all of them to participate. However, we must once more underline the very important distinction between the scientific fact and the opinion of the scientist. In the case of opinion it is imperative that we inquire if we deal with a scientific opinion, based on expertise, or with the opinion of a person who just happens to be a scientist.

A few years ago a news writer illustrated pseudo-science by referring to a talk by Irving Langmuir, a physicist and 1953 Nobel Laureate. Langmuir described "the science of the things that aren't so," and characterized it by the following features:

1 Maximum effect is caused by an agent of barely detectable intensity.
2 The size of the effect remains close to the limit of detectability.
3 Investigators claim great accuracy of measurement.
4 Fantastic theories contradict the experience.
5 Critics are given ad hoc explanations on the spur of the moment.
6 The ratio of supporters to critics rises to somewhere near 50 percent and then gradually falls into oblivion.[10]

A survey of much of the literature revealed to us an uncomfortably high level of pseudo-science or, as Jeremy Burgess called it, "imitations." Burgess referred to the literary "postmodernism" to criticize as unscientific the notion that "words have a meaning beyond that which most of us agree to give them." He illustrated the use of scientific terms in the pseudo-science debate by asking if "the next scare over *Streptococcus* should have gardeners watching for malevolent changes in their *Streptocarpus* plants." He then expressed his hopes that "postmodern *Streptomyces* remain willing to produce antibiotics."[11] We will add one more example to our collection.

Several studies have addressed an issue that has significant potential for an emotional response. There is growing concern that in the industrialized world, sharp drops in sperm counts are responsible for decreasing male fertility. Birth rates are dropping and testicular cancer is increasing. As was the case with cancer in general, the ready-made culprit is environmental pollution, more specifically organochlorine pollution.[12,13]

However, a much earlier observation offers another solution.[14] That earlier study found reduced sperm counts and testicular atrophy as the result of caffeine and its close relatives, theobromine and theophylline, in laboratory animals. Considering the substantial increase in coffee consumption, which supposedly is a significant factor in the destruction of the Brazilian rain forest, another explanation readily comes to mind. Tea and cocoa also contain significant amounts of caffeine and its relatives and caffeine is added in significant quantities to several popular beverages.

In other words, caffeine and its relatives are consumed in gram quantities while other pollutants occur in our food and water in sub-milligram amounts. Before we level yet another accusation at the organochlorines, some controlled studies on sperm counts relative to caffeine consumption and organochlorine exposure would seem to make eminent sense. The environment can almost certainly be blamed for a falling sperm count. That environment could well be the living room and the coffee shop. Although such a study is not available, we must avoid the jump to a pseudo-science conclusion that would designate a bona fide scientific observation to oblivion.

It is pseudo-science that has led to the rapid accumulation of factoids about the environment and about human health. This is unfortunate because it detracts from the real problems and it befuddles the already difficult process of setting priorities. Solving environmental problems requires large amounts of money, not just in terms of the cost of remedial action, but also in terms of social costs. Some solutions cause the closure of industries, unemployment, and other social upheaval. Simplistic assertions that chlorine-based materials can be replaced ignore the considerable energy costs of these replacements. Pseudo-science makes it difficult to allocate available resources according to priority.

Our discussions revealed two roads leading to the birth of factoids and we must underline these once more. Many health concerns came into being by studying the *real hazards* in the workplace and in the laboratory, and extrapolating to the *calculated hazards* in the indoor or outdoor environment. Asbestos, radiation, and organochlorines are some of the best known examples. A very large number of the dangers that lurk in our polluted environment are *calculated* dangers, and the interpretation of many of these calculations is trans-science.

The other road to factoids has to do with the corporate sinners, the "envirocriminals." The "polluter pays" mentality and the image of the ruthless corporations became an early part of our environmental discussions.[15] A brief look at the details should help us identify these envirocriminals. Corporations are in business to make profits. The people they employ for that purpose are paid out of these profits. The products and services these corporations provide are bought by people who need them, or think they need them, or are persuaded they need them. If we were to calculate how much of the waste and pollution comes from "big business" and how much comes from individuals that buy the products and services, the answer could well be a surprise. We could find that "the polluter is us."

Towards environmental sanity

The time has come to consider if it is really productive to ask if we can "do without" or if we can live "in harmony with nature" or if there is "an alternative way of life?" From a simple pragmatic point of view it may be too late for such questions. Short of a gigantic world revolution, the question is no longer how to find an alternative to chemical industries, but rather how to live with them, and how to control them. Instead of spending our efforts on the problem of replacing the steel axe with a stone axe, it would be more effective to manage the makers of the steel axe and their environmental impact. While we manage the steel industry and the paper industry and all the others, it is time to rethink our ideas of "zero." As long as industry operates, there will be waste, because that is a law of nature, the second law. Waste and pollution are never zero, because all we need is tomorrow's better instrument to find the next part per quadrillion. The only way to achieve true zero waste is to shut down.

Many scientists and engineers have probably viewed the evolution of the environmental movement with mixed feelings. Feelings of amusement or irritation for the pseudo-science, some indifference towards the emotional aspects, and perhaps anger at the attempts to discredit science and technology. However, few scientists may understand the idea of *moral outrage,* so well described by Roald Hoffmann.

Hoffmann is a professor of physical science at Cornell University. In 1981 Hoffmann shared the Nobel Prize for chemistry with Kenichi Fukui for the significant joint contribution these two scientists made to the theory of chemistry. A theory, incidentally, that is the basis for our observations on the close relationship between radiation and heat in food processing. A few years ago he was awarded the Priestley Medal, an award named after Joseph Priestley, a British born scientist who, after a colourful career, spent the last eight years of his life in the USA in the late 1700s.[16] Hoffmann's address to his peers of the American Chemical Society, on the occasion of this award, was reprinted in its entirety in *Chemical and Engineering News*.[17] We will highlight one small but important aspect. The controversy over alar, the growth regulator used for apples, is the illustration for the story.

Hoffmann urged his audience, nearly all of whom were chemists, to try and understand the basic reasons for chemophobia, by telling his own personal reaction to the news that his apples may have contained traces of a chemical that he did not know about. He attempted to place that experience in perspective by referring to a well-known factor in *risk perception* as distinct from *risk calculation*.

Experts in personal communication recognize several factors that influence a person's perception of risk. One example is the difference between the driver of an automobile and the passenger. The passenger perceives the risk of an accident quite differently than the driver does, because the driver is in control, the passenger *is dependent* on the driver. Hoffmann used the factor of voluntariness. We tend to perceive the risk associated with being pushed down a slope on two slippery planks very differently from the risk of going skiing. Voluntary risks and risky activities that are within our control are more acceptable than involuntary risks to which we are subjected without having control. There are several other factors that differentiate between acceptable and unacceptable risks. Widely scattered traffic fatalities are more "acceptable" than the same number of fatalities resulting from a plane crash.

For a chemist it is very easy to recognize alar and the related UDMH, both chemically and toxicologically, for what they are. The recognition of these chemicals as close relatives of the hydrazines in our mushrooms has a substantial effect on how the risk is perceived. The familiarity with the chemicals makes the minuscule risk of eating the apple not only acceptable, it makes it comparable to the risk of eating mushrooms in our salad. Without this recognition the threatening statement that the chemical can induce cancer in animals *and* the powerlessness associated with the situation, created the outrage underlying the protest.

Another example of the perception problem was told by Ottoboni.[18] In discussing the high expectations we have from science, she refers to the people who want to know "what is in there that makes me sick" when they refer to a chemical problem. When in the neighbourhood of a landfill, there appear to be more than the "normal" number of cancers, people do not understand why a scientist cannot just "go in and analyze things and tell us the truth." In reality most people have, by that time, accepted the *perception* that the leaking dump is the cause anyway, and all they, the scientists, have to do is confirm it.

An example of this approach was recently reported in *Nature*. The "Gardner hypothesis" suggested that an excess of childhood leukaemia near a nuclear reprocessing plant was caused by the exposure of the fathers of the children to radiation. The story involved a plant near the Northwest English town of Sellafield. Earlier investigations had ruled out waste or leakage from the plant and it was suggested that the men working in the plant were exposed to radiation that caused mutations in their sperm. It took much additional evidence, brought out by a court case, to show that the hypothesis was wrong. It was conclusively shown that parental irradiation could not have caused the problem. An alternative explanation was not presented in the report.[19]

A knowledge of the nature of the risk is an important factor in risk perception. That is why environmental factoids are so dangerous. They create a distorted picture of the real risks. Factoids about human health and the health of the environment are the biggest impediment to effective environmental action. They stand in the way of a realistic approach to sustainability and international agreement on how to manage the commons. We spend much of our resources on solving calculated dangers that have not yet happened, while many current problems, like poverty and traffic fatalities, remain unsolved.

Therefore this book must end with a plea. From the scientists we must ask that they not assume what Hoffmann called "a scientistic analytical stance." Instead they must try to understand the anxiety and moral outrage caused by agnosophobia, the fear of the unknown. Conversely, non-scientists must acquire what Commoner called many years ago,

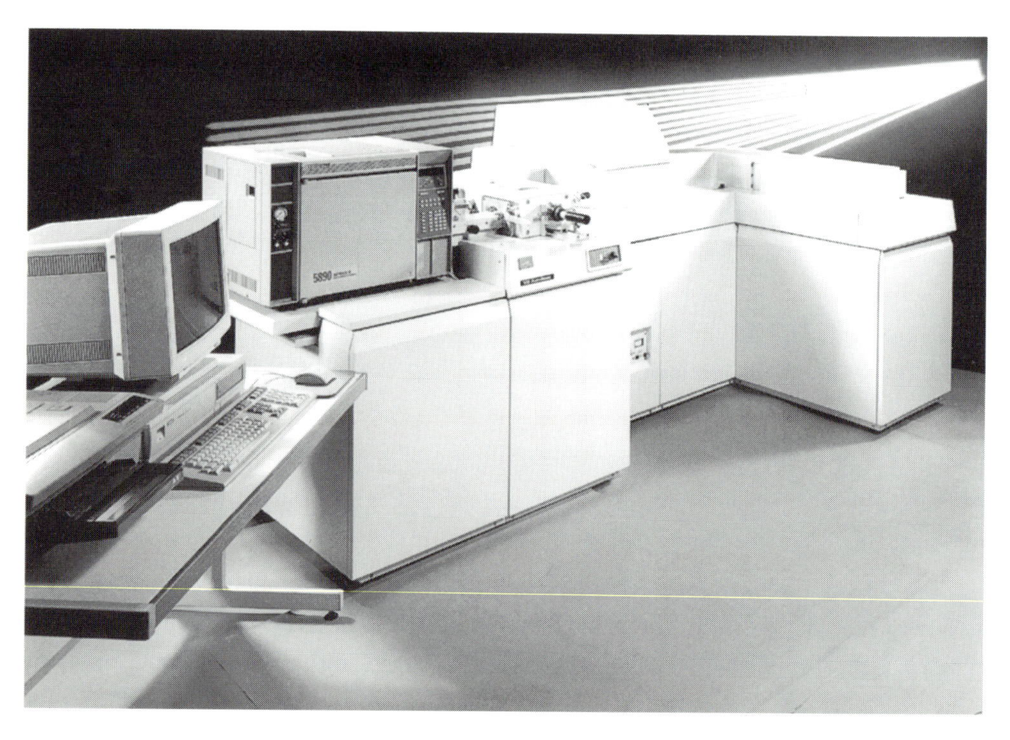

Figure 13.1 Modern instrumentation – a millionfold more sensitive than in the 1950s, but it cannot measure zero

> the new knowledge ... about the elements of the environmental crisis ... that must be the source of the grave new judgements.

We looked earlier at a fast and painless way to acquire some of that "new knowledge." A regular scan of the appropriate literature, the front sections of *Science, Nature,* and *Environmental Science and Technology,* is an essential part of the reading of anyone involved with environmental issues. The *New Scientist* contains many discussions and essays relating to environmental issues. *Chemical and Engineering News* is a news publication that frequently reports on events of environmental importance, and *Scientific American* translates the language of science into everyday English. These publications are available in almost all libraries in North America and in Europe. Some familiarity with this literature is a fast path to the language of the environment.

Then, with our newly acquired "new knowledge" we can lay the factoid of pollution-caused cancer to rest and we no longer need to spend our energy on the futile pursuit of zero. Zero toxicity, zero exposure and zero hazard are as unattainable as absolute safety. Instead, we must concentrate on those local and global issues that can be discussed in a more productive way. We started our discussions with a look at the multidisciplinary team that is needed to find our way to what Claus and Bolander called "ecological sanity."[20] It is imperative that all members of that team understand the language. Otherwise the "predicament of humanity" will continue to block the path towards sustainability.

Notes

Chapter 1 Introduction

1 Garrett Hardin, "Cassandra gets a hearing," in Garrett Hardin (ed.), *Population, Evolution and Birth Control, A Collage of Controversial Ideas* (2nd edn), San Francisco CA, W. H. Freeman and Company, 1969, p. 39.
2 Roderick Frazier Nash (ed.), *American Environmentalism* (3rd edn), New York, McGraw-Hill, 1990.
3 Paul R. Ehrlich, *The Population Bomb*, New York, Ballantine Books, 1968.
4 Nicholas Wade, "Sahelian drought: no victory for western aid," *Science*, 1974, vol. 185, July 19, pp. 234–7.
5 Donella H. Meadows *et al., Beyond the Limits: Confronting Global Collapse, Envisaging a Sustainable Future*, White River Junction VA, Chelsea Green Publishers, 1993.
6 Ralph Nader and Mark J. Green (eds), *Corporate Power in America, The Ralph Nader Conference on Corporate Accountability, Washington DC, 1971*, New York, Grossman Publishers, 1973.
7 Richard M. Salsman, "Corporate environmentalism and other suicidal tendencies," in J. Woiceshyn (ed.), *Environmentalism: What does it mean for Business? Proceedings of a Guest Lecture Series*, Calgary, The Faculty of Management, University of Calgary, 1992, pp. 117–41.
8 Vance Packard, *The Waste Makers*, Harmondsworth, UK, Penguin Books Ltd., 1963 (first published in USA 1960), p. 254.
9 Ayn Rand, *Atlas Shrugged*, New York, Random House, 1957.
10 Ayn Rand, *Capitalism, the Unknown Ideal*, New York, New American Library, 1967.
11 "The poisoning of America," *Time*, 1980, September 22, p. 58.
12 Dixy Lee Ray and Lou Guzzo, *Trashing the Planet*, Washington DC, Regnery Gateway, 1990, p. 163.
13 Aldo Leopold, *A Sand County Almanac*, New York, Oxford University Press, 1949.
14 As quoted by John Rodman, "Ecological sensibility," in Donald VanDeVeer and Christine Pierce, *People, Penguins, and Plastic Trees*, Belmont CA, Wadsworth Inc., 1986, p. 165.
15 Garrett Hardin and John Baden (eds), *Managing the Commons*, San Francisco CA, W. H. Freeman and Co., 1977.
16 J. Chowdhury, "The battle over the environment," *Chemical Engineering*, 1991, October, pp. 30–45.
17 Barry Commoner, "The environmental costs of economic growth," in Robert Dorfman and

Nancy Dorfman (eds), *Economics of the Environment, Selected Readings,* New York, W. W. Norton & Company Inc., 1972, p. 261.
18 Norman R. Ehmann, "Victims of gnosophobia," *Journal of Environmental Health,* 1990, vol. 53, p. 14.
19 Industrial Disease Standards Panel, *Report to the Workers Compensation Board of Ontario on the Occupational Exposure to PCBs and Various Cancers, Report No. 2,* Toronto, December 1987.
20 I. J. Selikoff, J. Church, and E. C. Hammond, "The occurrence of asbestosis among insulation workers in the United States," *Annals of the New York Academy of Science,* 1965, vol. 132, p. 139.
21 Barry Commoner, "Fundamental causes of the environmental crisis," in Roderick Frazier Nash (ed.), *American Environmentalism* (3rd edn), New York, McGraw-Hill, 1990, pp. 206–14.
22 Chowdhury, op. cit.
23 Ehrlich, op. cit.

Chapter 2 The three Es: earth, ecology, and the environment

1 Anne H. Ehrlich and Paul R. Ehrlich, *Earth,* New York, Franklin Watts, 1987.
2 Kenneth Boulding, "The economics of the coming spaceship earth," in Henry Jarrett (ed.), *Environmental Quality in a Growing Economy,* Baltimore MD, Johns Hopkins University Press, 1966, pp. 3–14.
3 Paul R. Ehrlich and Richard L. Harriman, *How to be a Survivor, A Plan to Save Spaceship Earth,* New York, Ballantine Books Inc., 1971.
4 Garrett Hardin, "Living in a lifeboat," *BioScience,* 1974, vol. 24, no. 10, pp. 561–8, reprinted in Garrett Hardin and John Baden (eds), *Managing the Commons,* San Francisco CA, W. H. Freeman and Co., 1977, p. 261.
5 Carl Sagan, *The Dragons of Eden – Speculations on the Evolution of Human Intelligence,* New York, Random House Inc., 1977.
6 T. S. Eliot (ed.), *Pascal's Pensées,* New York, E. P. Dutton & Co. Inc., 1958.
7 James Lovelock, *The Ages of Gaia, A Biography of Our Living Earth,* New York, W. W. Norton & Co, 1988.
8 James Lovelock, *Gaia: A New Look at Life on Earth,* Oxford, Oxford University Press, 1979.
9 Canadian Broadcasting Corporation, *Ideas – Religion and the New Science,* Montreal, CBC Transcripts, 1985.
10 Norman Myers (ed.), *Gaia, An Atlas of Planet Management,* New York, Anchor Books/ Doubleday, 1984.
11 Kees Zoeteman, *Gaiasophy, Another View of Evolution, Space, Order, and Environmental Management* (title translated), Deventer, The Netherlands, Ankh-Hermes, 1989.
12 Allan Hunt Badiner (ed.), *Dharma Gaia, A Harvest of Essays in Buddhism and Ecology,* Berkeley CA, Parallox Press, 1990.
13 Richard Kerr, "No longer willful, Gaia becomes respectable," *Science,* 1988, vol. 240, April 22, pp. 393–5.
14 James Lovelock, *Gaia: A New Look at Life on Earth* (2nd edn), Oxford, Oxford Paperbacks, 1995, Chapter 10.
15 Isaac Asimov, *Isaac Asimov's Biographical Encyclopedia of Science and Technology* (revised edn), New York, Avon Books, 1972.
16 H. S. Jennings, "Heredity and environment," *Science Monthly,* 1924, vol. 19, pp. 225–38.
17 William Paley, "Natural theology," in Garrett Hardin (ed.), *Population, Evolution and Birth Control, A Collage of Controversial Ideas* (2nd edn), San Francisco CA, W. H. Freeman and Company, 1969, p. 139.
18 A. E. Wilder Smith, *The Creation of Life, A Cybernetic Approach to Evolution,* Wheaton IL, Harold Shaw Publishers, 1970.
19 Stephen Clark, "A modern stoic takes on the reductionists," *New Scientist,* 1994, December 31, p. 62.
20 James Gleick, *Chaos, Making a New Science,* New York, Penguin Books, 1987.

21 Steven J. Schiff *et al.*, "Controlling chaos in the brain," *Nature*, 1994, vol. 370, August 25, pp. 615–20.
22 Isaac Asimov, *Asimov on Chemistry*, Garden City, Anchor Press/Doubleday, 1974, p. 218.
23 J. B. Sykes (ed.), *The Concise Oxford Dictionary of Current English*, Oxford, Oxford University Press, 1986.
24 John Maddox, *The Doomsday Syndrome*, London, Macmillan, 1972.
25 Rachel Carson, *Silent Spring*, Greenwich CO, Fawcett Publications Inc., 1962.
26 Robert H. MacArthur, *Geographical Ecology, Patterns in the Distribution of Species*, New York, Harper & Row, 1972.
27 E. G. Nisbet, *Leaving Eden, to Protect and Manage the Earth*, Cambridge MA, Cambridge University Press, 1991.
28 Donald VanDeVeer and Christine Pierce, *People, Penguins, and Plastic Trees*, Belmont CA, Wadsworth Inc., 1986, p. 70.
29 Roderick Frazier Nash, *American Environmentalism. Readings in Conservation History* (3rd edn), New York, McGraw-Hill Publishing Co., 1990.
30 Environment Canada, *Toxic Chemicals in the Great Lakes*, Ottawa, Supply and Services Canada, 1991.
31 C. S. Giam and H. J. M. Dou (eds), *Strategies and Advanced Techniques for Marine Pollution Studies: The Mediterranean Sea*, proceedings of a conference held in Beaulieu-sur-mer, France, 1984, Berlin, Springer Verlag, 1986.
32 David D. Kemp, *Global Environmental Issues, A Climatological Approach*, London, Routledge, 1990.
33 Andrew Goudie, *The Human Impact on the Natural Environment* (3rd edn), Cambridge MA, The MIT Press, 1990, p. 322.
34 Eugene Hargrove, *Foundations of Environmental Ethics*, Englewood Cliffs NJ, Prentice-Hall, 1989, p. 21.
35 Lynn White Jr., "The historical roots of our ecological crisis," *Science*, 1967, vol. 155, March 10, pp. 1203–7.
36 Lewis W. Moncrief, "The cultural basis for our environmental crisis." in Robert Dorfman and Nancy S. Dorfman (eds), *Economics of the Environment, Selected Readings*, New York, W. W. Norton & Company, Inc., 1972, pp. 284–93.
37 Garrett Hardin, "The tragedy of the commons," *Science*, 1968, vol. 162, pp. 1243–8, reprinted in Hardin and Baden, op. cit., pp. 16–33.
38 Peter Knudtson and David Suzuki, *The Wisdom of the Elders*, Toronto, Stoddart Publishing Co. Ltd., 1992.
39 Marvin Harris, *Culture, People, Nature. An Introduction to General Anthropology*, New York, Harper and Row, Publishers, 1985.
40 Ibid.
41 Calvin Martin, *Keepers of the Game*, Berkeley CA, University of California Press, 1978.
42 Nash, op. cit., p. xi.
43 Asimov, 1971, op. cit.
44 Clark Wissler, *Man and Culture*, New York, Thomas Y. Crowell & Co., 1923, p. 316.
45 Nash, op. cit., p. xi.
46 Paul R. Ehrlich, *The Population Bomb*, New York, Ballantine Books, 1968.
47 Garrett Hardin, "Malthus under attack," in Hardin (ed.), op. cit., p. 32.
48 Paul Ehrlich, "Paying the piper," *New Scientist*, 1967, vol. 36, pp. 652–5, reprinted in Hardin (ed.), op. cit., pp. 127–30.
49 Ehrlich and Harriman, op. cit.
50 Donella H. Meadows, Dennis L. Meadows, Jørgen Randers, and William W. Behrens III, *The Limits to Growth, A Report of the Club of Rome's Project on the Predicament of Mankind*, New York, Universe Books, 1972.
51 Jay H. Lehr (ed.), *Rational Readings on Environmental Concerns*, New York, Van Nostrand Reinhold, 1992, p. 13.
52 H. S. D. Cole, Christopher Freeman, Marie Jahoda, and K. L. R. Pavitt (eds), *Thinking About the Future, A Critique of The Limits to Growth*, London, Chatto & Windus for Sussex University Press, 1973.

53 World Commission on Environment and Development, *Our Common Future*, Oxford, Oxford University Press, 1987.
54 Gita Sen, "The world programme of action, a new paradigm for population policy," *Environment*, 1995, vol. 37, no. 1, pp. 10–15 and 34–7.
55 World Commision, op. cit., p. 46.
56 Rachel Carson, op. cit., p. 16.
57 Meadows *et al.*, op. cit., p. 19.
58 World Commission, op. cit., p. 6.
59 *Toronto Globe and Mail*, Report on Business, September 3, 1993.

Chapter 3 Science and the scientists

1 Smil Vaclav, "Acid rain: a critical review of monitoring baselines and analyses," *Power Engineering*, 1985, April, p. 59.
2 Thomas Hobbes, *Leviathan*, Harmondsworth UK, Penguin Books, 1968.
3 Rachel Carson, *Silent Spring*, Greenwich CO, Fawcett Publications Inc., 1962.
4 Barry Commoner, *The Closing Circle: Nature, Man, and Technology*, London, Jonathan Cape Ltd., 1972, p. 140.
5 Lynn Thorndike, *A History of Magic and Experimental Science, Volume 1*, New York, Columbia University Press, 1958.
6 Isaac Asimov, *Isaac Asimov's Biographical Encyclopedia of Science and Technology* (revised edn), New York, Avon Books, 1972, p. 55.
7 Ibid.
8 World Commission on Environment and Development, *Our Common Future*, New York, Oxford University Press, 1987, p. 103.
9 Lynn White Jr., "The historical roots of our ecological crisis," *Science*, 1967, vol. 155, March 10, pp. 1203–7.
10 J. B. Sykes (ed.), *The Concise Oxford Dictionary*, Oxford, Oxford University Press, 1982.
11 Henry I. Bolker, "Scaring ourselves to death," *Canadian Chemical News*, 1986, February pp. 18–22.
12 Elisabeth Whelan, *Toxic Terror*, Ottawa IL, Jameson Books, 1985, p. 291.
13 M. Perutz, *Is Science Necessary? Essays on Science and Scientists*, London UK, Barrie and Jenkins, 1988.
14 John Boslough, *Stephen Hawking's Universe*, New York, Quill/William Morrow, 1985.
15 Francis Crick, *What Mad Pursuit. A Personal View of Scientific Discovery*, New York, Basic Books Inc., 1988.
16 Farley Mowat, *Rescue the Earth! Conversations with the Green Crusaders*, Toronto, McClelland & Steward Inc., 1990.
17 Ibid., p. 206.
18 Mowat, op. cit., p. 15.
19 Mowat, op. cit., p. 175.
20 Subhash C. Basak and Gerald J. Niemi, "Predicting properties of molecules using graph invariants," *Journal of Mathematical Chemistry*, 1991, vol. 7, pp. 243–72.
21 Whelan, op. cit., p. 5.
22 White, op. cit.
23 Fernand Seguin, "Science's method misunderstood," *Science Dimension*, January 1968, p. 30.
24 Henry H. Bauer, *Scientific Literacy and the Myth of the Scientific Method*, Urbana IL, University of Illinois Press, 1992.
25 Albert Einstein, *Ideas and Opinions, With New Translations and Revisions by Sonja Bargmann*, New Jersey, Wings Books, 1988.
26 First announced simultaneously in *The Financial Times* and in *The Wall Street Journal*, March 23, 1989. The first report by Fleischer and Pons in a science publication appeared in the April 10 issue of *The Journal of Electro-Analytical Chemistry*.

27 Michael Heylin, "Sociology of cold fusion examined," *Chemical and Engineering News*, 1990, March 19, p. 24.
28 R. P. Turco, O. B. Toon, T. P. Ackerman, J. B. Pollack, and Carl Sagan, "Nuclear winter: global consequences of multiple nuclear explosions," *Science*, 1983, vol. 222, December 23, pp. 1283–92.
29 Bauer, op. cit.
30 Joseph Palca, "Get-the-lead-out guru challenged," *Science*, 1991, vol. 253, August 23, p. 842.
31 Joseph Palca, "Lead researcher confronts accusers in public hearing," *Science*, 1992, vol. 256, April 24, p. 437.
32 Arnold S. Relman, "Fraud in science: causes and remedies," *Scientific American*, 1989, April, p. 126.
33 Anon., "Cold fusion at Texas A&M: after a four month international review there is no direct evidence of fraud," *Science*, 1990, vol. 250, December 14, p. 1507.
34 George G. Reisman, "The growing abundance of natural resources and the wastefulness of recycling," in Jay H. Lehr (ed.), *Rational Readings on Environmental Concerns*, New York, Van Nostrand Reinhold, 1992, pp. 631–7.
35 Clark Wiseman, "Dumping: less wasteful than recycling," in Lehr, op. cit., pp. 637–9.
36 Thomas W. Culliney, David Pimentel, and Marcia H. Pimentel, "Pesticides and natural toxicants in foods," in David Pimentel and Hugh Lehman (eds), *The Pesticide Question, Environment, Economics, and Ethics*, New York, Chapman and Hall, 1993, pp. 126–50.
37 Whelan, op. cit.
38 John Maddox, *The Doomsday Syndrome*, London, Macmillan, 1972.
39 Jayadev Chowdhury, "The battle over the environment," *Chemical Engineering*, 1991, October, pp. 30–45.
40 M. Alice Ottoboni, *The Dose Makes the Poison. A Plain-Language Guide to Toxicology*, Berkeley CA, Vincente Books, 1984, p. 179.
41 Alvin Weinberg, "Science and trans-science," *Minerva*, 1972, vol. 10, p. 209.
42 Daniel D. Chiras, *Environmental Science, Action for a Sustainable Future* (3rd edn), Redwood City CA, Benjamin Cummings Publishing Co., 1991.
43 B. J. Nebel, and R. T. Wright, *Environmental Science, The Way the World Works* (4th edn), Englewood Cliffs NJ, Prentice-Hall, 1993, p. 315.
44 Nigel Bunce, *Environmental Chemistry* (2nd edn), Winnipeg, Wuerz Publishing, 1994.
45 T. VanDer Ziel, "Eco-tourism and the dance around the native," (translated title), *Nederlands Dagblad*, 1993, November 20, p. 76.
46 Günay Kocasoy, "Effects of tourist population pressure on pollution of coastal seas," *Environmental Management*, 1995, vol. 19, no. 1, pp. 75–9.

Chapter 4 Chemistry, chemicals, and the chemists

1 Victor B. Scheffer, *The Shaping of Environmentalism in America*, Seattle, University of Washington Press, 1991.
2 J. Cairns and D. I. Mount, "Aquatic toxicology," *Environmental Science and Technology*, 1990, vol. 24, no. 2, pp. 154–61.
3 S. Pope, M. Appleton, and E. Wheal, *The Green Book*, London, Hodder & Stoughton, 1991.
4 Colin Johnson, *Green Dictionary*, London, Macdonald & Co. Ltd., 1991.
5 G. Jones, A. Robertson, J. Forbes, and G. Hollier, *Collins Reference Dictionary of Environmental Science*, London, Collins, 1991.
6 Ibid.
7 John Holum, *Elements of General and Biological Chemistry* (9th edn), New York, John Wiley & Sons Inc., 1995.
8 M. Hein, L. R. Best, S. Pattison, and S. Arena, *College Chemistry, An Introduction to General, Organic and Biochemistry* (5th edn), Pacific Cove CA, Brooks/Cole Publishing Co., 1993.
9 F. A. Bettelheim and J. March, *Introduction to General, Organic and Biochemistry* (3rd edn), New

York, Saunders College Publishing, 1991.
10 P. Kruus and I. M. Valeriote (eds), *Controversial Chemicals, A Citizen's Guide* (2nd edn), Montreal, Multiscience Publications Ltd., 1984.
11 Jones *et al.*, op. cit.
12 Johnson, op. cit., p. 236.
13 F. J. Monkhouse and J. Small, *A Dictionary of the Natural Environment* (2nd edn), New York, John Wiley & Sons, 1978.
14 Martha Windholz (ed.), *The Merck Index* (10th edn), Rahway NJ, Merck & Co. Inc., 1983, Entry no. 9375.
15 Sybil Parker (ed.), *McGraw-Hill Dictionary of Scientific and Technical Terms* (4th edn), New York, McGraw-Hill, 1989.
16 Rachel Carson, *Silent Spring*, Greenwich CO, Fawcett Publications, Inc., 1962.
17 James B. Hendrickson, *Molecules of Nature*, New York, W. A. Benjamin Inc., 1965, p. 58.
18 W. B. Turner, *Fungal Metabolites*, New York, Academic Press, 1971.
19 Louis F. Fieser and Mary Fieser, *Steroids*, New York, Reinhold Publishing Corporation, 1959.
20 Timothy Johns, *With Bitter Herbs They Shall Eat It, Chemical Ecology and the Origins of Human Diet and Medicine*, Tucson AZ, University of Arizona Press, 1990, p. 9.
21 Margaret B. Kreig, *Green Medicine*, Chicago, Rand McNally & Co., 1964.
22 D. L. Illman, "Green technology presents challenge to chemists," *Chemical and Engineering News*, 1993, September 6, pp. 26–33.
23 Alan Newman, "Designer chemistry," *Environmental Science and Technology*, 1994, vol. 28, no. 11, p. 463A.

Chapter 5 How toxic is toxic?

1 W. A. N. Dorland, *Dorland's Illustrated Medical Dictionary* (27th edn), Philadelphia, W. B. Saunders Co., 1988.
2 Sybil P. Parker (ed.), *Dictionary of Scientific and Technical Terms*, New York, McGraw-Hill, 1989.
3 Ida Dix, John B. Melloni, and Gilbert M. Eisner, *Melloni's Illustrated Medical Dictionary* (2nd edn), Baltimore MD, Williams and Wilkins, 1985.
4 Alfonso R. Gennaro and George M. Gould, *Blakiston's Gould Medical Dictionary* (4th edn), New York, McGraw-Hill, 1979.
5 M. Alice Ottoboni, *The Dose Makes the Poison. A Plain-Language Guide to Toxicology*, Berkeley CA, Vincente Books, 1984.
6 Joseph V. Rodricks, *Calculated Risks: The Toxicity and Human Health Risks of Chemicals in Our Society*, Cambridge MA, Cambridge University Press, 1992.
7 David M. Halton and Christian Millet, *What Makes a Chemical Poisonous?*, Hamilton ON, Canadian Centre for Occupational Health and Safety, Publication P87-8E, 1988.
8 Paracelsus' real name was Philip Theophrastus Bombast von Hohenheim. He Latinized his name and published his medical observations, the "Drey Bücher" in Cologne in 1564. His famous statement in the original old German reads: "Was ist das mit gifft ist: alle ding sind gifft und nichts ohn gifft/ Allein die dozis macht das ein ding ein gifft ist."
9 John Cairns and Donald I. Mount, "Aquatic toxicology," *Environmental Science and Technology*, 1990, vol. 24, no. 2, pp. 154–61.
10 P. Kruus and I. M. Valeriote (eds), *Controversial Chemicals, A Citizen's Guide* (2nd edn), Montreal, Multiscience Publications Ltd., 1984, p. 77.
11 Elisabeth Whelan, *Toxic Terror*, Ottawa IL, Jameson Books, 1985, p. 36.
12 Alvin Weinberg, "Science and trans-science," *Minerva*, 1972, vol. 10, pp. 209–22.
13 H. F. Smyth, "Sufficient challenge," *Food and Cosmetics Toxicology*, 1967, vol. 5, pp. 51–8.
14 W. H. Baarschers, A. I. Bharath, M. Hazenberg, and J. E. Todd, "Fungitoxicity of methoxychlor and fenitrothion and the environmental impact of their metabolites," *Canadian Journal of Botany*, 1980, vol. 58, no. 4, pp. 426–31.
15 T. D. Luckey, *Radiation Hormesis*, Boca Raton CA, CRC Press, 1990.

16 A. R. D. Stebbing, "Hormesis – the stimulation of growth by low levels of inhibitors," *The Science of the Total Environment*, 1982, vol. 22, pp. 213–34.
17 Elizabeth A. Martin (ed.), *Concise Medical Dictionary* (2nd edn), Oxford, Oxford University Press, 1980.
18 E. Hodgson, R. B. Mailman, and J. E. Chambers, *Dictionary of Toxicology*, New York, Van Nostrand Reinhold Company, 1988.
19 Greta G. Fein, Joseph L. Jacobson, Sandra W. Jacobson, Pamela M. Schwarz, and Jeffrey K. Dowler, "Prenatal exposure to polychlorinated biphenyls: effects on birth size and gestational age," *Journal of Pediatrics*, 1984, vol. 105, no. 2, pp. 315–20.
20 Joseph L. Jacobson, Sandra W. Jacobson, and Harold E. B. Humphrey, "Effects of in utero exposure to polychlorinated biphenyls and related contaminants on cognitive functioning in young children," *Journal of Pediatrics*, 1990, vol. 116, no.1, pp. 38–45.
21 J. R. Allen, D. A. Barsotti, and L. A. Carstens, "Residual effects of polychlorinated biphenyls on adult non-human primates and their offspring," *Journal of Toxicology and Environmental Health*, 1980, vol. 6, pp. 55–66.
22 Lynda Dickinson, *Victims of Vanity*, Toronto, Summerhill Press, 1989, p. 30.
23 Daniel D. Chiras, *Environmental Science, Action for a Sustainable Future* (3rd edn), Redwood City CA, Benjamin Cummings Publishing Co., 1991, p. 323.
24 *A Prescription for Healthier Great Lakes, A Report on the Program for Zero Discharge*, A Joint Project of the National Wildlife Federation and the Canadian Institute for Environmental Law and Policy, Toronto, 1991, p. 10.
25 B. J. Feder, "Chemical firms abandoning medical equipment business," *Toronto Globe and Mail*, 1994, May 5, p. B18.
26 Mark A. Karaffa, Jean Ghesson, and James Russell, *Evaluation of Asbestos Levels in Two Schools Before and After Removal*, Office of Research and Development, Washington DC, US Environmental Protection Agency, 1989, Report on Contract no. 58-03-4006.
27 Donald VanDeVeer and Christine Pierce, *People, Penguins, and Plastic Trees, Basic Issues in Environmental Ethics*, Belmont CA, Wadsworth Inc., 1986.
28 Timothy Johns, *With Bitter Herbs They Shall Eat It*, Tucson AZ, University of Arizona Press, 1990.
29 Michael Bliss, *The Discovery of Insulin*, Chicago, University of Chicago Press, 1982.
30 Rebecca Kolberg, "Animal models point the way to human clinical trials," *Science*, 1992, vol. 256, May 8, pp. 772–3.
31 Constance Holden, "A pivotal year for lab animal welfare," *Science*, 1986, vol. 232, April 11, pp. 147–50.
32 Deborah M. Barnes, "Tight money squeezes out animal models," *Science*, 1986, vol. 232, April 18, pp. 309–11.
33 Rachel Carson, *Silent Spring*, Greenwich CO, Fawcett Publications Inc., 1962, p. 34.
34 Dickinson, op. cit.
35 Susan J. Ainsworth, "Cosmetics suppliers rally to meet consumers demand," *Chemical and Engineering News*, 1992, April 20, pp. 31–52.
36 Ibid, p. 44.
37 Alan M. Goldberg and John M. Frazier, "Alternatives to animals in toxicity testing," *Scientific American*, 1989, vol. 261, August, pp. 24–30.
38 Bruce N. Ames, W. E. Durston, E. Yamasaki, and F. D. Lee, "Carcinogens and mutagens: a simple test system combining liver homogenates for activation and bacteria for detection," *Proceedings of the National Academy of Sciences*, USA, 1970, pp. 2281–5.
39 Ronald A. House, "The Ames test – a critical appraisal," *Occupational Health in Ontario*, 1980, vol. 1, no. 2, pp. 2–11.
40 Subhash C. Basak and Gerald J. Niemi, "Predicting properties of molecules using graph invariants," *Journal of Mathematical Chemistry*, 1991, vol. 7, pp. 243–72.
41 Goldberg and Frazier, op. cit.
42 Whelan, op. cit., p. 31.

Chapter 6 Cancer everywhere?

1 Michael Jacobson, "Practical advice for cancerphobiacs," *Dangerous Properties of Industrial Materials Report*, New York, Van Nostrand Reinhold, November/December, 1980.
2 Anon., "Diagnosis," *Algemeen Dagblad*, 1994, December 30.
3 James D. Watson, *The Double Helix; A Personal Account of the Discovery of the Structure of DNA*, New York, Atheneum, 1968.
4 John Holum, *Elements of General, Organic, and Biochemistry* (9th edn), New York, John Wiley & Sons Inc., 1995.
5 Carl Sagan, *The Dragons of Eden – Speculations on the Evolution of Human Intelligence*, New York, Random House Inc., 1977.
6 B. J. Nebel and R. T. Wright, *Environmental Science, The Way the World Works* (4th edn), Englewood Cliffs NJ, Prentice-Hall, 1993, p. 314.
7 Daniel D. Chiras, *Environmental Science, Action for a Sustainable Future* (3rd edn), Redwood City CA, Benjamin Cummings Publishing Co., 1991, p. 318.
8 Ernest Hodgson, Richard B. Mailman, and Janice E. Chambers, *Dictionary of Toxicology*, New York, Van Nostrand Reinhold Company, 1988.
9 Ibid.
10 Samuel Epstein, "The political and economic basis of cancer," *Technology Review*, 1976, July/August, p. 35.
11 Thomas H. Maugh, "Cancer and the environment, Higginson speaks out," *Science*, 1979, vol. 205, September 28, pp. 1363–6.
12 Edith Efron, *The Apocalyptics, Cancer and the Big Lie*, New York, Simon and Schuster, 1984, pp. 96–119 and pp. 135–76.
13 Bruce N. Ames, W. E. Durston, E. Yamasaki, and F. D. Lee, "Carcinogens and mutagens: a simple test system combining liver homogenates for activation and bacteria for detection," *Proceedings of the National Academy of Sciences*, USA, 1970, pp. 2281–5.
14 Bruce N. Ames, "Dietary carcinogens and anticarcinogens," *Science*, 1983, vol. 221, September 23, pp. 1256–64.
15 Nancy Wertheimer and Ed Leeper, "Electrical wiring configurations and childhood cancer," *American Journal of Epidemiology*, 1979, vol. 109, no. 3, pp. 273–84.
16 W. E. Wright, J. M. Peters, and T. M. Mack, "Leukaemia in workers exposed to electrical and magnetic fields," *The Lancet*, 1982, November 20, p. 1160.
17 P. H. Ruey, S. Lin, Patricia C. Dischinger, Jose Conde, and Katherine P. Farrell, "Occupational exposure to electromagnetic fields and the occurrence of brain tumours," *Journal of Occupational Medicine*, 1985, vol. 27, no. 6, pp. 413–19.
18 W. E. Wright *et al.*, op. cit.
19 J. D. Bowman, D. H. Garaband, E. Sobel, and J. M. Peters, "Exposures to extremely low frequency (ELF) electromagnetic fields in occupations with elevated leukaemia rates," *Applied Industrial Hygiene*, 1988, vol. 3, no. 6, pp. 189–94.
20 Michel Coleman and Valerie Beral, "A review of epidemiological studies of the health effects of living near or working with electricity generation and transmission equipment." *International Journal of Epidemiology*, 1988, vol. 17, no. 1, pp. 1–13.
21 Bette Hileman, "Health effects of electromagnetic fields remain unresolved," *Chemical and Engineering News*, 1993, November 8, pp. 15–29.
22 Jeffrey D. Saffer and Sarah J. Thurston, "Cancer risks and electromagnetic fields," *Nature*, 1995, vol. 375, May 4, p. 22.
23 Adam Lacey-Hulbert, Roger C. Wilkins, T. Robin Hesketh, and James C. Metcalfe, "Cancer risks and electromagnetic fields," *Nature*, 1995, vol. 375, May 4, p. 23.
24 Rachel Carson, *Silent Spring*, Greenwich CO, Fawcett Publications Inc., 1962, p. 205.
25 Samuel Epstein, *The Politics of Cancer*, San Francisco CA, Sierra Club Books, 1978.
26 M. Alice Ottoboni, *The Dose Makes the Poison. A Plain-Language Guide to Toxicology*, Berkeley CA, Vincente Books, 1984, p. 145.
27 Efron, op. cit., p. 408.

28 A. R. D. Stebbing, "Hormesis – the stimulation of growth by low levels of inhibitors," *The Science of the Total Environment*, 1982, vol. 22, pp. 213–34.
29 Stephen Safe, "Synthesis and application of 1,3,6,8, and 2,4,6,8-tetrasubstituted dibenzofurans and -dibenzo-p-dioxins as antitumorigenic agents," US Patent Application no. 300,416, January 23, 1989, *Chemical Abstracts*, 1991, vol. 114, abstract no. 61914.
30 H. F. Smyth, "Sufficient challenge," *Food and Cosmetics Toxicology*, 1967, vol. 5, pp. 51–8.
31 Sohei Kondo, *Health Effects of Low-level Radiation*, Osaka, Japan, Kinki University Press, and Madison WI, Medical Physics Publishing, 1993, Chapter 3.
32 T. D. Luckey, *Radiation Hormesis*, Boca Raton CA, CRC Press, 1990.
33 Bruce N. Ames and Lois Swirsky Gold, "Environmental pollution and cancer: some misconceptions," in Jay H. Lehr (ed.), *Rational Readings on Environmental Concerns*, New York, Van Nostrand Reinhold, 1992, pp. 151–67.
34 Jean Marx, "Animal carcinogen testing challenged," *Science*, 1990, vol. 250, September 9, pp. 743–5.
35 The issue of naturally occurring organochlorine chemicals will be discussed in detail in Chapter 12.
36 Elizabeth Whelan, *Toxic Terror*, Ottawa IL, Jameson Books, 1985, p. 30.
37 D. M. Shankel, P. E. Hartman, T. Kada, and A. Hollaender (eds), *Antimutagenesis and Anticarcinogenesis Mechanisms*, New York, Plenum Press, 1986.
38 Ames, op. cit.
39 Timothy Johns, *With Bitter Herbs They Shall Eat It*, Tucson AZ, University of Arizona Press, 1990.
40 Aaron Novick and Leo Szilard, "Antimutagens," *Nature*, 1952, vol. 170, November 29, p. 926.
41 Shankel *et al.*, op. cit.
42 F. Xavier Bosch *et al.*, "Prevalence of human papillomavirus in cervical cancer: a worldwide perspective," *Journal of the National Cancer Institute*, 1995, June 7, pp. 796–802.
43 Peter Parham, "Politically incorrect viruses," *Nature*, 1994, vol. 368, April 7, pp. 495–6.
44 James A. Cleaver, "DNA repair and replication in *xeroderma pigmentosum* and related disorders," in D.E. Shankel *et al.*, op. cit., p. 425.
45 Jean Mark, "Zeroing in on individual cancer risk," *Science*, 1991, vol. 253, August 9, pp. 612–16.
46 Rachel Nowak, "Breast cancer gene offers surprises," *Science*, 1994, vol. 265, September 23, pp. 1796–9.
47 Yoshio Miki *et al.*, "A strong candidate for the breast and ovarian cancer susceptibility gene BRCA1," *Science*, 1994, vol. 266, October 7, pp. 66–71.
48 Joseph Levine and David Suzuki, *The Secret of Life*, Toronto, Stoddart Publishing, 1993.
49 H. Mussalo-Rauhamaa, E. Häsänen, H. Pyysalo, K. Antervo, R. Kauppila, and P. Pantzar, "Occurrence of beta-hexachlorocyclohexane in breast cancer patients," *Cancer*, 1990, vol. 66, pp. 2124–8.
50 G. N. Wogan and Wm. F. Busby, "Naturally occurring carcinogens," in I. E. Liener (ed.), *Toxic Constituents of Plant Foodstuffs*, New York, Academic Press, Inc., 1980, pp. 329–70.
51 Brian E. Henderson, Ronald K. Ross, and Malcolm C. Pike, "Toward the primary prevention of cancer," *Science*, 1991, vol. 254, September 22, pp. 1131–8.
52 Gio Batta Gori, "Regulation of cancer-causing substances: utopia or reality," *Chemical and Engineering News*, 1982, September 6, pp. 25–32.

Chapter 7 Energy and the laws of nature

1 Sybil Parker (ed.), *McGraw-Hill Dictionary of Scientific and Technical Terms*, (4th edn), New York, McGraw-Hill Book Co., 1989.
2 J. H. Harker and J. R. Backhurst, *Fuel and Energy*, London, Academic Press, 1981.
3 Tim Padmore, "Affluence for 10 billion," *The Vancouver Sun*, 1974, October 12.
4 Barry Commoner, *The Closing Circle: Nature, Man, and Technology*, London, Jonathan Cape Ltd., 1972, p. 45.

5 Fred Pearce, "First brick in China's colossal dam," *New Scientist*, 1994, no. 24, December 31, p. 6.
6 The Coal Association of Canada, *The Coal Kit*, Calgary AB, 1991.
7 Isaac Asimov, *Isaac Asimov's Biographical Encyclopedia of Science and Technology* (revised edn), New York, Avon Books, 1972, p. 466.
8 Barry Commoner, "The environmental costs of economic growth" in Robert Dorfman and Nancy S. Dorfman (eds), *Economics of the Environment, Selected Readings*, New York, W.W. Norton & Company Inc., 1972, pp. 261–83.
9 "Natural gas is gunning out ... now what?" (headline translated), Dutch Newspaper *De Telegraaf*, 1992, October 17, p. T23.
10 M. Eisenbud and G. Gleeson (eds), *Electric Power Generation and Thermal Discharges; Thermal Considerations in the Production of Electric Power*, New York, Gordon and Breach, 1969.
11 Alvin Weinberg, "Science and trans-science," *Minerva*, 1972, vol. 10, pp. 209–22.
12 John J. McKetta, "Acid rain – the whole story to date," in Jay H. Lehr (ed.), *Rational Readings on Environmental Concerns*, New York, Van Nostrand Reinhold, 1992, pp. 44–58.
13 The Coal Association of Canada, op. cit., Module 9.
14 Gunilla Asplund and Anders Grimvall, "Organohalogens in nature, more widespread than previously assumed," *Environmental Science and Technology*, 1991, vol. 25, no. 8, pp. 1346–50.
15 Anon., "The European Commission (EC) proposes to control evaporative emissions of gasoline from cars," *Chemical and Engineering News*, 1990, vol. 24, no. 8, p. 1113.
16 Anon., "Washington DC-area motor vehicles could contribute more than 40 tons per day of smog precursors to the atmosphere," *Chemical and Engineering News*, 1990, vol. 24, no. 11, p. 1609.
17 L. B. Lave, W. E. Wecker, W. S. Reis, and D. A. Ross, "Controlling emissions from motor vehicles," *Environmental Science and Technology*, 1990, vol. 24, no. 8, pp. 1128–35.
18 Ontario Ministry of the Environment, Air Resources Branch, *Polynuclear Aromatic Hydrocarbons*, Toronto, September 1979, ARB-TDA Report no. 58–79.
19 Paul J. Lioy and Arthur Greenberg, "Factors associated with human exposure to polycyclic aromatic hydrocarbons," *Toxicology and Industrial Health*, 1990, vol 6, no. 2, pp. 209–23.
20 Gene W. Bartholomew and Martin Alexander, "Soil as a sink for atmospheric carbon monoxide," *Science*, 1981, vol. 212, June 19, pp. 1389–91.
21 J. B. Kemeny (chair), *Report of the President's Commission on the Accident at Three Mile Island*, Washington DC, Superintendent of Documents, 1979.
22 Henry N. Wagner and Linda E. Ketchum, *Living with Radiation, The Risk, The Promise*, Baltimore MD, Johns Hopkins University Press, 1989, p. 114.
23 Ibid.
24 Daniel D. Chiras, *Environmental Science, Action for a Sustainable Future*, (3rd edn), Redwood City CA, Benjamin Cummings Publishing Co., 1991, p. 260.
25 Leo Yaffe, "The health hazards of not going nuclear," *Chemistry in Canada*, 1979, December, pp. 25–32.
26 Ibid.
27 C. Holden, "Psychiatrists study Three-Mile trauma," *Science*, 1980, vol. 208, p. 37.
28 Wagner and Ketchum, op. cit., p. 140.
29 Wagner and Ketchum, op. cit., p. 114.
30 T. D. Luckey, *Radiation Hormesis*, Boca Raton CA, CRC Press, 1991, p. 234.
31 Anon., "Another panel rejects Nevada disaster theory," *Science*, 1992, vol. 256, April 24, p. 434.
32 Robert G. Cochran and Nicholas Tsoulfanidis, *The Nuclear Fuel Cycle: Analysis and Management*, La Grange Park IL, American Nuclear Society, 1990, p. 220.
33 Cochran and Tsoulfonidis, op. cit., p. 315.
34 K. Schrader-Frechette, "Ethical dilemmas and radioactive waste," *Environmental Ethics*, 1991, vol. 13, no. 4, pp. 327–43.
35 P. Wuebben, A. C. Lloyd, and J. H. Leonard, "The future of electric vehicles in meeting the air quality challenges in southern California," *Electric Vehicle Technology*, Society of Automotive Engineers Publication, 1990, vol. SP-817, February, pp. 107–20.
36 Dieter W. Gruenwedel, "Industrial and environmental chemicals in the human food chain,"

Chain, New York, Van Nostrand Reinhold, 1990, p. 161.

37 The World Bank, *Alcohol from Biomass in the Developing Countries*, Washington DC, The World Bank, 1980.

38 Bern Keating, "The gasohol gamble, a bad bet for the future," *Next*, 1980, September/October, pp. 66–9.

39 Chiras, op. cit., p. 277.

40 The World Bank, op. cit.

41 V. D. Phillips and P. K. Takahashi, "Methanol from biomass," *Environmental Science and Technology*, 1990, vol. 24, no. 8, pp. 1136–7.

42 Julian Rose, "Biofuel benefits questioned," *Environmental Science and Technology*, 1994, vol. 28, no. 2, p. 63A.

43 W. H. Baarschers and H. S. Heitland, "Biodegradation of fenitrothion and fenitro-oxon by the fungus *Trichoderma viride*," *Journal of Agricultural and Food Chemistry*, 1986, vol. 34, pp. 707–9.

44 Charles L. Gray and Jeffrey A. Alson, *Moving America to Methanol: A Plan to Replace Oil Imports*, Ann Arbor MI, University of Michigan Press, 1985.

45 Daniel Sperling, *New Transportation Fuels: A Strategic Approach to Technological Change*, Berkeley CA, University of California Press, 1989.

46 Anon., "The great Berlin greenhouse compromise," *Nature*, 1995, vol. 374, April 27, pp. 749–50.

47 Anon., "Making nuclear power usable again," *Nature*, 1995, vol. 375, May 11, pp. 91–2.

48 C. Starr, M. F. Searl, and S. Alpert, "Energy sources, a realistic outlook," *Science*, 1992, vol. 256, May 15, pp. 981–7.

49 Philip Abelson, "Increased use of renewable energy," *Science*, 1991, vol. 253, September 6, p. 1073.

50 Starr *et al.*, op. cit.

51 Michael Heylin, "Sociology of cold fusion examined," *Chemical and Engineering News*, 1990, March 19, p. 24.

52 D. J. Freeman and F. C. R. Cattell, "Wood burning as a source of atmospheric polycyclic aromatic hydrocarbons," *Environmental Science and Technology*, 1990, vol. 24, no. 10, pp. 1581–5.

53 Julian L. Simon, *The Ultimate Resource*, Princeton, NJ, Princeton University Press, 1981, p. 90.

54 Starr *et al.*, op cit.

55 Harry Perry, "Coal in the United States: a status report," *Science*, 1983, vol. 221, August 19, p. 377.

56 M. King Hubbert, "Survey of the world's energy resources," *Canadian Mining and Metallurgical Bulletin*, 1973, vol. 66, no. 735, July, pp. 37–55.

Chapter 8 Air for breathing

1 James Lovelock, *The Ages of Gaia*, New York, W. W. Norton & Company, 1988.

2 Sherwood Idso, *Carbon Dioxide and Global Change: Earth in Transition*, Tempe AZ, IBR Press, Institute for Biospheric Research, 1989, pp. 1–10.

3 E. G. Nisbet, *Leaving Eden, To Protect and Manage the Earth*, Cambridge, Cambridge University Press, 1991.

4 Nigel Bunce, *Environmental Chemistry*, (2nd edn), Winnipeg, Wuerz Publishing, 1994.

5 David Kemp, *Global Environmental Issues, A Climatological Approach*, London, Routledge, 1990.

6 Lovelock, op. cit.

7 Isaac Asimov, *Asimov on Chemistry*, Garden City NY, Doubleday & Company Inc., 1974, pp. 160–70.

8 J. R. Kasting, O. B. Toon, and J. B. Pollack, "How climate evolved on the terrestrial planets," *Scientific American*, 1988, vol. 258, no. 2, pp. 90–7.

9 H.G. Hengeveld, "Global warming: why the controversy?," *Canadian Chemical News*, 1991, August, pp. 17–19.

10 Barry Commoner in "Balance and Biosphere", Toronto, Canadian Broadcasting Corporation, 1972, p. 81.

11 Sherwood Idso, "Do increases in atmospheric CO_2 have a cooling effect on surface air temperature?" *Climatological Bulletin*, 1983, vol. 17, p. 22.
12 J. D. Cure and B. Acock, "Crop response to CO_2 doubling," *Agricultural and Forestry Meteorology*, 1986, vol. 38, pp. 127–45.
13 Idso, 1989, op. cit., p. 9.
14 Bette Hileman, "Web of interactions makes it difficult to untangle global warming data," *Chemical and Engineering News*, 1992, April 27, pp. 7–19.
15 Leo Yaffe, "The health hazards of not going nuclear," *Chemistry in Canada*, 1979, December, pp. 25–32.
16 Dixy Lee Ray and Lou Guzzo, *Trashing the Planet*, Washington DC, Regnery Gateway, 1990, p. 57.
17 F. S. Rowland and M. J. Molina, "Ozone depletion, 20 years after the alarm," *Chemical and Engineering News*, 1994, August 15, p. 8–13.
18 Anon., "Antarctic spring of 1989 showed deepest ozone hole recorded so far," *Environmental Science and Technology*, 1990, vol. 24, no. 1, p. 6.
19 Lovelock, op. cit., p. 90.
20 Anon., "A 40 percent depletion of stratospheric ozone would reduce the yield of Loblolly Pine pulpwood by 30 percent," *Environmental Science and Technology*, 1990, vol. 24, no. 6, p. 766.
21 R. Douglas Evans and Frank H. Rigler, "Calculation of the total anthropogenic lead in the sediments of a rural Ontario lake," *Environmental Science and Technology*, 1980, vol. 14, no. 2, pp. 216–18.
22 K. S. Subramanian and J. W. Connor, "Lead contamination of drinking water, metals leached from soldered pipes may pose health hazard," *Journal of Environmental Health*, 1991, vol. 54, no. 2, pp. 29–32.
23 Lester B. Lave and Eugene P. Seskin, "Air pollution and human health," *Science*, 1970, vol. 169, August 21, pp. 723–33.
24 Jane V. Hall *et al.*, "Valuing the health benefits of clean air," *Science*, 1992, vol. 255, January 14, pp. 812–16.
25 Simon P. Wolff, "Air pollution and cancer risk," *Nature*, 1992, vol. 356, April 9, p. 471.
26 Henry N. Wagner and Linda E. Ketchum, *Living with Radiation, the Risk, The Promise*, Baltimore MD, Johns Hopkins University Press, 1989.
27 T. D. Luckey, *Radiation Hormesis*, Boca Raton CA, CRC Press, 1991, p. 234.
28 Luckey, op. cit., Chapter 1, pp. 1–31.
29 Sharon L. Roan, *Ozone Crisis*, New York, John Wiley and Sons, 1990.
30 Timothy J. Wallington *et al.*, "The environmental impact of CFC replacements – HFCs and HCFCs," *Environmental Science and Technology*, 1994, vol. 28, no. 7, pp. 320A–6A.
31 Julian Rose, "HCFCs may slow ozone layer recovery," *Environmental Science and Technology*, 1994, vol. 28, no. 3, p. 111A.
32 Ibid.
33 Gordon W. Gribble, "Natural organohalogens, many more than you think," *Journal of Chemical Education*, 1994, vol. 71, no. 11, pp. 907–11.
34 P. Wuebben, A. C. Lloyd, and J. H. Leonard, "The future of electric vehicles in meeting the air quality challenges in southern California," *Electric Vehicle Technology*, Society of Automotive Engineers Publication, 1990, vol. SP-817, February, pp. 107–20.
35 Julian Simon, *The Ultimate Resource*, Princeton NJ, Princeton University Press, 1981.
36 Barbara A. Plog (ed.), *Fundamentals of Industrial Hygiene* (3rd edn), Chicago IL, US National Safety Council, 1988.
37 Mary M. Agocs, *et al.*, "Mercury exposure from indoor latex paint," *New England Journal of Medicine*, 1990, vol. 323, no. 16, pp. 1096–101.
38 John D. Spengler and Ken Sexton, "Indoor air pollution: a public health perspective," *Science*, 1983, vol. 221, July 1, pp. 9–17.
39 L. R. Ember, "Survey finds high indoor levels of volatile organic chemicals," *Chemical and Engineering News*, 1988, December 5, pp. 23–5.
40 Bette Hileman, "Multiple chemical sensitivity," *Chemical and Engineering News*, 1991, July 22, pp. 26–42.

41 Elizabeth Whelan, *Toxic Terror*, Ottawa IL, Jameson Books, 1985, p. 159.
42 Alvin L. Alm, "There is no place like home – but is it safe?," *Environmental Science and Technology*, 1990, vol. 24, no. 5, p. 625.
43 US Environmental Protection Agency, *A Citizen's Guide to Radon*, Washington DC, Superintendent of Documents, *OPA-86-004*, 1986.
44 US Environmental Protection Agency, *Radon Reduction Methods*, Washington DC, Superintendent of Documents, *OPA-87-010*, 1987.
45 William W. Nazaroff and Kevin Teichman, "Indoor radon, exploring US federal policy for controlling human exposures," *Environmental Science and Technology*, 1990, vol. 24, no. 6, pp. 774–82.
46 Nazaroff and Teichman, op. cit.
47 Wm. W. Nazaroff and A. V. Nero (eds), *Radon and its Decay Products in Indoor Air*, New York, John Wiley & Sons, 1988.
48 Philip Abelson, "Mineral dusts and radon in uranium mines," *Science*, 1991, vol. 254, November 8, p. 777.
49 Ibid.
50 Keith J. Schiager, "Radon – risk and reason," in Jay H. Lehr (ed.), *Rational Readings on Environmental Concerns*, New York, Van Nostrand Reinhold, 1992, pp. 619–26.
51 Philip Abelson, "Uncertainties about the health effects of radon," *Science*, 1990, vol. 250, October 19, p. 353.
52 Richard J. Guimond, "Radon risk and EPA," *Science*, 1991, vol. 251, p. 251.
53 J. S. Neuberger, "Indoor radon and lung cancer," *Radiation Protection Dosimetry*, 1992, vol. 40, no. 3, p. 147.
54 Julian Peto and Sarah Darby, "Radon risk reassessed," *Nature*, 1994, vol. 368, March 10, pp. 97–8.
55 W. T. Vonstille and H. L. A. Sacarello, "Radon and cancer, Florida study finds no evidence of increased risk," *Journal of Environmental Health*, 1990, November/December, pp. 25–8.
56 Neuberger, op. cit.
57 Bernard L. Cohen, "Test of the linear no-threshold theory of radiation carcinogenesis for inhaled radon decay products," *Health Physics*, 1995, vol. 68, no. 2, pp. 157–74.
58 J. H. C. Miles, "Indoor radon and lung cancer: reality or myth?," *Radiological Protection Bulletin*, 1991, March, pp. 25–8.
59 Luckey, op. cit.
60 Irving J. Selikoff, "Occupational lung diseases," in Douglas H. K. Lee (ed.), *Environmental Factors in Occupational Lung Disease*, New York, Academic Press, 1972.
61 I. J. Selikoff, "Cancer risk of asbestos exposure," in H. H. Hyatt, J. D. Watson, and J. A. Winsten (eds), *Origins of Human Cancer, Cold Spring Harbor Conference on Cell Proliferation*, Cold Spring Harbor NY, Cold Spring Harbor Press, 1977.
62 Maria Albin MD, Lund University Hospital, Sweden, private communication.
63 Mark A. Karaffa, Jean Ghesson, and James Russell, "Evaluation of asbestos levels in two schools before and after asbestos removal," Cincinnati OH, Report on Contract no. 68-03-4006, US Environmental Protection Agency, 1989.
64 Philip H. Abelson, "The asbestos removal fiasco," *Science*, 1990, vol. 247, March 2, p. 1017.
65 Bill Denault, "Riding the wave, profiting from asbestos removal," *Occupational Health and Safety Canada*, 1990, vol. 6, no. 6, p. 110.
66 Richard Stone, "No meeting of the minds on asbestos," *Science*, 1991, vol. 254, pp. 928–31.

Chapter 9 Water for drinking

1 J. Lycklema and T. E. A. Van Hylckama, "Water, something peculiar," in David H. Speidel, Lon C. Ruedisili, and Allen F. Agnew (eds), *Perspectives on Water, Uses and Abuses*, Oxford, Oxford University Press, 1988.
2 J. W. Maurits la Rivière, "Threats to the world's water," *Scientific American*, 1989, September,

pp. 80–94.

3 Raymond F. Dasmann, "Can Egypt survive progress?," in Garret Hardin (ed.), *Population, Evolution and Birth Control, A Collage of Controversial Ideas* (2nd edn), San Francisco CA, W. H. Freeman and Company, 1969, pp. 52–6.

4 Nicholas Wade, "Sahelian drought: no victory for western aid," *Science*, 1974, vol. 185, July 19, pp. 234–7.

5 Malin Falkenmark and Gunnar Lindh, *Water for a Starving World*, Boulder CO, Westview Press, 1976.

6 Isaac Asimov, *Asimov on Chemistry*, Garden City NY, Doubleday & Co., Inc., 1974.

7 Falkenmark and Lindh, op. cit., p. 12.

8 Environment Canada, *Toxic Chemicals in the Great Lakes*, Ottawa, Supply and Services Canada, 1991.

9 John F. Rosen, "Effects of low levels of lead exposure," *Science*, 1992, vol. 256, April 17, p. 294.

10 Edward C. Krug, "The great acid rain flimflam," in Jay H. Lehr (ed.), *Rational Readings on Environmental Concerns*, New York, Van Nostrand Reinhold, 1992, pp. 35–43.

11 David D. Kemp, *Global Environmental Issues, A Climatological Approach*, London, Routledge, 1990, p. 94.

12 Dixy Lee Ray and Lou Guzzo, *Trashing the Planet*, Washington DC, Regnery Gateway, 1990, p. 64.

13 Stanton Miller, "Eisenreich and the transport of airborne chemicals in aquatic systems," *Environmental Science and Technology*, 1995, vol. 28, no. 2, pp. 92A–4A.

14 Bernhard Ulrich, "Waltsterben: forest decline in West Germany," *Environmental Science and Technology*, 1990, vol. 24, no. 5, pp. 436–41.

15 General Directorate for Environmental Hygiene, the Netherlands, *Interim Evaluation of Policies on Acidification*, Report no. 23D7012 (title translated), The Hague, 1987, December 23, p. 15.

16 "Wells polluted, tests show," *Toronto Globe and Mail*, 1992, November 5.

17 A. Bharath, C, Mallard, D. Orr, G. Ozburn, and A. Smith, "Problems in determining the water solubility of organic compounds," *Bulletin of Environmental Contamination and Toxicology*, 1984, vol. 33, pp. 133–7.

18 Logan A. Norris, "The movement, persistence and fate of the phenoxy herbicides and TCDD in the forest," *Residue Reviews*, 1981, vol. 80, pp. 65–135.

19 David A. Belluck and Sally L. Benjamin, "Pesticides and human health. Defining acceptable and unacceptable risk levels," *Journal of Environmental Health*, 1990, vol. 53, pp. 11–13.

20 Walter Klöpfer, "Photochemical degradation of pesticides and other chemicals in the environment: a critical assessment of the state of the art," *The Science of the Total Environment*, 1992, vol. 123/124, pp. 145–59.

21 M. Alexander and M. I. H. Aleem, "Effect of chemical structure on microbial decomposition of aromatic herbicides," *Agricultural and Food Chemistry*, 1961, vol. 9, no. 1, January/February, pp. 44–7.

22 William C. White and Thomas R. Sweeney, "The metabolism of DDT, a metabolite from rabbit urine, di(*p*-chlorodiphenyl)acetic acid; its isolation, identification and synthesis," *Public Health Reports*, 1945, vol. 60, January 19, pp. 66–71.

23 M. K. Ahmed and J. E. Casida, "Metabolism of some organophosphorus insecticides by micro-organisms," *Journal of Economic Entomology*, 1958, vol. 51, pp. 59–63.

24 S. Dagley, W. C. Evans, and D. W. Ribbons, "New pathways in the oxidative metabolism of aromatic compounds by micro-organisms," *Nature*, 1960, vol. 188, November, p. 560.

25 B. J. Nebel and R. T. Wright, *Environmental Science, The Way the World Works* (4th edn), Englewood Cliffs NJ, Prentice-Hall, 1993, pp. 314–15.

26 Barry Commoner, "The environmental costs of economic growth," in Robert Dorfman and Nancy Dorfman (eds), *Economics of the Environment, Selected Readings*, New York, W. W. Norton & Company, Inc., 1972, p. 261.

27 John A. Bumpus, Ming Tien, David Wright, and Steven D. Aust, "Oxidation of persistent environmental pollutants by a white rot fungus," *Science*, 1985, vol. 228, June 28, pp. 1434–6.

28 D. Ghosal, I.-S. You, D. K. Chatterjee, and A. M. Chakrabarty, "Microbial degradation of

halogenated compounds," *Science*, 1985, vol. 228, April 12, pp. 135–42.
29 Gerhard Gottschalk and Hans-Joachim Knackmuss, "Bacteria and the biodegradation of chemicals achieved naturally, by combination, or by construction," *Angewandte Chemie, Int. English Edn*, 1993, vol. 32, pp. 1398–408.
30 F. Matasamura, K. C. Petil, and G. M. Boush, "DDT metabolized by micro-organisms from Lake Michigan," *Nature*, 1971, vol. 230, pp. 325–6.
31 Norris, op. cit., p. 113.
32 Jack Manno, Sheila Meyers, Dieter Riedel, and Neil Tremblay, "Effects of Great Lakes basin environmental contaminants on human health," *State of the Great Lakes Ecosystem Conference Report*, Ottawa, Environment Canada, July 1994.
33 Julian Simon, *The Ultimate Resource*, Princeton NJ, Princeton University Press, 1981, p. 249.
34 Graeme Rodden, "When it comes to water, Canada does most things right," *Canadian Chemical News*, 1994, July/August, p. 11.
35 W. H. Baarschers, J. Elvish, and S. P. Ryan, "Adsorption of fenitrothion and 3-methyl-4-nitrophenol on soils and sediments," *Bulletin of Environmental Contamination and Toxicology*, 1983, vol. 30, pp. 621–7.
36 The International Bank for Reconstruction and Development/The World Bank, *World Development Report 1992, Development and the Environment*, Oxford, Oxford University Press, 1992, p. 46.
37 The International Bank for Reconstruction and Development/The World Bank, op. cit., p. 49.
38 Jocelyn Kaiser, "Alzheimer's: could there be a zinc link?" *Science*, 1994, vol. 265, September 2, p. 1365.
39 Nebel and Wright, op. cit., p. 255.
40 M. Hamilton, "Risk perspective on water fluoridation," *Journal of Environmental Health*, 1992, vol. 54, no. 6, p. 27.
41 J. Seiber, "Industrial and environmental chemicals in the human food chain – Part 2, Organic chemicals," in Carl K. Winter, James N. Seiber, and Carole F. Nuckton (eds), *Chemicals in the Human Food Chain*, New York, Van Nostrand Reinhold, 1990, p. 193.
42 R. R. Christman, "A perspective on chlorination," *Environmental Science and Technology*, 1980, vol. 14, No, 1, p. 5.
43 Robert D. Morris, Anne-Marie Audet, Italo F. Angelillo, Thomas C. Chalmers, and Frederick Mosteller, "Chlorination, chlorination by-products and cancer: a meta-analysis," *American Journal of Public Health*, 1992, vol. 82, no. 7, pp. 955–63.
44 Roy L. Wolfe, "Ultraviolet disinfection of potable water," *Environmental Science and Technology*, 1990, vol. 24, no. 6, pp. 768–73.
45 Anon., "Tucson, AZ, will use ozone to disinfect water at its new 150 million gal/day potable water treatment plant," *Environmental Science and Technology*, 1990, vol. 24, no. 1, p. 6.
46 Enric Volante, "Pouring CAP water on reservation weighed," *The Arizona Daily Star*, 1995, March 28, p. 1.
47 Wolfe, op. cit.
48 The International Bank for Reconstruction and Development/World Bank, op. cit., p. 197.
49 Maurits la Rivière, op. cit.
50 Rodden, op. cit.
51 Dieter W. Gruenwedel, "Industrial and environmental chemicals in the human food chain – Part 1, Inorganic chemicals," in Winter *et al.*, op. cit., p. 166.
52 "System would cost $1.7 billion," *Toronto Globe and Mail*, July 6, 1994.
53 Julian Rose, "And the detergent 'eco-label' goes to....," *Environmental Science and Technology*, 1994, vol. 28, no. 4, p. 179A.
54 Maurits la Rivière, op. cit.
55 Gruenwedel, op. cit. p. 177.
56 D. Wheeler, M. L. Richardson, and J. Bridges (eds), *Watershed 89. The Future for Water Quality in Europe, vol. II*, Oxford, Pergamon Press, 1989.
57 Nebel and Wright, op. cit., p. 250.

Chapter 10 Food for eating

1 Garret Hardin, "The tragedy of the commons," in Garrett Hardin and John Baden (eds), *Managing the Commons*, San Francisico CA, W. H. Freeman and Co., 1977, pp. 16–30.
2 Melvin A. Benarde, *The Chemicals We Eat*, New York, American Heritage Press, 1971.
3 J. N. Seiber, "Introduction" in Carl K. Winter, James N. Seiber, and Carole F. Nuckton (eds), *Chemicals in the Human Food Chain*, New York, Van Nostrand Reinhold, 1990.
4 T. Johns, *With Bitter Herbs They Shall Eat It*, Tucson AZ, University of Arizona Press, 1990.
5 Bruce N. Ames, "Dietary carcinogens and anticarcinogens," *Science*, 1983, vol. 221, September 23, pp.1256–62.
6 Margaret Kreig, *Green Medicine, The Search for Plants that Heal*, Chicago, Rand McNally & Co. 1964.
7 J. M. Watt and M. G. Breyer-Brandwijk, *The Medicinal And Poisonous Plants of Southern and Eastern Africa* (2nd edn), London, Livingstone, 1962.
8 Johns, op. cit., p. 71.
9 The Bible, 2 Kings 4:40.
10 The Bible, Numbers 11:33.
11 Nicholas Wade, "Sahelian drought: no victory for western aid," *Science*, 1974, vol. 185, July 19, pp. 234–7.
12 Donella Meadows *et al.*, *Beyond the Limits: Confronting Global Collapse, Envisaging a Sustainable Future*, White River Junction VA, Chelsea Green Publishers, 1993.
13 John Maurice, "The rise and rise of food poisoning," *New Scientist*, 1994, December 17, pp. 28–33.
14 S. E. Epstein, *The Politics of Cancer*, San Francisco, Sierra Club Books, 1978, p. 178.
15 Epstein, op. cit., p. 181.
16 G. K. York and S. H. O. Gruenwedel, "Food additives in the human food chain" in Winter *et al.*, op. cit., pp. 87–128.
17 Zachary A. Smith, *The Environmental Policy Paradox*, Englewood Cliffs NJ, Prentice-Hall, 1992, p. 27.
18 York and Gruenwedel, op. cit.
19 G. J. Marco, R. M. Hollingworth, and W. Durham (eds), *Silent Spring Revisited*, Washington DC, American Chemical Society, 1987.
20 J. E. Davies and R. Doon in Marco *et al.*, op. cit., pp. 113–24.
21 Sandra O. Archibald and Carl K. Winter, "Pesticides in our food, assessing the risks," in Winter *et al.*, op. cit., pp. 1–50.
22 E. L. Price, *Pest Control with Nature's Chemicals*, Norman OK, University of Oklahoma Press, 1983, Chapter 4.
23 Jeffrey Miller, "Effects of a microbial insecticide, *Bacillus thuringiensis*, on non-target Lepidoptera in a spruce budworm-infested forest," *Journal of Research on the Lepidoptera*, 1990 (92), vol. 29, no. 4, pp. 267–76.
24 David Pimentel *et al.*, "Environmental and economic impacts of reducing US agricultural pesticide use," in David Pimentel and Hugh Lehman (eds), *The Pesticide Question, Environment, Economics, and Ethics*, New York, Chapman and Hall, 1993, pp. 223–78.
25 B. J. Nebel and R. T. Wright, *Environmental Science, The Way the World Works* (4th edn), Englewood Cliffs NJ, Prentice-Hall, 1993, op. cit., p. 235.
26 Daniel E. Koshland, "Scare of the week," *Science*, 1989, vol. 244, April 7, p. 9.
27 Eliot Marshall, "A is for apple, Alar and ... alarmist?," *Science*, 1991, vol. 254, October 4, pp. 20–2.
28 Edward Groth, "Alar in apples," *Science*, 1989, vol. 244, May 19, p. 755.
29 Bruce Ames, "Pesticides, risk, and applesauce," *Science*, 1989, vol. 244, May 19, p. 755.
30 Eric Dewailly *et al.*, in Santé Québec, Mireille Jetté (ed.), *A Health Profile of the Inuit, Report of the Santé Québec Health Survey Among the Inuit of Nunavik, 1992*, Montréal, Gouvernement du Québec, 1994, pp. 73–107.
31 Swann Committee Report, *Joint Committee on the Use of Antibiotics in Animal Husbandry and*

Veterinary Medicine, London, Her Majesty's Stationery Office, 1969.
32 B. B. Norman and A. J. Edmondson, "Sources of chemicals in animal products," in Winter *et al.*, op. cit., pp. 51–85.
33 D. W. Gruenwedel, "Industrial and environmental chemicals in the human food chain," in Winter *et al.*, op. cit., pp. 129–219.
34 S. Yannai, "Toxic factors induced by processing," in I. E. Liener (ed.), *Toxic Constituents of Plant Foodstuffs*, New York, Academic Press, 1980, pp. 371–427.
35 Yannai, op. cit., p. 413.
36 Yannai, op. cit., p. 371.
37 Ames, 1983, op. cit.
38 R. R. Christman, "A perspective on chlorination," *Environmental Science and Technology*, 1980, vol. 14, no. 1, p. 5.
39 Wei Chen *et al.*, "Unexpected mutagen in fish," *Nature*, 1995, April 13, p. 599.
40 Maurice, op. cit.
41 Edith Efron, *The Apocalyptics, Cancer and the Big Lie*, New York, Simon and Schuster, 1984, p. 22.
42 Winter *et al.*, op. cit.
43 Liener, op. cit.
44 National Research Council (US) Committee on Food Protection, *Toxicants Occurring Naturally in Food*, Washington DC, National Academy of Sciences, 1973.
45 Ames, 1983, op. cit.
46 C. K. Winter, Dennis P. H. Hsieh, and Stefan H. O. Gruenwedel, "Natural toxins in the human food chain," in Winter *et al.*, op. cit., pp. 221–67.
47 Rachel Carson, *Silent Spring*, Greenwich CO, Fawcett Publications Inc., 1962, p. 17.
48 Gordon W. Gribble, "The natural production of chlorinated compounds," *Environmental Science and Technology*, 1994, vol. 28, no. 7, pp. 310A–19A.
49 Peter S. Spencer, *et al.*, "Guam amyotropic lateral sclerosis-Parkinsonism-dementia linked to a plant exicant neurotoxin," *Science*, 1987, vol. 237, p. 517.
50 Thomas W. Culliney, David Pimentel, and Marcia H. Pimentel, "Pesticides and natural toxicants in foods," in Pimentel and Lehman, op. cit., p. 131.
51 E. Cameron, L. Pauling, and B. Leibovitz, "Ascorbic acid and cancer: a review," *Cancer Research*, 1979, vol. 39, pp. 663–81.
52 Maryce M. Jacobs (ed.), *Vitamins and Minerals in the Prevention and Treatment of Cancer*, Boca Raton CA, CRC Press, 1991.
53 Joseph Palca, "Vitamin C gets a little respect," *Science*, 1991, vol. 254, pp. 374–6.
54 *Journal of Clinical Nutrition*, 1991, January, Supplement to vol. 53.
55 Gladys Block, "Vitamin C and cancer prevention: the epidemiological evidence," *American Journal of Clinical Nutrition*, 1991, Supplement to vol. 53, no. 1, p. 270S.
56 Steven R. Tannenbaum, John S. Wishnok, and Cynthia D. Leaf, "Inhibition of nitrosamine formation by ascorbic acid," *American Journal of Clinical Nutrition*, 1991, Supplement to vol. 53, no. 1, January, pp. 247S–50S.
57 James McD. Robertson, Allan P. Donner, and John R. Trevihick, "A possible role for vitamins C and E in cataract prevention," *American Journal of Clinical Nutrition*, 1991, vol. 53, pp. 346S–51S.
58 Lawrie Mott and Karen Snyder, *Pesticide Alert, A Guide to Pesticides in Fruits and Vegetables*, San Francisco CA, Sierra Club Books, 1987.
59 Thomas D. Wyllie, Lawrence G. Moorehouse (eds), *Mycotoxic Fungi, Mycotoxins, Mycotoxicoses, An Encyclopedic Handbook, vol. 1*, New York, Marcel Dekker, 1978.
60 Ann Gibbons, "Does war on cancer equal war on poverty?," *Science*, 1991, vol. 253, p. 260.
61 Winter *et al.*, op. cit., p. 224.
62 M. Grieve, *A Modern Herbal*, New York, Hafner Publishing Co., 1959.
63 M. R. Malinow, E. J. Bardana, and P. McLaughlin, "Systemic lupus erythematosus-like syndrome in monkeys fed alfalfa sprouts: role of a non-protein amino acid," *Science*, 1994, vol. 216, April 23, pp. 415–17.
64 Barbara J. Culliton, "Eat your broccoli (and Brussels sprouts)," *Nature*, 1992, vol. 356, April 2,

p. 377.
65 M. Windholz (ed.), *The Merck Index* (10th edn), Rahway NJ, Merck & Co., 1983, Entry no. 8843.
66 Maryce M. Jacobs and Roman J. Pienta, "Relationships between potassium and cancer," in Jacobs, op cit., pp. 227–46.

Chapter 11 Waste not, want not

1 Vance Packard, *The Waste Makers*, Harmondsworth, UK, Penguin Books Ltd., 1963 (first published in USA 1960).
2 G. H. Eduljee, D. H. F. Atkins, and A. E. Eggleton, "Observations and assessment relating to incineration of chlorinated waste chemicals," *Chemosphere*, 1986, vol 15, nos. 9–12, pp. 1577–84.
3 L. G. Oeberg, N. Wagman, R. Anderson, and C. Rappe, "De novo formation of PCDD/Fs in compost and sewage sludge – a status report," *Organohalogen Compounds*, 1993, vol. 1, pp. 297–302.
4 Gina Bara Kolata, "Love Canal: false alarm caused by botched study," *Science*, 1980, vol. 208, June 13, pp. 1239–42.
5 Dwight T. Janerch, William S. Burnett, Gerald Feck, Margaret Hoff, Philip Nasca, Peter Greenwald, and Nicholas Vianna, "Cancer incidence in the Love Canal area," *Science*, 1981, vol. 212, June 19, pp. 1404–7.
6 Elizabeth Whelan, *Toxic Terror*, Ottawa IL, Jameson Books, 1985, p. 87.
7 Constance Holden, "Love Canal residents under stress, psychological effects may be greater than physical harm," *Science*, 1980, vol. 208, June 13, p. 1242.
8 P. Costner and J. Thornton, *Playing with Fire, Hazardous Waste Incineration*, Washington DC, Greenpeace USA, 1990.
9 Clement International Corporation, *Scientific Peer Review of Greenpeace's Position on Hazardous Waste Incineration Impacts in its "Report on the Hazardous Waste Incineration Crisis" and "Playing with Fire,"* Fairfax VI, Clement International Corporation, 1992, January 29.
10 Julian Simon, *The Ultimate Resource*, Princeton NJ, Princeton University Press, 1981.
11 David M. Wiles, "Degradable plastics – fact or myth?" *Canadian Chemical News*, 1990, January, p. 20–1.
12 Ontario Ministry of Municipal Affairs, *Municipal Waste Management Powers in Ontario, A Discussion Paper*, Toronto, Queen's Printer for Ontario, 1992.
13 *Community Workshops for the Environment*, Ottawa, Harmony Foundation of Canada, 1991.
14 H. E. Goeller and Alvin M. Weinberg, "The age of substitutability," *Science*, 1976, vol. 191, February 20, pp. 683–9.
15 Lester B. Lave, Chris Hendrickson and Francis C. McMichael, "Recycling decisions and green design," *Environmental Science and Technology*, 1994, vol. 28, no. 1, pp. 19A–24A.
16 Francesco L. Galassi, "Is the blue box a red herring?" *Toronto Globe and Mail*, 1992, April 10.
17 C. K. Brown, F. J. Hopton, R. G. W. Laughlin, and S. E. Sax, *Energy Analysis of Resource Recovery Options*, Report no. P-2032/G, Mississauga ON, Ortech International (formerly The Ontario Research Foundation), 1976.
18 Brown *et al.*, op. cit., p. 70.
19 Lave *et al.*, op. cit.
20 W. R. J. Sutton, "The world's need for wood," *Environmental Science and Engineering*, 1995, January, pp. 87–8.
21 Barry Commoner, *The Closing Circle: Nature, Man, and Technology*, London, Jonathan Cape Ltd., 1972, p. 153.
22 Rachel Carson, *Silent Spring*, Greenwich CO, Fawcett Publications Inc., 1962, p. 60.
23 Wm. Rathje and Cullen Murphy, *Rubbish! The Archeology of Garbage*, New York, Harper Collins, 1993.
24 Wiles, op. cit.
25 Michael Judge, "Plastics in the green age: science, fact, and fiction," *Canadian Chemical News*, 1993, July/August, p. 5.

26 Marlene Simmons, "Disintegrating plastics," *Science Affairs*, 1978, vol. 8, no. 3, p. 11.
27 Brown *et al.*, op. cit.
28 Goeller and Weinberg, op. cit.
29 Malin Falkenmark and Gunnar Lindh, *Water for a Starving World*, Boulder CO, Westview Press, 1976.
30 Donella H. Meadows, Dennis L. Meadows, Jørgen Randers, and William W. Behrens III, *The Limits to Growth, A Report of the Club of Rome's Project on the Predicament of Mankind*, New York, Universe Books, 1972.
31 Simon, op. cit., p. 286.
32 Simon, op. cit., p. 287.
33 Philip Abelson, "Limits to Growth," *Science*, 1972, vol. 175, March 17, p. 1197.
34 Meadows *et al.*, *Beyond the Limits: Confronting Global Collapse, Envisaging a Sustainable Future*, White River Junction VA, Chelsea Green Publishers, 1993.
35 Goeller and Weinberg, op. cit.
36 Simon, op. cit., p. 6.
37 D. L. Ray and L. Guzzo, *Trashing the Planet*, Washington DC, Regnery Gateway, 1990.
38 Alan Newman, "Designer chemistry," *Environmental Science and Technology*, 1994, vol. 28, no. 11, p. 463A.
39 Lave *et al.*, op. cit.
40 Stuart Pimm, "Cassandra versus Pangloss," *Nature*, 1994, vol. 372, December 8, p. 512.
41 Anon., "Europeans generate only half as much garbage per person as do US residents," *Environmental Science and Technology*, 1994, vol. 28, no. 4, p. 171A.
42 Anon., "Innovative recycling," *Canadian Chemical News*, 1990, February, p. 8.
43 "Provinces stalled on tire disposal," *Toronto Globe and Mail*, 1993, January 2.
44 Michael Chejlava, "Paper versus plastic," *Chemical and Engineering News*, 1990, July 2, p. 2.
45 Martin B. Hocking, "Paper versus polystyrene, a complex choice," *Science*, 1991, February 1, pp. 504–5.
46 Anon., "Michigan and Dart Container have a program to reclaim more than 2 million polystyrene cups a year," *Environmental Science and Technology*, 1990, vol. 24, no. 9, p. 1276.

Chapter 12 Organochlorines – killer chemicals?

1 Science Advisory Board of the International Joint Commission on the Great Lakes, 1989 Report, Hamilton, Ontario, as quoted by Gordon W. Gribble, "Naturally occurring organohalogen compounds – a survey," *Journal of Natural Products*, 1992, vol. 55, no. 10, pp. 1353–95.
2 Ivan Amato, "The crusade against chlorine," *Science* 1993, vol. 261, July 9, pp. 152–4.
3 Paul Ehrlich, *The Science of Ecology*, New York, Macmillan, 1987, p. 616.
4 Martha Windholz (ed.), *The Merck Index* (11th edn), Rahway NJ, Merck & Co. Inc., 1989.
5 Irving N. Sax, *Dangerous Properties of Industrial Materials* (6th edn), New York, Van Nostrand Reinhold Co., 1984.
6 E. Hodgson, R. B. Mailman, and J. E. Chambers, *Dictionary of Toxicology*, New York, Van Nostrand Reinhold Co., 1988.
7 Shirley A. Briggs, *Basic Guide to Pesticides, Their Characteristics and Hazards*, Washington DC, Hemisphere Publishing Corp., 1992.
8 Sandra O. Archibald and Carl K. Winter, "Pesticides in our food, assessing the risks," in Carl K. Winter, James N. Seiber, and Carole F. Nuckton (eds), *Chemicals in the Human Food Chain*, New York, Van Nostrand Reinhold, 1990, pp. 1–50.
9 William C. White and Thomas R. Sweeney, "The metabolism of DDT, a metabolite from rabbit urine, Di(*p*-chlorodiphenyl)acetic acid; its isolation, identification and synthesis," *Public Health Reports*, 1945, vol. 60, January 19, pp. 66–71.
10 S. Dagley, W. C. Evans, and D. W. Ribbons, "New pathways in the oxidative metabolism of aromatic compounds by micro-organisms," *Nature*, 1960, vol. 188, November, p. 560.
11 John A. Bumpus, Ming Tien, David Wright, and Steven D. Aust, "Oxidation of persistent

environmental pollutants by a white rot fungus," *Science*, 1985, vol. 228, June 28, pp. 1434–6.
12 D. Ghosal, I.-S. You, D. K. Chatterjee, and A. M. Chakrabarty, "Microbial degradation of halogenated compounds," *Science*, 1985, vol. 228, April 12, pp. 135–42.
13 M. Alexander and M. I. H. Aleem, "Effect of chemical structure on microbial decomposition of aromatic herbicides," *Agricultural and Food Chemistry*, 1961, vol. 9, no. 1, January/February, pp. 44–7.
14 B. B. Norman and A. J. Edmondson, "Sources of chemicals in animal products," in Winter *et al.*, op. cit., pp. 51–85.
15 Anon., "Federal approach outlined for chlorinated substances," *Environmental Science and Engineering*, 1995, January, p. 46.
16 Windholz, op. cit., and Sax, op. cit.
17 Rachel Carson, *Silent Spring*, Greenwich CO, Fawcett Publications, Inc., 1962.
18 H. Raistrick and G. Smith, "The metabolic products of *Aspergillus terreus*, two new chlorine containing mold products," *The Biochemical Journal*, 1936, vol. 30, pp. 1315–22.
19 F. Kavanagh, A. Hervey and W. J. Robbins, "Antibiotic substances from Basidiomycetes (VIII), *Pleurotus multilus* and *P. passeckerianus*," Proceedings of the National Academy of Science of the USA, 1952, vol. 38, p. 555.
20 W. B. Turner, *Fungal Metabolites*, London, Academic Press, 1971.
21 J. S. M. Marais, "Monofluoroacetic acid, the toxic principle of 'Gifblaar' *Dichapetalum cymosum* (Hook)," *Onderstepoort Journal of Veterinary Science and Animal Industry*, 1944, vol. 20, pp. 67–73, (*Chemical Abstracts*, 1944, vol. 39, p. 4116).
22 K. C. Engvilt, "Chlorine containing natural compounds in higher plants," *Phytochemistry*, 1986, vol. 25, no. 5, p. 781–91.
23 Gribble, op. cit.
24 J. Done, P. H. Mortimer, A. Taylor, and D. W. Russel, "Production of sporidesmin and sporidesmolides by *Pithomyces chartarum*," *Journal of General Microbiology*, 1961, vol. 26, pp. 207–22.
25 Gribble, op. cit., p. 1372.
26 Elizabeth Whelan, *Toxic Terror*, Ottawa IL, Jameson Books, 1985, p. 146.
27 Deborah L. Illman, "Green technology presents challenge to chemists," *Chemical and Engineering News*, 1993, September 6, pp. 26–33.
28 Michael Gough, *Dioxin, Agent Orange. The Facts*, New York, Plenum Press, 1986.
29 M. P. Esposito, T. O. Tierman, and F. E. Dryden, *Dioxins*, Washington DC, US Environmental Protection Agency, Report No. EPA-600/2-80-197, 1980.
30 Ontario Ministry of the Environment *Scientific Criteria Document for Standard Development No. 4-84*, Toronto, 1985, September (reprinted December 1988).
31 L. L. Lamparski, R. H. Stehl and R. L. Johnson, "Photolysis of pentachlorophenol-treated wood. Chlorinated dibenzo-*p*-dioxin formation," *Environmental Science and Technology*, 1980, vol. 14, no. 2, pp. 196–200.
32 J. M. Czuczwa, B. D. McVeety, and R. A. Hites, "Polychlorinated dibenzo-*p*-dioxins and dibenzofurans in sediments from Siskiwit Lake, Isle Royale," *Science* 1986, vol. 226, pp. 568–9.
33 M. J. Blumich, M. V. D. Berg, and K. Olie, "Sources and fate of PCDDs and PCDFs, an overview," *Chemosphere*, 1985, vol. 14, no. 6/7, p. 581.
34 R. M. Smith, P. O'Keefe, K. Aldous, R. Briggs, D. Hilker, and S. Connor, "Measurements of PCDFs and PCDDs in samples and lake sediments at several locations in upstate New York," *Chemosphere*, 1992, vol. 25, nos. 1–2, pp. 95–8.
35 Shunji Hashimoto, Tadaaki Wakimoto, and Ryo Tatsukawa, "PCDDs in the sediments accumulated about 8120 years ago from Japanese coastal areas," *Chemosphere*, 1990, vol. 21, no. 7, pp. 825–35.
36 Kavanagh *et al.*, op. cit.
37 Esposito *et al.*, op. cit.
38 Ontario Ministry of the Environment, op. cit., p. 3130.
39 T. J. Nestrick and L. L. Lamparski, "Isomer-specific determination of chlorinated dioxins for assessment of formation and potential environmental emission from wood combustion," *Analytical Chemistry*, 1982, vol. 54, no. 13, pp. 2292–9.

40 R. E. Clement, H. M. Tosine, and B. Ali, "Levels of polychlorinated dibenzo-*p*-dioxins and dibenzofurans in wood burning stoves and fireplaces," *Chemosphere,* 1985, vol. 14, no. 6/7, pp. 815–19.

41 Gribble, op. cit.

42 Gordon W. Gribble, "The natural production of chlorinated compounds," *Environmental Science and Technology,* 1994, vol. 28, no. 7, pp. 310A–19A.

43 C. E. VanWagner, "Forest fire statistics and the timber supply," in D.G. Brand (ed.), *Canada's Timber Resources, Proceedings, Natl. Conference, Canadian Forestry Service, Victoria BC, June 1990,* Ottawa, Canadian Forestry Service, pp. 111–18.

44 Richard M. Berry, Corinne E. Luthe, and Donald H. Voss, "Ubiquitous nature of dioxins: a comparison of the dioxins content of common everyday materials with that of pulps and papers," *Environmental Science and Technology,* 1993, vol. 27, no. 6, pp. 1164–8.

45 Anders Svenson, Lars-Owe Kjeller, and Christopher Rappe, "Enzyme-mediated formation of 2,3,7,8-tetrasubstituted chlorinated dibenzodioxins and dibenzofurans," *Environmental Science and Technology,* 1989, vol. 23, no. 7, pp. 900-2.

46 Gribble, 1994, op. cit.

47 L. G. Oeberg, N. Wagman, R. Anderson, and C. Rappe, "De novo formation of PCDD/Fs in compost and sewage sludge – a status report," *Organohalogen Compounds,* 1993, vol. 1, pp. 297–302.

48 R. L. Dillman, *Secondary Metabolites from the Bermudian Sponge Tedania ignis,* Bozeman MT, Ph.D. Dissertation, Montana State University, 1990.

49 Curtis C. Travis and Holly A. Hattemer-Frey, *Human Exposure to Dioxin,* US Dept. of Energy contract No. DE-AC-o5-840R 21400 (Oak Ridge National Laboratories), 1990.

50 Tony Reichhardt, "EPA rebuffs challenge to its assessment of dioxin data," *Nature,* 1994, September 22, p. 272.

51 Briggs, op. cit., p. 202.

52 Windholz, op. cit.

53 National Research Council of Canada, Report no. 14094, *Chlordane: its Effects on Canadian Ecosystems and its Chemistry,* Ottawa, National Research Council of Canada, 1973, p. 90.

54 A. Bharath, C, Mallard, D. Orr, G. Ozburn, and A. Smith, "Problems in determining the water solubility of organic compounds," *Bulletin of Environmental Contamination and Toxicology,* 1984, vol. 33, pp. 133–7.

55 D. Liu, "Biodegradation of Aroclor 1221 type PCBs in sewage waste water," *Bulletin of Environmental Contamination and Toxicology,* 1981, vol. 27, pp. 695–703.

56 D. Liu, K. Thomson, and W. M. J. Strachan, "Biodegradation of pentachlorophenol in a simulated aquatic environment," *Bulletin of Environmental Contamination and Toxicology,* 1981, vol. 26, pp. 85–90.

57 Bumpus *et al.,* op. cit.

58 G. Asplund, A. Grimvall, and C. Pettersson, "Naturally produced adsorbable organic halogens (AOX) in humic substances from soil and water," *The Science of the Total Environment,* 1989, vol. 81/82, pp. 239–48.

59 B. Wigilius, B. Allard, H. Borén, and A. Grimvall, "Determination of adsorbable organic halogens (AOX) and their molecular weight distribution in surface water samples," *Chemosphere,* 1988, vol. 17, no. 10, pp. 1985–8.

60 Asplund *et al.,* op. cit.

61 Gunilla Asplund and Anders Grimvall, "Organohalogens in nature, more widespread than previously assumed," *Environmental Science and Technology,* 1991, vol. 25, no. 8, pp. 1346–50.

62 Olaf Dahlman, Roland Morck, Pierre Ljungquist, Anders Relmann, Carina Johansson, Hans Boren, and Anders Grimvall, "Chlorinated structural elements in high molecular weight organic matter from unpolluted waters and bleached-kraft mill effluents," *Environmental Science and Technology,* 1993, vol. 27, no. 8, pp. 1616–20.

63 David P. Barr and Steven D. Aust, "Mechanisms white rot fungi use to degrade pollutants," *Environmental Science and Technology,* 1994, vol. 28, no. 2, pp. 78A–87A.

64 K. D. Crow, "Chloracne (Halogenacne)," in F. N. Marzulli and H. I. Maibach (eds), *Dermatotoxicology,* Washington DC, Hemisphere Publishing Corp., 1987, pp. 515–34.

65 Vic Peck and Ralph Daley, "Toward a 'Greener' Pulp and Paper Industry," *Environmental Science and Technology*, 1994, vol. 28, no. 12, pp. 524A–527A.
66 Ibid.
67 H. Mussalo-Rauhamaa, E. Häsänen, H. Pyysalo, K. Antervo, R. Kauppila, and P. Pantzar, "Occurrence of beta-hexachlorocyclohexane in breast cancer patients," *Cancer*, 1990, vol. 66, pp. 2053–128.
68 Franc Falck, Andrew Ricci, Mary S. Wolff, James Gobold, and Peter Dekkers, "Pesticides and polychlorinated biphenyl residues in human breast lipids and their relation to breast cancer," *Archives of Environmental Health*, 1992, vol. 47, no. 2, pp. 143–6.
69 Stephen Safe, "Synthesis and application of 1,3,6,8, and 2,4,6,8-tetrasubstituted dibenzofurans and -dibenzo-*p*-dioxins as antitumorigenic agents," US Patent Application, 1989, January 23, no. 300,416, (Chemical Abstracts, 1991, vol. 114, abstract no. 61914).
70 Gribble, 1992, op. cit., p. 1378.
71 Dietrich Henschler, "Toxicity of chlorinated organic compounds: effects of the introduction of chlorine in organic molecules," *Angewandte Chemie, Internl. English Edition*, 1994, vol. 33, pp. 1920–35.
72 R. D. Kimbrough (ed.), *Halogenated Biphenyls, Terphenyls, Naphthalenes, Dibenzodioxins, and Related Products*, Amsterdam, Elsevier, 1980.
73 D. A. Barsotti, R. J. Marlar, and J. R. Allen, "Reproductive dysfunction in Rhesus monkeys exposed to low levels of polychlorinated biphenyls (Aroclor 1248)," *Food and Cosmetics Toxicology*, 1976, vol. 14, pp. 99–103.
74 James A. Allen, Deborah A. Barsotti, and Laurine A. Carstens, "Residual effects of polychlorinated biphenyls on adult non-human primates and their offspring," *Journal of Toxicology and Environmental Health*, 1980, vol. 6, pp. 55–66.
75 J. R. Allen and D. H. Norback, "Carcinogenic potential of the polychlorinated biphenyls," in H. H. Hyatt. J. D. Watson, and J. A. Winsten (eds), *Origins of Human Cancer, Cold Spring Harbor Conference on Cell Proliferation*, Cold Spring Harbor NY, Cold Spring Harbor Laboratory Press, 1977.
76 G. G. Fein, J. L. Jacobson, S. W. Jacobson, P. M. Schwarz, and J. K. Dowler, "Prenatal exposure to polychlorinated biphenyls: effects on birth size and gestational age," *Journal of Pediatrics*, 1984, vol. 105, no. 2, pp. 315–20.
77 Joseph L. Jacobson, Sandra W. Jacobson, and Harold E. B. Humphrey, "Effects of in utero exposure to polychlorinated biphenyls and related contaminants on cognitive functioning in young children," *Journal of Pediatrics*, 1990, vol. 116, no. 1, pp. 38–45.
78 Michael A. Kamrin and Lawrence J. Fisher, "Workshop on human health impacts of halogenated biphenyls and related compounds," *Environmental Health Perspectives*, 1991, vol. 91, pp. 157–64.
79 Wayland R. Swain, "Human health consequences of consumption of fish contaminated with organochlorine compounds," *Aquatic Toxicology*, 1988, vol. 11, pp. 357–77.
80 *Report to the Ontario Workers' Compensation Board on Occupational Exposure to PCBs*, Toronto, Industrial Disease Standards Panel, IDSP Report No. 2, 1987.
81 Proceedings Dioxin Workshop Québec City, May 7–8, 1987, Canadian Council of Ministers of the Environment, ISBN 0-919074-46-4.
82 Gangadhar Choudhary, "Occupational exposure to polychlorinated dibenzo-*p*-dioxins and dibenzofurans: a perspective," in Gangadhar Choudhary, Lawrence H. Keith, and Christoffer Rappe (eds), *Chlorinated Dioxins and Furans in the Total Environment*, Boston MA, Butterworth Publishers, 1983, pp. 335–54.
83 R. J. Kociba and O. Cabey, "Comparative toxicity and biologic activity of chlorinated dibenzo-p-dioxins and furans relative to 2,3,7,8-tetrachlorodibenzo-p-dioxin (TCDD)," *Chemosphere*, 1985, vol. 14, no. 6/7, pp. 649–60.
84 Crow, op. cit.
85 Kociba and Cabey, op. cit.
86 A C&EN Special Issue, "Dioxin Report," *Chemical and Engineering News*, 1983, June 6, pp. 20–64.
87 Rebecca L. Rawis, "Dioxin's human toxicity is most difficult problem," *Chemical and*

Engineering News, 1983, June 6, pp. 44–8.
88 Bette Hileman, "Dioxin toxicity research, studies show cancer, reproductive risks," *Chemical and Engineering News*, 1993, September 6, p. 5.
89 Marilyn Fingerhut *et al.*, "Cancer mortality in workers exposed to 2,3,7,8-tetrachlorodibenzo-*p*-dioxin," *New England Journal of Medicine*, 1991, vol. 324, no. 4, pp. 212–18.
90 J. J. Collins, J. F. Acquavella, and B. R. Friedlander, "Reconciling old and new findings on dioxin," *Epidemiology*, 1992, vol. 3, no. 1, pp. 65–8.
91 M. A. Fingerhut *et al.*, "Old and new reflections on dioxin," *Epidemiology*, 1992, vol. 3, no. 1, pp. 69–71.
92 Reichhardt, op. cit.
93 G. Mason and S. Safe, "Synthesis, biologic and toxic effects of the major 2,3,7,8-tetrachlorodibenzo-*p*-dioxin metabolites in the rat," *Toxicology*, 1986, vol. 41, no. 2, pp. 153–9.
94 Eric Dewailly *et al.*, "High levels of PCBs in breast milk of Inuit women from arctic Québec," *Bulletin of Environmental Contamination and Toxicology*, 1989, vol. 43, no. 5, pp. 641–6.
95 Eric Dewailly *et al.*, in Santé Québec, Mireille Jetté (ed.) *A Health Profile of the Inuit, Report of the Santé Québec Health Survey Among the Inuit of Nunavik, 1992, Volume 1*, Montréal, Gouvernement du Québec, 1994, pp. 73–107.
96 Mireille Jetté (ed.) *A Health Profile of the Inuit, Report of the Santé Québec Health Survey Among the Inuit of Nunavik, 1992, Volume 1*, Montréal, Gouvernement du Québec, 1994, p. 247.
97 Anon., "Blood samples taken by Greenpeace from nine workers at an English chemical plant," *Environmental Science and Technology*, 1994, vol. 28, no. 12, p. 506A.
98 John P. Giesy, James P. Ludwig, and Donald E. Tillitt, "Deformities in birds of the Great Lakes region, assigning causality," *Environmental Science and Technology*, 1994, vol. 28, no. 3, pp. 128A–34A.

Chapter 13 Conclusion

1 H. F. Smyth, "Sufficient challenge," *Food and Cosmetics Toxicology*, 1967, vol. 5, pp. 51–8.
2 Rael J. Isaac and Erich Isaac, *The Coercive Utopians*, Chicago IL, Regnery Gateway Inc., 1985.
3 The World Bank, *Averting the Old Age Crisis, Policies to Protect the Old and Promote Growth*, Oxford, Oxford University Press, 1994.
4 Dixy Lee Ray and Lou Guzzo, *Trashing the Planet*, Washington DC, Regnery Gateway, 1990.
5 Tim Padmore, "Affluence for 10 billion," *The Vancouver Sun*, 1974, October 12.
6 Garrett Hardin, "The tragedy of the commons," *Science*, 1968, vol. 162, pp. 1243–8.
7 J. H. Dales, "Land water and ownership," in Robert Dorfman and Nancy S. Dorfman (eds), *Economics of the Environment, Selected Readings*, New York, W. W. Norton & Company Inc., 1972, p. 171.
8 James Lovelock, *Gaia: A New Look at Life on Earth* (2nd edn), Oxford, Oxford Paperbacks, 1995.
9 Helen B. Hiscoe, "Does being natural make it good?" *Journal of the American Medical Association*, 1983, June 16, p. 1474.
10 "Managing planet earth," Special edn of *Scientific American*, 1989, September, p. 17.
11 Jeremy Burgess, "Beware of imitations," *New Scientist*, December 31, 1995, p. 47.
12 E. Carlsen, A. Giwercman, N. Keiding, and N. Skakkebek, "Evidence for decreasing quality of semen during the past fifty years," *British Medical Journal*, 1992, vol. 305, pp. 605–13.
13 A. Giwercman and A. Skakkebek, "The human testis: an organ at risk?," *International Journal of Andrology*, 1992, vol. 15, pp. 373–5.
14 M. A. Weinberger, L. Friedman, T. M. Farer, F. M. Moorland, E. L. Peters, C. E. Gilmore, and M. A. Kahn, "Testicular atrophy and impaired spermatogenesis in rats fed high levels of the methyl xanthines, caffeine, theobromine or theophylline," *Journal of Environmental Pathology and Toxicology*, 1978, vol. 1, no. 5, pp. 669–88, (*Carcinogenesis Abstracts*, 1979, vol. 16, abstract no. 78-6123).
15 Ralph Nader and Mark J. Green (eds), *Corporate Power in America, The Ralph Nader Conference on Corporate Accountability, Washington DC, 1971*, New York, Grossman Publishers, 1973.

Publishers, 1973.

16 Isaac Asimov, *Isaac Asimov's Biographical Encyclopedia of Science and Technology* (Revised edn), New York, Avon Books, 1972, p. 182.

17 Roald Hoffmann, "Chemistry, democracy, and a response to the environment," *Chemical and Engineering News*, 1990, April 23, pp. 25–9.

18 M. Alice Ottoboni, *The Dose Makes the Poison*, Berkeley CA, Vincente Books, 1984.

19 R. Doll, H. J. Evans, and S. C. Darby, "Paternal exposure not to blame," *Nature*, 1994, vol. 367, February 24, pp. 678–80.

20 G. Claus and K. Bolander, *Ecological Sanity*, New York, David McKay, 1977.

Name index

Subject index